OUR TRIBAL FUTURE

OUR TRIBAL FUTURE

HOW TO CHANNEL OUR
FOUNDATIONAL HUMAN INSTINCTS
INTO A FORCE FOR GOOD

DAVID R. SAMSON

ST. MARTIN'S PRESS
NEW YORK

 For information, address St. Martin's Publishing Group, 120 Broadway, New York, NY 10271.

www.stmartins.com

The Library of Congress Cataloging-in-Publication Data is available upon request.

ISBN 978-1-250-27224-9 (hardcover)
ISBN 978-1-250-27225-6 (ebook)

Our books may be purchased in bulk for promotional, educational, or business use. Please contact your local bookseller or the Macmillan Corporate and Premium Sales Department at 1-800-221-7945, extension 5442, or by email at MacmillanSpecialMarkets@macmillan.com.

First Edition: 2023

10 9 8 7 6 5 4 3 2 1

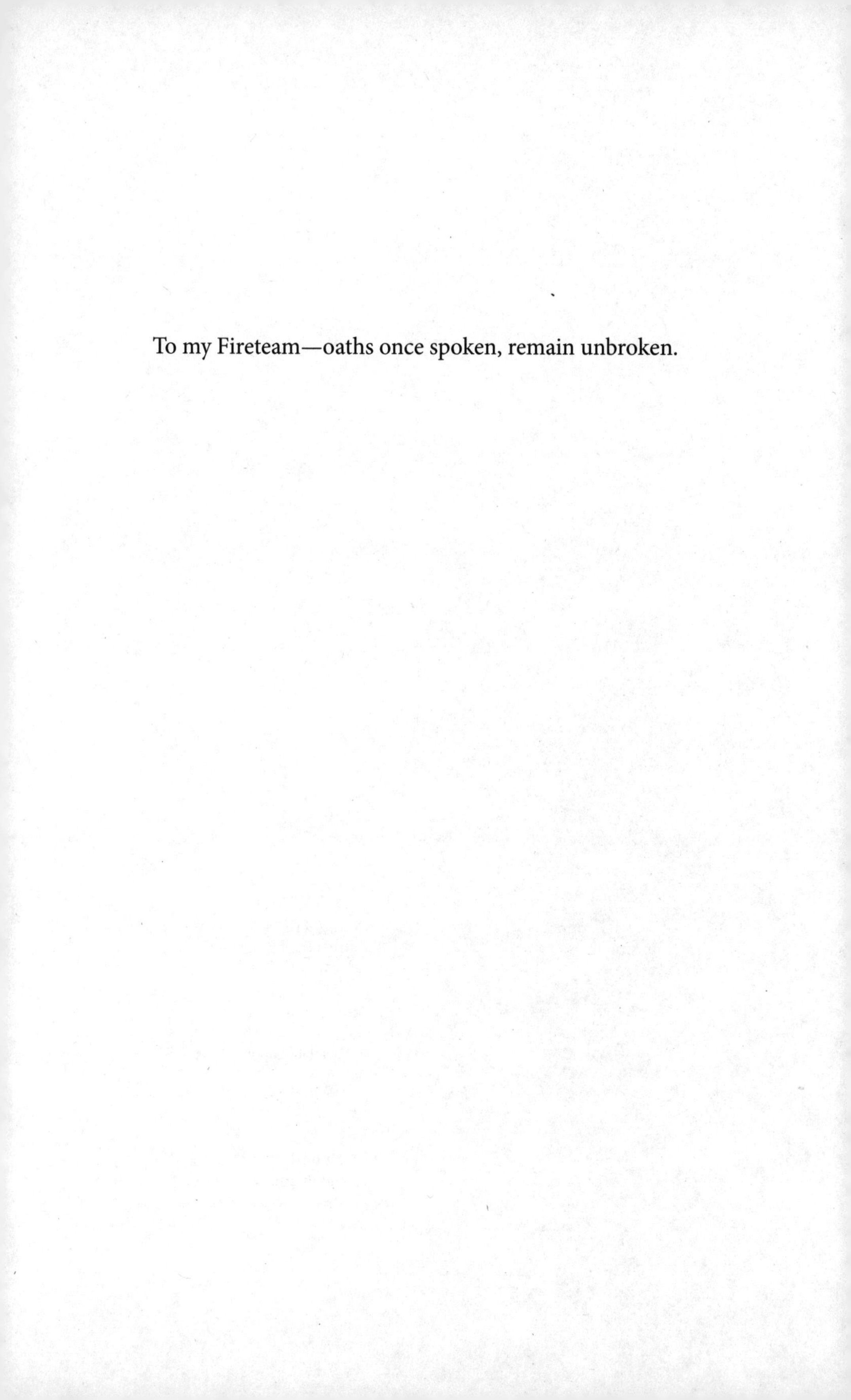

To my Fireteam—oaths once spoken, remain unbroken.

CONTENTS

PROLOGUE

In Whom Do You Put Your Trust?

Without trust . . . civilization collapses.

–Yuval Noah Harari, 2022

For all social species, one of the most intractable problems is *whom to trust.* I call this the Trust Paradox. A paradox can be defined as the contemplation of a seemingly self-contradictory statement that can help to illuminate a larger truth. As the philosopher G. K. Chesterton quipped, "A paradox is simply Truth standing on its head to get our attention." At first glance, the question "In whom do you put your trust?" does not seem controversial, never mind paradoxical; the answer is usually agreed upon by scientist and layman alike. You trust family.

Your kin shares your blood. On a foundational level, you and your kin are one and the same. The genomes imprinted in the cells that instruct who and what you are closely resemble those of your immediate family and cousins. Kin selection—the preferential bias for those who are genetically related—was the *first* answer concocted by evolution several hundred million years ago and has faithfully served Earthlings ever since. The challenge is scale. When considering plants, social insects, and even naked mole rats, kin selection works predictably. However, when you scale up complexity to species that rely on meaty brains, long-term memory, and special kinds of social arrangements, then something else is needed.

The thought experiment of the Trolley Problem is a touchstone example of this challenge. Over the history of philosophy, it has evolved into a number of iterations, but is distilled to something like this: You are riding in a trolley without functioning brakes. On the current track stand five people who are certain to be killed if the trolley continues on its path. You have access to a switch that would divert the trolley to another track, but another individual stands there. That person will be killed if the switch is activated. So do you switch tracks or not?

When we are confronted with this thought experiment, we face an ethical dilemma. That's because our nervous system, housing our massive brains, has multiple—sometimes competing—internal structures; these brain regions evolved for different functions, and thus compete for neural resources over what is the moral (or "ethically optimized") outcome. While the high-minded "forebrain" seeks idealistic outcomes based on logic, reason, and numerical *utilitarianism*, the more ancient limbic system seeks to maximize *principled** outcomes that preserve harmony for *your* group, and the most core biological system seeks only outcomes that maximize your own odds of survival.

This is where the Trust Paradox comes to light. How do we solve the problem of whom to trust when different parts of our brain seek different solutions to the same problem? This is a paradox worth investigating, because the fate of our species depends on a scientifically robust solution to the contradictions we inherited.

The Trust Paradox is a challenge faced by all life, but answers to the question have varied depending on evolutionary pressures. Even within an individual species, the answers can change over time as it gains a foothold over its environment. Once our ancestors got the hang of survival and proliferated, their success, *paradoxically*, came back to haunt them. With so many people living in ever-expanding larger groups, how did we begin to scale trust to individuals who weren't family? As we will explore in this book, humans innovated novel ways to solve this problem. The next solution, after kin, was friendship. Rare in the animal kingdom, friendships worked keenly for humans living in face-to-face social worlds. Humans are not the only species with friendship, but its not typ-

---

* "Deontological," for you philosophy nerds out there.

ical in the animal world and the way human friendship is expressed is special. Friendship serves as our species' crowning ethical achievement and has been suggested to be the origin point for the evolutionary emergence of morality. As opined by Nicholas Christakis: "Our assembly into networks of friendships sets the stage for the emergence of moral sentiments. At their core, moral compunctions relate to how people interact with others, especially those who are not kin and for whom the bonds of kinship and the inexorable workings of inclusive fitness are not enough of a guide . . . friendship lays the foundation for morality."[1] Perhaps the most noble and virtuous quality ever produced by natural selection is the transcendence of kin selection to a truly moral sensibility expressed by way of friendship. After all, friends don't need to have sex with each other or share in childrearing in order to feel that they have a special relationship.[2] This too was a novel, innovative answer to the question "In whom do you put your trust?"

But at this point in the evolutionary story our ancestors were multiplying in droves. Friendship and kin selection gave us the new and unprecedented heroes of the Paleolithic, the avengers of kin *and* kith. Myths and legends remain of their exploits—a mother laying down her life for her daughter, a blood brother giving his life for a friend. These were all canonized in the stories that our ancestors passed down by word of mouth, which were later written down by scribes. But this success came with another round of costs incurred by increases to scale. We were so successful in reproducing that we became encroached by strangers. Humanity needed a new answer to the trust dilemma.

The answer was to become tribal.

As we explore the natural history of tribalism, we will see that some three hundred thousand years ago humans chanced upon a revolutionary adaptation that led to the encoding of the Tribe Drive in our DNA. This was the evolution of nested groups, each with their own particular symbols—and enshrined shared myths and values—that bound participants together in trusting relationships. All tribes are, in effect, a type of secret society, and the passwords to unlocking full rights and responsibilities of membership reside in the symbols used to verify that one is part of the tribe. Religion is one such signal. The ancient Israelites recorded it in their sacred texts, in Psalms 16:1: "Preserve me, O God, for in You I put my trust." The Mesopotamian goat herders of three thousand years ago were putting to scale the tribal solution our African ancestors had innovated three hundred thousand years beforehand. *If we all believe in the same tribal God, we can trust each other even though we may have never before met.* If your signal is not received as honest, you gain no entrance into the social inner sanctum. But if others recognize your signals as honest, you pass the test and are treated with a positive bias and, buttressed by your shared *identity protective cognition*, given tribal privileges. Humanity had a new way to promote cooperation . . . but at a terrible, horrific cost. Once in-groups exist, by definition, so too do *out-groups*. It was both feature and bug, *curse and blessing*. This book is an exploration of the natural history of how our species innovated tribalism as a novel answer to the Trust Paradox. I believe a deep, scientifically robust understanding of the contours of this adaptation will be critical to the survival of our species.

IN THE OPENING CHAPTER OF Part I, "The Science of Tribalism," we confront one of the greatest mysteries in science: How do we get from a gene to a behavior? To get to the root of this problem, we'll begin by exploring the ways in which the drive to be tribal—henceforth, the Tribe Drive—significantly (and mostly imperceptibly) influenced my life. But to do this, we'll have to pose the fundamental question "What is a drive?" What does it *feel* like to be influenced by your genes, hormones, brain, or culture? The naked answer is that most of the time it doesn't feel like anything—the *imperceptibility* of the drive is one of the keys to its evolutionary success. Importantly, elevating a drive from imperceptibility to perceptibility will allow us to begin gaining control over it, and not be controlled by it.

In chapter two we will look at one of the most important scientific concepts for understanding the modern human condition—evolutionary mismatch. Mismatch is the idea that humans are adapted for how things *were*, not how they *are*. The way evolution usually works is that when an environment rapidly changes, an animal must change with it or go extinct; modern humans have found a way around this, in a sense, because our culture shapes our environments to our needs.

Once we have a better understanding of the drives that influence our behavior, and of the problems that a modern, mismatched society cause in detriment to the healthy expression of those behaviors, we will move on to a scientifically robust definition of *the tribe concept*. In chapter three, by seeking the answers to the question "Why are we tribal?" we will explore the archaeology of past human societies to uncover the natural history of tribalism in our species, and survey the latest theories as to why evolution favored the tribal organization of societies.

In chapter four, we will intimately engage the new behavioral science of the Tribe Drive. Drawing inspiration from Robert Sapolsky's book *Behave*, we investigate the causal levels of tribal in-group "us" and out-group "them" behavior. In the quest to be better equipped at predicting our own Tribe Drives, we'll consider some of the most powerful factors that lead to the ultimate expression of a tribal behavior.

Unexpectedly, standard evolutionary models fail to address one critical aspect of the Trust Paradox—evolution hides the truth from our eyes! In chapter five, we will hear a revolutionary and startling new idea called the Fitness-Beats-Truth (FBT) theorem. The claim is that evolution by natural selection does not favor true perceptions; in reality, it drives them to extinction. If true, this lays waste to common intuitions, since the very language of our perceptions—including shape, hue, brightness, texture, sound, smell, motion, and even space-time—cannot describe objective reality. If this is true, how do we know that our group identities aren't fundamentally distorting our perception of reality?

In Part II, "The Practice of Tribalism," we will embark on a radical journey into the new practice of applied evolutionary anthropology. Standing on the firm scientific foundation that we built in Part I, we will dare to use the science of the Tribe Drive in praxis, to empower us to build thriving friendships, strong pairbonds (romantic relationships), and protective honor groups. We will call these important groups *camps*, denoting a tight social network of people sharing resources and dwelling in geophysical proximity with shared identity and superordinate goals. Ideally, these camps are embedded in healthy, prosocial communities.

Knowing the natural history of tribalism is one thing, but in the twenty-first century should we still try to build and belong to distinct groups with strong identities? What are the costs and benefits to doing so? In chapter six, we will consider the case for doing just that by looking at the tangible rewards of a life spent in active pursuit of a cultivated social network with a shared identity—an honor or sympathy group.

The next several chapters will lay down a roadmap, a "DIY How-To Guide" that teaches us how to *find the others*. We will look at an evolutionary guide to forging a family and crafting a camp. We'll also take a deep dive into the latest scientific understanding of the Big Five personality traits so that we can understand the underlying factors that determine whether two people become and stay friends and succeed or fail as a team within a community.

But if tribalism can lead to some of the worst of human behaviors, how do we guard against these dangers? In the final chapters, we will wade into the philosophical and ethical implications of the Trust Paradox. We survey how many thinkers in both camps of philosophy and science have been collaborating in transdisciplinary ways to probe one of the greatest ontological* challenges of both fields: If humans are the product of evolution, is morality objective or the subjective by-product of our evolutionary history?

If tribalism is such a powerful force biasing the human mind to prefer one's own tribe over another, then what does it mean for the future of humanity? I don't know the answer. But the new science of cognitive immunology may point us in the right direction. The twenty-first century has been waylaid by ideological derangement. Extremist culture warriors dot the landscape in a winner-take-all attack on each of their enemies. Mass shootings (spurred by loneliness and social isolation), terror bombings (spurred by religious tribalism), and hate crimes (spurred by racist tribalism) are endemic; tribalist ideology is at the very center of the modern epidemic of bad ideas. How do we inoculate ourselves against the tribe virus? In the concluding chapters, I will use the new science of cognitive immunology to craft a ready-to-use tribe vaccine that will serve as an immune booster to combat the tribalist ways of thinking that plague our species.

The ultimate goal of this book is to foster personal and collective empowerment. The Tribe Drive, characterized by its own flawless, internal

* Ontological: relating to the branch of metaphysics dealing with the nature of being.

logic—*obey tribal consensus of your in-group at the expense of the out-group*—took hold because it got our ancestors out of a lot of scrapes. But now, as our species scales in numbers beyond comprehension, its risks are no less than existential. The following pages will help you to learn its language and know its secrets so that when the day comes, you will not fall prey to its curse but instead elevate yourself with its blessings. The fate of our species hangs in the balance.

PART I

THE SCIENCE OF TRIBALISM

1

THE TRIBE DRIVE

What Is Tribalism and Why Does It Matter?

So next time you hear a raving demagogue counseling hatred for other, slightly different groups of humans, for a moment at least see if you can understand his problem: He is heeding an ancient call that–however dangerous, obsolete, and maladaptive it may be today–once benefited our species.

–Carl Sagan and Ann Druyan, 1993

The science is out. Every citizen has an obligation to understand our drives.

–Geoffrey Miller, 2019

Until you make the unconscious conscious, it will direct your life, and you will call it fate.

–Carl Jung, 1951

What do you think of when you hear the word *tribalism*?

You may have a hunch, but the word is now used in so many varying contexts by media talking heads, podcasters, and everyday people that it has almost lost its meaning. For many, it conjures images of racism and sectarian violence, playing in a constant loop on the twenty-four-hour news cycle. They may see leaders mobilizing their supporters by pitting the differences of one group against another in a zero-sum political game. For others, though, the first thoughts may be of their own group. They may envision community, family, and faith. It may remind them of the bonds of loyalty that bind us all together, or a true friend demonstrating personal sacrifice and benevolence in a way that blood relations would be pressed to match. There is an element of tribalism in all of these things, but let me tell you what I see from my perspective as an evolutionary anthropologist.

I see a band of Paleolithic humans dancing and singing in a ring around a fire. I see a mother nursing her infant and handing it off to another camp member so that she can attend other business. I see a hunter sharing his prize of meat with an infirm grandmother. Juxtaposing these images of altruism, I also see the very first moment where a human murdered another human not of "their kind" in cold blood. I see a band of men with sharpened wooden spears and flint-knapped handaxes circling the foraging camp of a rival group in the night, waiting to ambush, kill, or abduct any who may wander out alone.

Beyond these visions of our prehistoric past, I see a twenty-first-century single mother, lonely and exhausted from another sleepless night with her newborn; a Wednesday-night bowling team in matching jerseys celebrating after making a tournament-winning strike; a socially isolated teenage boy entering a school with a loaded semiautomatic rifle. I see churchgoers, soccer hooligans, and suicide bombers. I see Donald Trump and the populism that fueled his rise to presidency, and his followers cheering at rallies and storming the U.S. Capitol.

Tribalism is governed by a force so motivationally powerful that it predicts more of your behavior than does your race, class, nationality, or religion. The formal analysis of this incredible phenomenon has only just begun, but the emerging science reveals that these factors are mere subjugates to our primal instinct to be a member of a tribe. This "Tribe Drive" is an ancient adaptation that has been a prerequisite for survival for 99.9 percent of our species' evolutionary history. It is a critical piece of cognitive machinery—honed by millions of years of evolution—that gave us the ability to navigate, both cooperatively and competitively, increasingly complex social landscapes. But now that our species spans billions across the globe, does this adaptation continue to serve us, or is it mismatched to its environment? In other words, *what happens when humans become either tribeless or destructively consumed by tribalism?*

The Tribe Drive is *coalitionary instinct*. Specifically, it is the instinct to belong to *a nested group—a tribe—that uses symbols that represent a shared reality to identify membership.* For good or ill, we all have it. This instinct ignores your political affiliation and cares little for the color of your skin or the gender you identify with. It stays with you from the time you are born to the time that you die. It disregards whether you are rich or poor. It scoffs at your intelligence—and may even use your formal education against you, allowing you to craft more convincing narratives to justify your actions. Surprisingly, neither expertise, intelligence, numeracy, nor political ideology serve as an inoculation to being tribal.[1] Studies show that people with specialized training become better at deceiving themselves and others when truth conflicts with their prior beliefs. In other words, the Tribe Drive is indiscriminate.

Tribalism is the worst kind of manifestation from the Tribe Drive. It is the belief that different identity-based coalitions possess distinct characteristics, abilities, or qualities, especially so as to distinguish them as inferior or superior to one another. Sound familiar? It should. If racism is the belief that skin color can distigiush people as superior or inferior, then tribalism—using skin color as a predictive factor in group identity—is the root code of this phenomenon. This is profound because if we want to solve racism, we have to understand tribalism. You and every person you interact with emit and receive signals embedded with coalitionary information that shape our behavior, and this behavior spills into our social networks and societies. On a daily basis, I take note and register tribalism's influence on myself and others. This book is about why this matters and why it should matter to you.

In the coming chapters, we will explore the science of how various human drives manifest in distinct behaviors. We will also investigate the anthropological definition of a tribe, and how belonging to one is not always a matter of personal choice. Nor is it only the purview of a handful of people, to be determined by legislators and judicial systems or social justice advocates. Belonging to a tribe is, in fact, the birthright of every human being. One of the most pressing challenges to becoming masters of our drives—not mastered by them—is *knowing they exist in the first place*. In this chapter, we will uncover the nature of the imperceptibility of the dispositions that influence our behavior. Ultimately, this knowledge will be essential in improving our moral landscapes.

Deconstructing My Tribalism

Looking back now, over the span of my lifetime, the Tribe Drive has been the unseen force lurking in the shadows of not only my origin story, but also some of its most critical, identity-shaping moments. Now I see these moments in a new light; they carry with them through time important lessons, harbored in secret for years until I learned the code to unlock their deeper meanings.

My Conception

My French-Canadian grandfather, Roland Samson, was born in 1928. My Catholic great-grandparents—Emilio and Laura Samson—had twelve children in all, and Roland was the eighth. In 1979, my mother, Dana Monroe, fell in love with my father, Daniel Samson, son of Roland, in part because they were both attending the same religious college. Religious devotion is a theme in the Samson family tree. In fact, my great-grandfather Emilio's faith at the time forbade the use of contraceptives, which explains in part why they had twelve children, a number unheard of nowadays. My father, in seeking a partner, was hoping to marry one who shared his devotion and religious affiliation. Group affiliation—in this case, religion—brought me into the world as a Samson.

Lesson: Group affiliation guides and drives reproduction.

My First Memory

I watch with some measure of awe as my father stands upon the dais and speaks to hundreds of his parishioners. He leans on a podium embossed with a golden icon of a lion, a boy, and a lamb, symbolizing the millennial age and the coming of Christ. The parishioners look up to him and listen in rapt attention. Moreover, this is an international affair. We are in Martinique, a Caribbean island that's part of the Lesser Antilles, for a church "feast." Even as a young boy, I am struck by the human diversity of the crowd. As with all good shamans, my father nourishes the souls of those in attendance with homiletical concepts of group unity under one ultimate God. The final goal (highlighted from sermon to sermon) is that of worldwide peace when evil will be destroyed, and the "lamb" (Christ) will come back to rule, paradoxically, with a rod of iron as all nations submit to his will.

Lesson: Religion is a powerful group adhesive and tribal signal.

My First Fight

My family is transferred by the church from New Brunswick, Canada, a primarily English-speaking province, to French Québec. There, as a newly minted fourth grader in a Quebecois French–dominant elementary school, I am standing in defiance of an older, stronger fifth-grade bully. Being one of the only English-speaking kids at my school, my poor French language skills make me an easy target on the playground. I am being goaded into a fight. I have yet to back down, but I don't dare throw a punch. "Are you a pussy, you little English piece of shit!?" He launches a punch and I take it in the stomach, doubling over. His friends laugh as he continues to throw blows on my hunched-over body.

Lesson: Language is a powerful coalitional identifier.

My First Friendships

I am kneeling in the dusty basement of my best friend's home. Metallica's heavy metal ballad "Nothing Else Matters" plays in the background. A stainless steel sword tip touches my left shoulder, then my right, and then after an ascending arc, comes to rest on a lock of unkempt teenage hair. "Arise, Sir Davers, a [redacted] of the order of [redacted]." We share a sacred, secret handshake known only to our group. The ritual complete, I stand and am embraced by my fictive kin—family by choice, not genealogy. I feel a love and acceptance identical to familial bonds. These first true, time- and stress-tested friends remain by my side and to this day call me "Davers." Our secret society, I would learn only as an anthropologist much later, is called a *sodality* and serves a critical survival function on the band level of tribes.

Lesson: Ritual, symbols, music, codes, and nicknames sum up to make a group's creed. This creed functions as a proof of group membership.

My Early Awakening

I am twelve years old. My father leaves a book on my nightstand: Carl Sagan and Ann Druyan's *Shadows of Forgotten Ancestors*. After several days of it collecting dust, I take it with me to school and begin reading the thrilling saga that starts with the origin of the Earth and ends with the evolution of our species. Little do I know that this book is destined to alter the course of both my life and my father's. For my father, the book served as the catalyst to the ontologically shocking realization that his fundamentalist church was in error in denying the fact of evolution. Despite the trepidation that came with having to retrain for another career in order to support our family, this would lead to his decision in 1999 to resign his ministry.

Lesson: Even the strongest of tribal allegiances, grounded in the most profound sense of fundamentalist faith, are not immutable. If we no longer identify with a creed, we can relinquish its identity and find a new one.

My First Love

Tears well up in my eyes as the first woman I ever loved tells me: "David, it can't work. We can't be together. You believe in evolution. I don't. Our children would be confused. We can't live a life together . . ." After the encounter, I am driving home in the darkness of Indiana on a backwoods road. It is raining hard, making it difficult to see. The weeping becomes so uncontrollable that I have to pull over for fear of veering off into a ditch. I spend the night alone in the car awash in the anguish of loss.

Lesson: Your ideological belief system, a cornerstone of tribal identity, will influence whom you will love.

My First Win

It is the Southern Indiana football sectional championship, and we are playing on our home turf—Lidey Field at Castle High School in Newburgh, Indiana. I am the starting defensive end for the Knights; the jersey I am wearing features a proud blue-and-gold motif. There are only a few seconds left on the clock and we are up by three points. It is fourth down, and only several yards to the end zone. Harrison High School's dangerous and agile quarterback snaps the ball; I speed off the line of scrimmage far to the outside of the offensive tackle I had been fighting all game, and as he overcompensates his speed to catch up, I throw an uppercut move to his torso that uses his own momentum against him, sending him crashing to the ground. It is just me and the elusive quarterback now, and every sense in my body is tingling. I am completely focused on him. I send him crashing to the ground with the game-winning sack. I am lifted up by my teammates, and ten thousand screaming fans in attendance share in the sense of victory and conquest that we, "*the us-es,*" are having at the expense of "*the thems,*" our fallen enemy. Until that moment in life, I had never felt such glory, never felt such power or sense of purpose.

Lesson: Enacting socially sanctioned physical violence on behalf of one group over another can be intoxicating, thereby reinforcing an addictive quality to the behavior.

My First Political Debate with a Friend

I am shocked! One of my closest, most cherished friends is attacking me with conservative talking points about fiscal responsibility and barbs against the welfare state. I am in a debate with someone whom I love like a brother and whose intellect I respect, yet I feel—to the very core of my being—that he couldn't possibly be more wrong. I think to myself: *How can someone so smart be on the wrong side of history?* Being an anthropology graduate student, I have, up to this point, been baptized as a hardcore card-carrying liberal, and I viciously, with no small measure of righteousness, counter with my side's political talking points.

He looks disgusted, an expression I had never witnessed coming from him before. For the first time in my life, I feel a fracture in a cherished relationship and leave the debate angry and confused.

Lesson: If identity is in question, group ideology has the power to fracture historically strong relationships.

My First Hobby

I am on a field of battle, with a thousand other warriors dressed in full plate and leather armor, in Slippery Rock, Pennsylvania. Every year a crowd of twenty thousand congregate for two weeks in what is the equivalent to the Burning Man of medieval martial arts, called Pennsic.* Savage blows rain down on me. I am in the midst of an assault the likes of which I have never experienced. Sweat and dust sting my eyes. I am now exhausted and my shield arm begins faltering inch by inch with the constant drum of rattan swords and polearms beating across my upturned shield. An enemy column plunges deep within our ranks, revealing an open flank. My blood brother† is at my right; he pushes into my exposed left flank, his shield replacing mine. He absorbs the brunt of a charge from two huge adversaries in full plate armor who tower above us and who wish us a violent end. At this point, the enemy push has forced me down onto one knee. With a rage ranging from frenzied to berserk, my blood brother reveals his flank vulnerably and expels both enemy combatants with killing blows. I promise myself never to forget this moment, when a friend puts his well-being and chance of real bodily harm on the line for my sake.

Lesson: Modern humans will spend incredible sums of resources, including risk of bodily harm, to simulate in-group cooperation and competition with out-groups.

---

* The Society of Creative Anachronism, which sponsors Pennsic, was founded in 1966. It is an international living history group with the aim of studying and recreating medieval European cultures and their histories before the seventeenth century.

† Through ceremony and ritual, as in many cultures through space-time, we formally awknowledged each other as fictive kin. The power of ritual to such ends will be expounded upon in Part II.

My First Year of Field Work

I am in Uganda, Africa. I have been in the field for months now, observing, following, and studying a community of wild chimpanzees. I have followed them to the periphery of their common home range, and I can sense the group of males I am following and observing closely is agitated and anxious. The hair on the alpha male's body stands erect—a phenomenon primatologists call *piloerection*—as a pant hoot "oooowwaa waaaaaaaai waaaaaaaaai" from a rival community echoes across the "no man's land" between the two different, warring communities. The males closer to me pant hoot, fear defecate, and run back to the safety of their inner community home range.

Lesson: Ways of delineating "us" and "them" are ancient and predate the existence of our species.

My First Depression

In the fall of 2016 I have it all—professional, financial, romantic, and scientific success—yet I am more miserable than I have ever been. Through hard work, grit, and a fair amount of luck, I find myself in a prestigious postdoctoral fellowship that quickly morphs into a multi-year senior research scientist position at Duke University. I work in the lab of Charles Nunn, one of the brightest and most prolific minds in evolutionary anthropology. The work is intellectually challenging and deeply satisfying. We publish innovative, theoretically relevant work at a fantastic rate in respected journals. Durham, North Carolina, equidistant from the coastal beaches and Appalachian Mountains, is a beautiful and charming place to live. Moreover, for the first time in my life, I have financial stability. Despite all this, and having never been prone to bouts of melancholy, I am teetering on the edge of depression. It has been three years since I left Indiana for the pursuit of my career in academia. I left my best friends and my family and now I find myself tribeless and profoundly alone.

Lesson: Being too far removed from our kith and kin increases our vulnerability to a host of mental and physical ailments despite having all other extrinsic needs met.

My First Days Living with African Hunter-Gatherers

I speak with a Hadza elder. The Hadza are a group of indigenous people who have lived here, in north-central Tanzania within the central Rift Valley, for thousands of years. It has been days since I explained in broken field-Swahili the purpose of my research and the camp agreed to collaborate with me. This moment feels significant though, as it is the first time I am alone with the elder. He asks me why I am here. I tell him, "Because there are important lessons that your people can teach me and us Westerners about how to live better lives." He nods, and ponders my request. He responds: "We live together and depend on each other to survive. It is not always easy, and things between people can become heated, but we know we can rely on each other to survive."

Lesson: In environments analogous to the kinds our ancestors survived in, hunter-gatherers' only insurance policy against an ever-changing environment is each other.

My Exposure to the Power of Belonging

It is my first year of university. In the hopes of expanding our social network in this new place, my long-standing high school friend and I have pledged one of the local fraternities. As part of the pledging process, we have been working overtime for the fraternity for weeks. Another friend, an African American I befriended in my dorm, recounts to me a recent experience where he was trying to go to a party at this fraternity and they denied his entrance with a "Your kind aren't welcome here" denial of entry. When I hear about this, I confront the president of the frat and end up quitting after being dismissed with no meaningful response to the incident. When I compel my friend to leave with me, he, too, dismisses it. This is shocking to me on several fronts, but the one point that has always stuck with me is my high school friend was himself a minority. How could he want to stay?

Lesson: The drive to belong to a group can overshadow our personal identities. Ultimately, we are not born racist, but we are born coalitionist.

My Realization of Ignorance

It is the night of the 2016 U.S. presidential election and Donald Trump is winning. I am stunned, as I am realizing, along with a clearly shaken media, that their prognostications of imminent Democratic victory were to be unrealized. The next morning I wake up to find the final results. My mind reels as I contemplate the alternate reality I now live in. My previous certainty and hubris of what I thought I knew about the world crumbles before me. I feel utterly humbled. *My models of reality are inaccurate and corrupt*, I think to myself. I know in that moment there are forces in the world I do not yet comprehend and they have the power to affect nations.

Lesson: Entire societies can be driven by the power of tribalism.

WHETHER WE KNOW IT OR not, the Tribe Drive is part of us. In fact, its imperceptibility is part of its power, and it works best when you remain blissfully ignorant of its existence. In the next section, our aim is to derive a deeper understanding of one of science's greatest investigations: How do we get from something like life's blueprint—the ATCG(U) nucleotides that make up DNA and RNA—all the way to a behavior?

Reverse Engineering the Drive

The brain is the final conduit, the ultimate pathway, that mediates behavior. And yes, different scientific disciplines will attempt to explain the ever-elusive *why* behind behavior with their pet "*because*." *Because* of childhood experience; *because* of this or that hormone; *because* of some cue in their immediate environment; *because* of diet during gestation; *because* they had a bad night's sleep; *because* of the culture they were born into; (here is an infamous one) *because* of their genetics. These are all drives. The result of a drive is a *disposition* that is realized in the process of certain behaviors. Evolution crafts, for every species, the suite of drives that parameterize the dispositions that give life to our behaviors. If the set of drives is successful, it is passed along and the species thrives. If a set of drives fails, the species may dwindle and go extinct.

Fascinatingly, every specialist exploring any aspect of the way drives influence behavior is, more often than not, only a bit right; these drives are all intimately, exquisitely intertwined, and they all contribute to the final thoughts, feelings, and perceptions that ultimately produce the behavior.

The topic of drives carries a few burdens, several of which are stained with ideological dye. Let's cover them up front so that we can move forward more productively. The first is the common consternation of people who are troubled by any link between genes and behavior. Specifically, the claim that individuals with gene "x" deterministically exhibit behavior "y." Much of this concern is justified, as many discoveries produced by a plethora of scientific fields have been maliciously distorted, and pseudoscientific jargon has been used to justify atrocious ends; think here of all the bad *isms* (e.g., racism, sexism, and many more) that have plagued our species. On the opposite side of the spectrum, with exponential computational powers decoding genomic sequences of countless species with ease, splicing genes therapeutically, and outright editing the blueprint of life, we are living in a golden age of molecular discovery. This leads reductionists to lean heavily on genes as the irreducible part to a complex whole that carries ultimate predictive power.

Robert Sapolsky, a professor of neurology at Stanford University, notes the tension between these two extremes of reductionism and antireductionism.[2] On the one side, "such pseudoscience has fostered racism and sexism, birthed eugenics and forced sterilizations, allowed scientifically meaningless versions of words like 'innate' to justify the neglect of have-nots. And monstrous distortions of genetics have fueled those who lynch, ethnically cleanse, or march children into gas chambers." And on the other side, "overenthusiasm for genes can reflect a sense that people possess an immutable, distinctive essence . . . people see essentialism embedded in bloodlines—i.e., genes." As with most extremes, it is best to hedge bets in the middle.

A gene can drive a behavior *and* at the same time rarely, if ever, determine it. This fact is profound and it devastates a belief in genetic determinism. Making the claim that genes "decide" is like saying that a dinner recipe decides *when* to cook a meal. It's not genes but the environment that decides the *when.* Importantly, "genes aren't about inevitability. Instead, they're about context-dependent tendencies, propensities, potential, and vulnerabilities. All embedded in the fabric of the other factors, biological and otherwise."[3] Perhaps we can distill all these context-dependent factors down to a single concept—*dispositions.* An immediate environment triggers a vast network of (often competing and mostly unconscious) dispositions.

Relatively small changes to your genetic makeup can combine to create radical types of beings, including new species—yet rarely forecast behavior. Richard Wrangham, professor of evolutionary anthropology at Harvard University, writes: "Mostly, individual gene differences are only weakly predictive. . . . Genes can influence behavior; they rarely determine it."[4] The natural world gives us hints when we attempt to figure out the extent to which genes have the power to influence or alter our behavior.

Generally, genes prime traits to be expressed in certain contexts, but a good clue that genes are responsible for a human trait is when the trait is found in most cultures. Anthropologists call these *cultural universals.* Take, for example, a trait we will revisit often in this book—the disposition for hunter-gatherer societies to exhibit egalitarian social norms. Roughly speaking, foragers live in societies with mild hierarchies and context-dependent distribution of social powers between genders. On the topic, anthropologist Christopher Boehm writes: "The egalitarian syndrome . . . whenever we find a behavior that is universal among the fifty foraging societies . . . we can appropriately ask if it is likely to have some substantial (if less than wholly 'determinative') genetic preparation."[5] *Genetic preparation* is a keen way of describing a context-sensitive system that improves the chances of a species surviving a constantly changing environment.

Scientific reductionism (the attempt to explain a phenomenon by reducing it to smaller parts) is the running foil for thinkers who wish to insulate human free will from determinism. In the context of our discussion on drives, take one common anti-reductionist counter: "If evolutionists say aggression is adaptive, then logically war is an inevitability." Yet, this simplification overlooks the context sensitivity of the evolved system. Violent behaviors are not unstoppable dictates from tyrannical genes. Behaviors, both "good" and "bad," respond to circumstances: "Genes affect the size and sensitivity of different brain regions, the nature and activity of the physiological stress systems, the production and fate of neurotransmitters, and on and on. . . . A primate that invariably produced aggression as predictably as it went to sleep or felt hungry or pulled away from a smelly cadaver would quickly fail in the evolutionary game. The secret to successful aggression is appropriate behavioral flexibility."[6] In other words, context is king. For example, two people are in a fistfight. Objectively, both are engaged in violence. One is the aggressor and the other the defender. We often label the former a jerk and the latter a hero. One behavior is "good" and the other "bad," despite both people lighting up identical brain regions and activating their muscles to produce nearly identical behaviors.

On the Imperceptibility of Drives

Another difficulty that clouds the discussion of drives is that, for the most part, only a sliver of them are registered consciously as forces that cultivate our (enacted or suppressed) behavior.* Steven Pinker, professor of psychology at Harvard University, writes in his book *The Language Instinct*: "Our flexible intelligence comes from the interplay of many competing instincts. Indeed, the instinctive nature of human thought is just what makes it so hard for us to see that it is an instinct. . . . The effortlessness [with which we behave], the transparency, the automaticity are illusions, masking a system of great richness and beauty." Here is the challenge: since we don't often *feel* like we are driven by a disposition, it is that much easier to say we *know* we are not driven. Yet, neuroscientist David Eagleman states: "To know oneself may require a change of the definition 'to know.' Knowing yourself now requires the understanding that the conscious *you* [emphasis his] occupies only a small room in the mansion of the brain, and that it has little control over the reality constructed for you."[7] To make matters more difficult, we do not even have the capacity to *feel* a gene in action. We didn't even know they existed until a nineteenth-century monk, Gregor Mendel, meticulously recorded the results of his experiments on the inheritance of breed traits in pea plants. Donald Hoffman, a professor at the University of California, observes: "For the most part, these machinations of genes fly under the radar of conscious experience and foster, but do not force, a

* Many argue this is the basis of the illusory feeling of free will.

choice of action."[8] If genes don't determine behavior, can anything? What chance is there, then, that any other single drive *determines* any behavior? Likely zero. At any given instant, there are many (often competing) drives that ultimately influence the behavior.*

It is important to observe that many of these drives only work effectively as a consequence of being unconscious. Consider the suite of dispositions we call "love" and its role in finding and keeping a mate. UMass Boston professor Patrick Clarkin writes: "Love would not be very good at its job if it was left to rational choice, or if we knew where the off switch was. Instead, it is much more effective because it seems to grab us by the throat."[9] Critically, the property of a drive being unconscious is often a crucial component of its function. In fact, there is growing evidence that evolution purposefully crafted our (and every other organism's) perception system to hide reality from us, so that mercifully we would only have to focus on fitness-relevant stimuli; in other words, it protected us from a deluge of incoming data that would distract us from our prime directive—to survive and reproduce.[10]

Some readers may cringe at the thought that the only purpose of humanity is to live long enough to make copies of and spread our genes. To this point, Carl Sagan, former planetary scientist at Cornell University, and award-winning science writer Ann Druyan state: "Passion for life and sex are built into us, hardwired, pre-programmed. . . . So we are the mostly unconscious tools of natural selection, indeed its willing instruments." Because evolution has endowed us with intelligence, we possess the ability to probe and assess our behaviors after the fact. We tend to ascribe an (often flattering) narrative to our actions, but much of this is done as a type of after-action report. "As deeply as we can go in assessing our own feelings, we do not recognize any underlying purpose. All that is added later (if at all!). All the social and political and theological justifications are attempts to rationalize, after the fact, human feelings that are at the same time utterly obvious and profoundly

* The last burden to the topic of drives is one worth mentioning—but we won't broach it in any serious way, given it is one of the longest-lasting ongoing debates in the history of science—free will versus determinism. If behavior is the product of drives, where then is there room for free will for people to choose their behavior? It turns out, people have opinions about this topic.

mysterious."[11] In the same vein, former professor of biology at Harvard University Stephen Jay Gould wrote in his book *The Mismeasure of Man*: "It strikes us that a skeptical and dispassionate extraterrestrial ethologist studying our species might reasonably conclude that *Homo sapiens* are, for the most part, automatons with overactive and highly verbal public relations departments to apologize for and cover up our foibles." Put another way, we are fantastic at justifying our behaviors but bad at truly understanding their underlying causes.

As an evolutionary biologist who researches human sleep, circadian physiology, and cognition, let me provide an example from my scientific wheelhouse to further elucidate this concept. Did you know that if you are born between the months of January and March, you are more likely to suffer from a host of diseases throughout development and as an adult—including general pathologies, psychiatric disorders, and neurological illness?[12] This mystery sounds something akin to astrological signs predicting future life events, but the underlying biology of these patterns is fantastically more fascinating. Scientific observation has uncovered a circannual birth rhythm in humans. Human birth rates in preindustrial societies were tethered to the seasons, with one in particular standing out. If you copulate and conceive in the earliest part of the new year, your gestating fetus will be afforded all the lush, calorically valuable, nutrient- and mineral-dense nutrition of the spring season.

Perhaps this fundamental rhythm informs the feelings of restlessness we get when cooped up over a long, dark winter. We have a term for this phenomenon: *spring fever*.* We *feel* like we are choosing to get out of the homestead and explore our environment of our own volition, but could it not be that a flurry of impulses and interconnected physiological processes—a dash of hormone here, a cascading firing of neurons there—results in something that feels like the exercise of choice but in actuality is the product of subroutines driving your behavior? Sagan and Druyan poetically contemplate this impulse: "But from the inside it may well feel as though the weather is intoxicating, life is tempestuous,

* Apparently, the aphorism "the birds and the bees" is a springtime phenomenon that has real scientific backing.

and moonlight becomes you."[13] It's not like you are consciously thinking through the optimized parameters through which conception and gestation can be timed to provide the perfect birth window to maximize the likelihood of the preservation of your DNA . . . is it?

The Brain Wave

The very instant a behavior occurs, there is one pathway all information must travel along that informs how the behavior manifests—the nervous system. Thus, it is by using the language of neurobiology that we can come to some deeper appreciation for the ways in which the brain commands or silences the muscles that expel the behavior. Together let's explore an analogy that may help expert and novice alike appreciate the beauty and majesty of the brain wave action. We will liken the brain to an ocean, filled with water and swells of energy, that when combined produce experiences in consciousness we will call waves.

The nervous system has two main parts: the central nervous system, which is made up of our brains and spinal cord, and the peripheral nervous system, which is the nerves that jut and root off from the spinal cord throughout our bodies. Now think of this system as you would an ocean. Oceanographers call the ocean walls and floors the abyssal plain (the central nervous system), and the channels and deeper roots of this plain they call ocean trenches (the peripheral nervous system). Now, let's fill our ocean up with water. Just like a real ocean (filled with water molecules), so, too, our nervous system is comprised of atoms that when charged become ionic energy. Without these molecules, neither ocean nor brain would be filled with substance. In a real ocean, currents shape the direction that, by force, energy flows. Thus, the crucial cells of the nervous system—the neurons—guide molecular ionic energy that proves to be the currency of all sensation, perception, thought, and experience.

The nervous system (our ocean) is comprised of hundreds of billions of "brain cells" neurobiologists like to call neurons (our currents of water). As far as cells go, neurons are idiosyncratic and unique. Cells are usually tiny and relatively independent; neurons, however, are hyper-connected and can be bizarrely long with their special subparts;

neurons speak and listen to each other.* When a neuron talks to another neuron, they drive each other to excitation. If we analogize and anthropomorphize the neuron to be a little guy, the dendrites would be the ears (receiving information from other neurons) and the mouth would be the terminal nerve ending that talks to the next neuron.

The beginning of behavioral swell starts with the axon (which would be the torso housed under the cell nucleus that looks like a human head). The swell itself is a flow of information that if broken down to chemistry, Sapolsky would call a "*wave* [emphasis mine] of electrical excitation." This flow of information occurs when a neuron has gotten excited enough (stimulated by some input) to spread the charge toward the axon's terminal by allowing ions to flow simultaneously *in* and *out* with the sum result producing one of two conditions: (i) a positively or (ii) negatively charged dendrite. This interplay produces a stark contrast in binary states of the neuron. The neuron, in essence, is yelling at the top of its metaphorical lungs, "I have something to say!" (positively charged), or it is completely mute (negatively charged). Neuroscientists call the former *the action potential* and the latter *the resting potential.* Excitation—that is, the yelling of a neuron—hinges on the swell of the ionic wave having enough energy to push through the *axon hillock*, a specialized part of the cell body at the base of the axon. If all the summated inputs from the dendrites produce an ionic ripple (measuring in millivolts as between -70 mV and -40 mV), the tiny swell that started the ripple becomes a tsunami wave, opening up a massive migration of ionic flow (to 30 mV). The threshold is broken and the action potential realized. The neuron's digital system with its speak-or-don't-speak signaling yells: "It's go time!"

Thus, deep in the ocean's trenches, in the literal roots and shoots of our nervous system's spinal cord, you have neurons projecting neatly one by one in a line. But in the ocean proper (our brain), where a torrent of currents interplay, your typical neuron sends projections into thousands of others that receive inputs—each with their own neuron's axon hillock determining action potentials of their own. One finer detail to note with

* Spinal cord neurons in humans project several feet long; in blue whales, they can unravel to almost one hundred feet.

respect to the action potential of a neuron: for excitatory signals to have a global cascade, you need a host of neuromodulators that encourage the charge to break a threshold. Again, by way of analogy, a neuromodulator does not have the ability to activate or start a current, but it does have the capacity to excite (speed up) or inhibit (slow down) a current. The neuromodulators that inhibit ionic charges (succinctly called neuroinhibitors) work to silence the energy in the ionic waters.

Sapolsky notes it is the interplay between modulators that determines the break of the axon hillock by way of a swell of ionic energy: "You've taken a nice smooth calm lake, in its resting state, and tossed a little pebble in. It causes a bit of a ripple right there, which spreads outward, getting smaller in its magnitude, until it dissipates not far from where the pebble hit. . . . You can't get a wave, rather than just a ripple, unless you throw in a lot of pebbles." It's all a beautiful, interconnected network of convergent and divergent signaling; each current (neuron) impacts the (ionic) flow of the next current (neuron).

Next, we consider a swell. Swells are really important because, in this analogy, they are the suite of drives that ultimately influence a behavior. A swell, in the context of bodies of water like oceans, seas, or lakes, is a series of mechanical forces that propagate along the interface between water and air, resulting in surface gravity waves. You can't have waves without swells. Oceans would be perfectly placid without them. Swells manifest by way of internal (depth of the water) and external (force of the wind) competing energies.

If the swell of energy is powerful enough to make not just a single neuron yell loud enough to be in an action potential state but many neurons to yell in concert, the resulting excitatory cries produce a drive on a behavior—a wave in the ocean. Finally, and an important detail to consider with our discussion on drives, the threshold of the axon hillock can be cultivated over time to be either less or more excitable—in other words, more or less predisposed to yell in one direction or the other. What kinds of things have such neuronal priming power? Other *becauses* to the *why* of our behavior, such as nutritional state, plasticity gained by a good night's sleep, hormones, cultural information, and all the other myriad competing factors that inhibit or encourage an impulse to behave. Thus, as a matter of experience, when the swell of ionic energy breaks through the axon hillock, a sensation, thought, and/or feeling wells up as a wave object in consciousness. This is a point that may ring true for experienced meditators, as some meditation experts have argued that if there is a goal to meditation, it is not to identify with but to observe the object within consciousness for what it is—something that was compelled to your attention by your conduit, the brain.

Bear with me as we wade into old waters, because we need to take a brief moment to talk about a question that has been analyzed and debated by philosophers and scientists for millennia: What is consciousness? Sam Harris, a neuroscientist and meditation expert, likens consciousness to water and the wave to an object that appears within it: "Everything in some sense, is made of consciousness as a matter of experience. Everything you notice is like a wave in the ocean inseparable from water. To speak of water without waves or waves without water is to ignore the actual character of what's happening. Everything that appears by virtue of it appearing is a kind of wave upon the surface of consciousness."[14] To extend the metaphor, the swells and waves are made of the water, so the swell is like the drive producing the behavior, while the wave is the act of perceiving a feeling or sensation that is directly tied to the excitation of the brain. Thus, the sum of the ocean—the water, the swell, and the wave—is that which permits both conscious experience and the ability to act on, or inhibit, a behavioral drive.

RECALL THAT WE OPENED THIS chapter with a goal: to gain greater understanding of the Tribe Drive by exploring the nature of its imperceptibility. As a reminder, there are two reasons to do this. The first is that scientific exploration is a just pursuit in and of itself. Knowing more about the natural world is awesome! But we want to go beyond the joy of understanding. As Eagleman notes in *Incognito: The Secret Lives of the Brain*, "We are not conscious of most things until we ask ourselves questions about them." Thus, we want to ask questions about tribalism to ultimately become aware of its properties so that we can successfully apply knowledge in a way that improves our moral landscapes. We want to use that understanding to better our lives and brighten our societies. The only way to do that is to align incentives to drive the best, over our worst, behaviors. We've now circled back to the prophylactic. Perhaps most important for the survival of our species, the only real defense we have against being manipulated by our drives is knowledge of their existence. As observed by a professor of psychology at Simmons University, Sarah Cavanaugh:

> How much more of human behavior is determined unconsciously, based on instinct and on motivations outside of our awareness, than we may have ever thought before. So probably the largest difference between humans and honeybees isn't whether unconscious impulses drive a lot of behavior—which is likely true of both bees and humans—but rather that we humans have evolved to understand these impulses and act to override some of them. . . . We can model Ulysses and metaphorically latch our more impulsive selves to the ship's mast so that we do not harken to the call of the sirens.[15]

Some of science's greatest accomplishments have been the discoveries that permit us deeper understanding of our drives. Without these revelations, we remain slaves to their commands. Understanding a new drive is like finding a mental key to a lock of an invisible shackle we didn't even know confined us. With this new knowledge, we can take a bit more ownership of our individual and collective destiny.

The next chapters of the book will give you the key to unlocking the shackles of the Tribe Drive. Put the key in the lock and turn.

2

TRIBALISM MISMATCHED

The Sea Turtle Brigade

A substantial proportion of human misery is probably due to genetic and cultural mismatch with our current environments.
–Elisabeth Lloyd, David Wilson, and Elliott Sober, 2011

Moths are unprepared for a world with windows.
–Carl Sagan and Ann Druyan, 1993

A Beacon of Deadly Light

An egg begins to stir. The fault lines of a shell shear against each other in tension. Intermittent cracking sounds reverberate through the cool, salty air, punctuating the ebb and flow of the waves on the seashore. Several tiny hatchlings scramble over each other, some lazily, others in earnest. Freshly minted one-and-a-half-ounce leatherback turtles are mostly flipper, with tiny heads and paddle-shaped appendages protruding from a smooth, striped shell. The flippers span an impressive four inches, twice the length of their body. They are awkward as they stumble over each other, seeking something, compelled to move forward. By what streams of information do these "lutes"* channel the energy of their new bodies to drive them to the place nature intends for them to thrive and flourish? What is the signal? What is the environmental cue on which their very survival depends? That signal is light. The "Sea Drive" is an ancient call for this kind of animal.

* So called because of the seven ridges that run the length of their backs that resemble the musical instrument of the same name.

A few of the early hatchlings move stridently toward the sea, as the brightest beacon of light reflects off the cresting waves on the horizon. These fortunate little lutes will grow to live remarkable lives as one of earth's most charismatic and enchanting animals. When they reach adulthood, some will span six to eight feet in length and weigh close to two thousand pounds. Evolution will have provided them with a perfectly adapted teardrop-shaped body, sporting the most hydrodynamic design in nature. The swimming sea turtles will outpace all of the planet's reptiles in speed by traveling up to twenty-two miles per hour. Sadly, most of this batch of hatchlings will not be so lucky. They are about to encounter a perilous trap. A trap that has happened to countless species across eons of time.

They encounter a new environment, disrupted by rapid change. As hours pass, and dusk approaches, the moon shrouds behind cloud cover, casting darkness across the seashore. With disoriented squeaks, the host of remaining hatchlings begin scrambling in circles. In the blink of a small reptilian eye, the lights on the seashore, in buildings, in homes and streetlamps are turned on in tandem. The hatchlings, perhaps with some trepidation, feel the pull . . . the drive toward the lights. Something deep within them is manifesting—an impulse—they must move toward the light. If the lights on the seashore overpower the light from the moon, or if the moon is hidden behind the clouds, then the sea turtle can only do what evolution, over millions of years, has prepared it to do. This tribe of lutes must follow the only cue that has served their ilk for countless generations—they follow the light.

The false light compels them forward and as they move deeper and deeper inland, they find themselves in dire straits. The brothers and sisters straggle together, and blindly follow the lights across a street where cars crush many of their siblings. Several survive the pernicious crossing, but their numbers mean little on land. Here, the fox, the snake, the owl, perceive the world in high fidelity and with few errors of perception. The disoriented and mismatched squeaks of the little troupe of sea turtle hatchlings do nothing but alert predators to their presence. Those baby turtles are now preyed upon from sky and land until the last of the ill-fated hatchlings are dragged into the darkness of a forest glen and eaten alive.

Sea turtles of the species *Dermochelys coriacea* are critically endangered. It turns out that there is an instinctual behavior, honed by vast swaths of time, that leaves them vulnerable to self-annihilation. In a twist of evolutionary fate, the species is turning into its own greatest enemy. Thousands of sea turtles on Florida's Atlantic coast prematurely die due to artificial, human-constructed light. The instinct that was a strength in the past dooms many of them to death in the present. Clearly, the species finds itself in a prickly *mismatched* scenario* for which evolution did little to prepare it.[1] The species appears destined to go extinct.

Or is it?

What Does It Mean to Be Mismatched?

How many of all of the species of life that have ever existed have gone extinct?

The profound and unnerving answer to this question is *almost every species that has come into existence has gone extinct.* If we were to tally the pool of living (extant) species studied by biologists and compare it to the number of dead (extinct) species dug up by paleontologists, the latter would be approximately 99.9 percent of the total.[2] It is no wonder that evolutionary biologists have been engaged in a spirited investigation over what makes species go extinct. Yes, they die off, but why? What are the mechanisms? Recently, the field has been injected with new vigor as an idea has emerged to help us make sense of this complex question.

Evolutionary mismatch is a crucial scientific concept that helps inform our understanding, not only of all life on this planet, but of the human condition. Mismatch is the idea that organisms are adapted for how things *were* in the past, not how they *are* in the present. Elisabeth Lloyd, the Arnold and Maxine Chair of History and Philosophy of Science at Indiana University, and her colleagues define evolutionary mismatch as "a negative consequence that results from a trait that evolved in one

---

* I first became aware of this example of mismatch from The Evolution Institute's *This View of Life*, edited by the experienced science educator, evolutionary biologist, and author David Sloan Wilson.

environment being placed in another environment. . . . Traits that were originally neutral or even deleterious can potentially acquire new harmful consequences in an altered environment."[3] In other words, when a species finds itself in a mismatched state, it is the result of a trait that evolved in one environment becoming maladaptive in another environment and therefore reaches a state of disequilibrium.*

In our baby sea turtle example, the hatchlings on the beach that have evolved to use the light reflecting off the water's surface (the original environment) are becoming disoriented by the lights from the beach houses and buildings (the new environment), causing them to fatally head inland. Thus, evolution had crafted a finely tuned perceptual system for the environment at the time, but in the new environment there is no guarantee that the population will be able to respond in a way that outmaneuvers extinction. As Carl Sagan once wrote: "Extinction is the rule. Survival is the exception."[4] But there are exceptions; to save the turtles from themselves, an intervention is required.

Evolutionary mismatch, conceptually, is a relatively new, but critically important theory for scientists who study patterns of change through space and time; it is an increasingly common framework not only for natural systems in general, but also within which to study species living in human-altered environments. If it can happen in turtles, can it happen in humans too? Absolutely. One of the great challenges of the twenty-first century is to understand mismatch in relation to our own species. This chapter is an outline of the emerging scientific understanding of the several ways that humans are in new, unexpected environments of our own making.

Humans are unique for many reasons. One of our greatest strengths is the rapidity with which we can shape our technologies, cultures, and societies. We have drastically altered our own environment in the quest for survival, but this has left us particularly vulnerable to mismatches that can occur on many levels. We are not the only species to face the

* To put it in a scientific framework, the original environment (E1) commonly called by evolutionary anthropologists the "environment of evolutionary adaptedness" or "ancestral environment," was where there was strong selection on trait (T). A species can find itself in mismatch if a novel environment (E2) comes along in such an abrupt manner that it requires too many generations to solve the problem.

challenges of evolutionary mismatch—but our circumstances are somewhat unique given the pace and scope with which humans can alter their environments. Highlighting the extent of this change, in 2008 a transition ten thousand years in the making came to a head when more *Homo sapiens* dwelled in major cities than any other type of human settlement.[5] Is it possible that we humans find ourselves similarly, and unwittingly, mismatched to a modern-day beachhead of our own making? I believe the answer is yes, and that an understanding of the Tribe Drive is at the core of an approach to get us unmoored.

Finding the Mismatch Signature—The Case of WEIRD Diets

Twelve percent of the world's population is argued to be some of the most psychologically unusual people on Earth. You count among them if you happen to live in a Western, Educated, Industrialized, Rich, and Democratic (WEIRD) country.[6] Joseph Henrich, professor of human evolutionary biology at Harvard University, and his colleagues have pointed out that a vast majority of research studies recruit WEIRD subjects. It makes sense. It's where most science is performed. Psychology is especially prone to the WEIRD bias with 96 percent of its subjects recruited from WEIRD populations—most of which are a uniquely strange cohort: university undergraduates. Importantly, this is the group most likely to be influenced by rapid changes in social and technological innovation. It stands to reason that this group also suffers more than most from mismatched states. If we will find any evidence for mismatch in our species, it will be in WEIRD humans.

What is the signature of mismatch? What data manifests as a smoking gun for us to target our potential intervention? Fortunately, the evidence required to demonstrate a case of evolutionary mismatch, in some cases, can be documented to the level of any scientific fact. Our trait of concern should be shown to positively influence a species in the past ancestral environment yet to no longer confer an advantage—and thus is out of sync—in the present changed environment. To demonstrate the former with an example, conservation biologists observed the fatal attraction to the glassy surfaces of solar panels by aquatic insects.[7]

These insects find themselves utterly attracted toward the bright polarized light of the reflective human-made surfaces—even more so than their ultimate goal of getting safely to the water's surface. In choice study experiments, scientists documented that the solar panels were more reflective than water, acting as a type of insect super-stimulus that overrode their senses. The scientists hypothesized a mismatch mechanism. They reasoned that the insects' senses had evolved to find large bodies of water and predicted the insects would not be triggered by reflective surfaces enclosed by small non-reflective boundaries. With a simple and cheap solution, the scientists intervened by adding white non-polarizing borders, resulting in a ten- to twenty-six-fold decrease in the insects' attraction to the solar panels. Previously dying in droves, the insects were restored to equilibrium. *This is the mismatch signature*: find the cue that causes the mismatch and test to see if alternative cues reduce the problematic behaviors.

Yet, it is important to consider that an out-of-sync trait doesn't need to be immediately fatal to be considered mismatch. For example, think about a human disease that only manifests late in life; it may not affect evolutionary success because later in life we have already had children, yet still be well worth an early intervention to ultimately reduce human suffering. Let's take a look at such a disease.

Shockingly, among the six thousand species of toothed mammals who eat their natural diet, there are no examples of tooth decay! None. If only one group of toothed mammals in six thousand experiences a pathology, it takes no Darwinian Sherlock Holmes to deduce that something has gone awry. Clues like the human outlier of tooth decay point to the conclusion that diet may be among the strongest contemporary cases of evolutionary mismatch.[8]

The obesity epidemic can be explained as a consequence of throwing a primate that evolved in environments without ready access to sugars, fats, and complex carbohydrates into a postindustrial world where such resources are available cheaply, and on every street corner. Our ancestral diet was rich in whole foods. In the ancient food pyramid, the bulky base of our diet consisted of wild tubers, game, greens, and fruits that were accessible only during seasonal windows of availability. For those of our lucky ancestors who were coastal-dwelling people, seafood and sea vegetables abounded. For more than three hundred thousand years, our teeth, jaws, and guts processed these foods. It is worth reiterating the importance of teeth (given their excellent preservation in the fossil record) as they are an archaeological smoking gun. Teeth are minerals and thus already made of the stuff of fossils. In essence, you are walking around with fossils in your mouth, and what you put into it determines the patterns of wear and tear your teeth will accrue over a lifetime's use. The notable times you find other non-human mammals with decay (such as gorillas, baboons, and bears) are almost always associated with the consumption of human trash. As noted by the dentist John Sorrentino: "This screams at anyone paying attention that tooth decay is a disease of environmental mismatch." So, too, is the modern obesity epidemic.

The cravings that drive overconsumption result in poor downstream health outcomes such as type 2 diabetes, stroke, gallbladder disease, osteoarthritis, mental disorders, cancer, and these outcomes increase to all-cause mortality. But it turns out the kinds of food we overconsume are the most powerful predictors of whether or not they are bad for our health. Hunter-gatherers consume an incredible breadth of varying proportions of carbohydrates, fats, and proteins and still show a near-absence of non-communicable disease.[9] The overarching pattern revealed from these global reviews of food consumption is that "food should be alive." That is, all ancestral diets have two things in common: they are whole, meaning no milled grain flours or refined sugars, and the foodstuff—be they plants or animal—were recently living things. It is the *degree of processing* that appears to be the point of mismatch with our physiology.[10]

We can point to an added dimension of the dietary mismatch example: our guts. Plant cells differ from processed foods (derived from sugar and flour) in one dramatic and essential way—they have built-in defenses. Indeed, by mass, plant cells top out at carbohydrate storage around 25 percent, whereas flour and sugar are super-dense nutrients, which, as a trade-off to the caloric watershed they provide us, unfortunately also serves as a growth medium for nasty microbes that can harm our bodies' natural biome. Just like our ancestors (and unlike artificially processed foodstuff) plants have battled and survived through hundreds of millions of years to keep bacteria from invading their treasury of carbs. Bowels that are regularly fed a whole foods diet that emphasizes plants have bacterial loads that reflect a well-defended environment where pathogens cannot prosper. It stands to reason that high-processed diets may be causally linked to autoimmune conditions that create a pro-inflammatory response, given that the lymphatic tissue that houses the body's immune cells are in constant, chronic exposure to a pathogen-facilitating biome.

Analogous to reflective surfaces for aquatic insects or artificial light for sea turtles, in the case of the human diet, the super-stimulus is tasty, sweet, calorically dense processed food. In essence, by having easy access to any kind of food our imaginations can innovate, WEIRD societies provide daily food choice tests in a grand experiment that no human was either prepared for or consented to: Do we eat crappy or healthy foods? McDonald's or vegan? Walmart or Whole Foods? Pizza or veggies from the garden? Every meal is a chance to be in or out of sync with our evolved bodies. In the words of philosophers Pierrick Bourrat and Paul Griffiths: "Organisms track environments across space and time on multiple scales in order to maintain an adaptive match. . . . Failures of adaptive tracking lead to disease."[11] It appears we've lost track.

To describe the delicate interlocking nature of evolution's dietary balancing act, the metaphor that Ian Spreadbury, a Canadian neuroscientist at the Gastrointestinal Diseases Research Unit at Queen's University in Kingston, Ontario, provides us is that of three dancers moving synchronically together through space and time: "We've long suspected that natural foods are better for health, but there are now clues that living foods might play a crucial part in a delicately evolved dance between host, microbe, and food. This dance needs all three dancers to be up to speed on the steps. The sudden replacement of one element with a slurry of flour might trigger a cascade of microbial changes that lead to overweight or disease in the susceptible." This illustrates an important concept for anyone interested in realigning themselves away from any state of mismatch and thus creating a healthier life—all dancers need to be in step. There needs to be balance. States of health and well-being rest in this balance. We need to find equilibrium in the dance of life. We need to be *in sync.*

"For Sale: A New Way of Life"—The Corporate Invention of the Modern Nuclear Family

One humid spring day in 1950, the twenty-square-mile strip of land in Hempstead, Long Island, became the birthplace of suburbia. This was a housing project unlike any the world had ever witnessed. In an assembly-line construction method, men organized as specialized units followed each other with antlike precision from house to house, completing incremental steps in construction. Each home was reduced to its minimal parts, only taking a rapid twenty-seven steps to complete. Literally thousands of houses were being built, not in sequential order, but at the same time. At the construction's frenzied peak, a suburban house was being completed every eleven minutes. The development was eponymously named Levittown after its conceiver William Jaird Levitt, known by many as "The King of Suburbia."[12] He and his brothers built a sprawling suburban empire that changed the very fabric of American life.

The Levittown development heralded a massive transformation of America's social and physical infrastructure. It did for homes what General Motors had done for cars—turn their construction into a factory assembly line. The Levitt brothers were proud to say they did not *build* homes, they *manufactured* them. And they did it ruthlessly. To lower costs, they froze out union labor—a move that provoked picket lines. They also cleverly cut out the construction middlemen of the day by purchasing lumber and spray paint directly from manufacturers. Starting in 1947 and ending in 1951, they manufactured more than seventeen thousand homes, which became the cookie-cutter model for every future postwar planned development. A review from *Time* magazine in 1950 remarked: "The community has an almost antiseptic air." And Levitt took steps to ensure the "purity" of his developments by way of segregation. Levitt, of Jewish descent, refused to integrate his developments, barring Jews (for a time) and African Americans from his homes, and even forbade resale properties to Blacks by using restrictive contracts. This, too, provoked picket lines, but this time from civil rights activists.

There is a hint of conspiracy to the cabal that got together to create an intentional *anti-community* community: during the war President Franklin D. Roosevelt and a think tank of psychologists helped lay the groundwork for the Levitt brothers to scheme up a way to deal with a serious threat to public order—the men returning from the war. Afraid of the destabilizing force of men congregating together—be it by forming unions or painting the town red—they dreamed up Levittown to draw men into a more controllable space. And thus, the suburb was born. Douglas Rushkoff, professor of media theory and digital economics at the City University of New York, highlighted the importance and deep-running implications of the creation of the first suburban tract housing areas on Long Island: "They were specifically concerned about men congregating. They wanted to prevent veterans returning from WWII from organizing labor or drinking together—they invented an intentionally de-social community." According to Rushkoff, this was a way to reduce the opportunity of assembly.[13] There may be something to this assessment, as Levitt said of his project: "No one who owns a house and a lot can be a communist. He has too much to do."

By 1950, 80 percent of Levittown's men commuted daily to Manhattan. With the passing Housing Act of 1949 that allotted credit toward every American with 5 percent down (0 percent if you were a veteran), a thirty-year mortgage was now available for anyone who wanted their very own "modern" home. A white man could walk up to the table with $400 and leave with a Levittown home. There was a great pent-up demand, and the landscape around the average American city was transformed within the decade to fit the suburban commuter model pioneered in Levittown.[14] On July 3, 1950, a cover story in *Time* magazine on Levitt was subtitled "For Sale: a new way of life."[15]

But by now, we know better than to assume every rapid environmental change is good. Many, in fact, can be downright dangerous.

The Social Mismatch Hypothesis—Death by Loneliness

In the same way that moonlight is overwhelmed by roadside lamps, light reflecting from solar panels is brighter and more alluring than light bouncing off of a body of water, and the craving of another bite of processed cheesecake overwhelms the need to finish eating your broccoli, so, too, does dwelling in domiciles that are large, independent, security system–protected, socially disconnected, and isolated put WEIRD humans in a mismatched state. This is the social mismatch hypothesis.

But if this is the hypothesis, what is the supporting evidence? The signatures of social mismatch in those of us that dwell in WEIRD societies are myriad. Our physical and mental health and well-being are affected. Even the meaning and purpose we derive from life is influenced. But all these threads intertwine with a fundamental factor that mediates the others—*loneliness by isolation*.

Let's start with happiness. The data from WEIRD societies indicates something astonishing—the safest,* wealthiest, most privileged people of any group of humans that has existed since the dawn of our species are experiencing a *happiness deficit*. How does that make any sense?

Steven Pinker authored the 2018 book *Enlightenment Now*,[16] a data-driven paean to the gains our species has made, which offers a cutting-edge analysis of how Enlightenment values (reason, science, and humanism) have ushered in unprecedented progress.† His thesis has stirred

* I believe the alienating effects of wealth, safety, and security that modernity has brought us have slowly percolated and spilled out into our culture. The number of horror films with zombie hordes obliterating modern society by blasting it into a postapocalyptic survivalist dystopian world has substantially increased since the mid-1990s. In the year 2000 there were 358, and by 2006, no fewer than 874 fear flicks per year were being produced. The reason why zombies are so compelling is because the disaster that hurtles the protagonists and their allies into a post-apocalypse leveling of modern life returns them to having a natural predator—the zombie (or whatever monster). By watching these people's stories we are actually running a thought experiment of having to rely on our own abilities for survival in an environment that mirrors the environment of our paleolithic ancestors. It is notable that most resolutions to these plot lines end with humans mastering strategies to counter their zombie predators and getting on to the business of crafting small-scale social order that recapitulates hunter-gatherer lifestyle—camps, bands, and tribes.

† The seventy-five graphs within meticulously detail a counternarrative to the commonly held public perception that the world is in poor shape. In fact, the data show the exact opposite. Pinker uses fifteen different measures of human well-being to bolster his argument that life is quantitatively getting better for most people in most places. The least controversial among these arguments is the statistically undeniable fact that compared to previous generations, people live longer and healthier lives.

up quite the controversy, and the most common attack against him is that he is a Pollyanna that sees the world in excessively, and unjustifiably, optimistic ways. I do not agree with his critics, but for the sake of argument let us assume he is being foolishly or blindly optimistic. It makes the assessment he presents in his chapter titled "Happiness" even more striking. For all our gains, the typically optimistic Pinker notes that after correcting for increases to GDP, Americans "punch under their weight" when it comes to happiness:

> But are we any happier? If we have a shred of cosmic gratitude, we ought to be. An American in 2015, compared with his or her counterpart a half-century earlier, will live nine years longer, have had three more years of education, earn an additional $33,000 a year per family member (only a third of which, rather than half, will go to necessities), and have an additional eight hours of leisure each week. If popular impressions are a guide, today's Americans are not one and a half times happier (as they would be in happiness tracked income), or a third happier (if it tracked education), or even an eighth happier (if it tracked longevity). People seem to bitch, moan, whine, carp, and kvetch as much as ever.

The United States and United Kingdom are outliers from the global trend in subjective well-being. Factoring in life satisfaction and income, the wealthier individuals are, the happier they tend to be. In a 2015 survey, despite ranking third for average income (behind Norway and Switzerland) the United States came in thirteenth place in happiness. The U.S. trailed three Commonwealth countries, eight countries in Western Europe, and Israel. The United Kingdom, comparable in average wealth per person (and arguably the most culturally similar Western European country to the U.S.) ranked twenty-third in happiness.[17] It begs the question—why are the countries that have seen the greatest amount of technological and financial prosperity (with concomitant rises in safety and security) not as happy as they should be? Is it, perhaps, something we are doing?

According to *self-determination theory,* humans need three foundational elements to live satisfactory lives. They need to feel that they are competent in their work, they need to feel that their life is being lived authentically, and they need to feel that they are socially connected. These are what positive psychologists call *intrinsic* values; the emerging science of happiness and well-being shows us that they are significantly more powerful predictors of a life well lived than *extrinsic* values such as money, status, and beauty. In an insightful and revealing commentary on the human condition, war journalist Sebastian Junger, in his 2016 book *Tribe: On Homecoming and Belonging*[18] noted a curious historical trend of seventeenth-century America. White captives of indigenous tribes, if returned to Western pioneer society, would often escape and return to the Native American lifestyle to which they had originally been taken captive. In contrast, there was no counter example of a Native American taken captive into early pioneer society deciding to remain in any colonial society after being freed.

Benjamin Franklin pointed out there were numerous settlers who were captured as adults and still seemed to prefer Native American society to their own. "And what about people who *voluntarily* (emphasis his) joined the Indians? What about men who walked off into the tree line and never came home? The frontier was full of men who joined Indian tribes, married Indian women, and lived their lives completely outside civilization." As pointed out by Junger, a French immigrant named Hector de Crèvecoeur offered an answer: "There must be in their social bond something singularly captivating and far superior to anything to be boasted of among us." Likely, by this stage of America's cultural evolution, the allure of extrinsic values of material wealth, individualism, and self-determination had taken root to the point where some were willing to abandon them to better align their intrinsic values with their lives.

Laurie Santos, professor of psychology at Yale University and director of the Comparative Cognition Laboratory, teaches the most popular course on this subject offered to date, titled The Science of Well-Being. In an interview, Santos expressed concern for national-level trends in Gen Z's (born roughly between 1995 and 2015) well-being: "I look at young people today who mortgage their happiness to a future state: *when they get into the perfect college, when they get married, when they get the perfect job.* It is the case that we are dealing with a generation of incredibly fragile students in terms of mental health.[19] The most recent 2018 national college health survey[20] had the following stats:

1. Over 40 percent of college students report being too depressed to function most days.
2. Over 50 percent say they regularly feel hopeless most of the time.
3. Two-thirds sáy they are overwhelmingly anxious.
4. Over 10 percent of college students today have seriously considered suicide in the last year.
5. The rates of depression have doubled since a 2009 study.

It is deeply sad and ironic that the modern cohorts of *Homo sapiens* who have the most material wealth and safety of any humans who have ever lived are the least happy, and they are unravelling into anxiety and depression. Santos continues her assessment: "When young people hear the science of well-being, that you will hedonically adapt to the perfect job and perfect marriage, they are angry because of how much they have sacrificed to get here."* For those students, how much of that sacrifice is levied from building social capital as an opportunity cost? Probably most of it—why make a new friend when what you really need is a nice car and a high-paying job?

Compare this to stunning reports of extremely high rates of happiness in individuals who experience a dangerous life of privation and join terrorist groups; one must wonder what a debilitating state we are in when an ISIS combatant has greater perceived happiness than a sophomore at Northwestern University. Clues emerge from their sophisticated and effective recruitment video *There Is No Life Without Jihad*, which reports they are a strong social collective, fighting in unison to protect Muslim children and women from being disgraced by the out-group Westerners and liberalizing Arabs within their own countries. The youthful Jihadist males are portrayed as heroic. The video pans over them, sitting together

* The hedonic treadmill, also known as hedonic adaptation, is the observed tendency of humans to quickly return to a relatively stable level of happiness despite major positive or negative events or life changes.

as a collective group of comrades, and it is clear they are enjoying being in each other's presence. Their purpose is clearly stated: "Oh you who believes, answer the call of what gives you life. And what gives you life is Jihad. Allah does not need you, you need Allah, and to sacrifice for him. . . . You will feel the *happiness* [emphasis mine] we feel." Another youth describes how he felt that he was nothing before he took on Jihad and began reading the Qur'an. He states that he knows that those living in Western society feel depressed and that if you only come to Jihad with ISIS, you will be welcomed and happy.*

THE LONELINESS PANDEMIC OF THE twenty-first century can be traced in origin to the first day of 2020; with the closing of the Huanan seafood market in China due to the novel coronavirus, the human species began one of the greatest experiments in social isolation the world has ever seen. By February 11, 2020, the World Health Organization gave the mysterious novel coronavirus its official name: COVID-19. The director general, Tedros Adhanom Ghebreyesus, called the outbreak "a very grave threat for the rest of the world."[21] China responded with the largest quarantine ever enacted in human history. Eleven million people were put on lockdown to contain the spread of the virus. The quarantine lasted seventy-six days. By March 16, 2020, the infections and deaths outside China surpassed those within. By mid-March, the world's countries began enforcing and adopting significant social distancing measures to "flatten the curve" of the virus's spread. By April, North Americans were feeling the reality of the situation with the U.S.–Canadian border closed, and the enforcement of a temporary closure of all non-essential business and institutions. By this point, the stock market had collapsed into a bear market, and the global economic impact reached tragic proportions. In a grand experiment, public health officials asked us to do something that does not come naturally to an

* Picturing this scene, I cannot help but think that when a terrorist is giving us advice on how to live a happier life, we should at least acknowledge that something strange is going on.

ultra-social species—avoid large gatherings and close contact with others.

Self-quarantine appears to be an acute version of what has been happening by a trickle since the industrial era—Western society has been slowly, but methodically, eroding the power of the small social group in favor of the individual. As noted by Jacqueline Olds and Richard Schwartz in *The Lonely American: Drifting Apart in the Twenty-First Century*: "The movement in our country toward greater social isolation is subtle. It is especially easy to miss in everyday life because we, as a people, try to let others live their own lives. We don't believe that we are our brothers' keepers. If neighbors seem to have disappeared into their houses or apartments, we treat it as their own business. Who are we to interfere?"[22] Since the advent of the postindustrial era and the spread of capitalism throughout the developed world, norms have been changing in favor of greater individual self-isolation; thus, I saw the COVID-19 pandemic as an opening of the isolation floodgates. This worldwide experiment in social distancing diminished our already eroded capacity to experience face-to-face contact in meaningful social groups. The consequences will surely impact our society in deep and lasting ways.

A CHIMPANZEE IN ISOLATION IS a pitiful thing. Unfortunately, due to the shortage of regulation of non-human animals in captivity, the twentieth century provided many opportunities to observe chimpanzees ripped from the social fabric of their lives. Mostly, this was the result of research chimpanzees being housed in woeful conditions that resembled a solitary confinement cell more than a home within which to live some semblance of a life. The signs of a socially isolated chimpanzee are telltale: they rock in place, pull out their hair in patches, grotesquely pick at self-inflicted wounds, and become coprophagous.* These are all behaviors that indicate a loss of sanity in humans, and worst of all, they are what some humans also do during social isolation. In its worst manifestations, a human being in true isolation is just as

* Layman's translation: they eat their own shit.

pitiful. There is a reason why severe punishment for prisoners is to be sent to solitary confinement, isolated from other. Yet, throughout the twentieth and twenty-first centuries, Americans have been moving in that direction, inch by social inch, becoming more and more self-isolated. Robert Putnam, a political scientist and professor at Harvard University and author of the 2000 book* *Bowling Alone: The Collapse and Revival of American Community*, analyzed data that suggests community structures in the United States had been atrophying throughout the twentieth century.[23] He discovered that virtually all forms of "family togetherness" (including eating, watching TV, and vacationing together) decreased throughout the last quarter of the twentieth century. Strikingly, in the ten-year period between 1985 and 1994, active participation in community-level institutions decreased by 45 percent.

Even the way couples geospatially orient themselves has taken a more isolated bent. By the 1990s, the acronym *LATs* was coined to describe couples "living apart together." By analyzing data from the 2004 General Social Survey (GSS), Miller McPherson and his team at Duke University unearthed trend lines of loneliness in the United States. Between 1985 and 2004, the number of Americans that said they had no one to discuss important matters with tripled, with the overall average dropping to two people whom they considered confidants.[24] Shockingly, nearly a quarter of those surveyed recorded they had no one.

In addition, the U.S. census has observed a continual rise in single-person homes since the 1940s at 7 percent, in the 1960s at 13 percent, in 2000 at 25 percent, and in 2018 at 28 percent, approaching nearly one in three and totaling 35.7 million single-person households. This trend is exacerbated by metropolitan living. In the city I dwell in, Toronto, 32.3 percent of households are one person (compared to the Canadian national average of 28.2 percent)[25]; in New York, specifically the island of Manhattan, nearly one in two households are one person.[26] Even starker than the LAT phenomenon is the practice of "sologamy," which

* Cited more than an astonishing sixty-five thousand times; somebody thinks social isolation in the United States is a problem!

has been coined as a term of "self-empowerment" to express to others that you do not need a partner to live a fulfilled life.[27]

It is not just an American phenomenon. Sweden was highlighted in the OZY article by Fiona Zublin "How Do You Stop a Plague of Loneliness?" Despite being one of the strongest welfare states in the world, Sweden reports a population-wide level of loneliness of around 40 percent. One in ten people die without being buried by their family. The Swedish Theory of Love was attributed for this rise in loneliness, given women were not wanting to cohabitate to reduce the likelihood that they would end up doing the housework and emotional labor for two. Historian Lars Trägårdh coined the above term for the idea that all relationships should be chosen, not maintained out of cultural loyalty, duty, or need.

It is not just Sweden, either; this is a postmodern idea that has swept many economically developed countries, as the number of people worldwide living on their own has increased after the turn of the century by 33 percent.[28]

In 1970, Philip Slater published his book, *The Pursuit of Loneliness,* in which he encapsulates the American ideal of privacy and individual freedom over social connection:

> We seek a private house, a private means of transportation, a private garden, a private laundry, self-service stores, and do-it-yourself skills of every kind. . . . Even within the family Americans are unique in their feeling that each member should have a separate room, and even a separate telephone, television, and car when economically possible. We seek more and more privacy, and feel more and more alienated and lonely when we get it.[29]

Slater has his finger on the pulse of an emergent phenomenon that has been encroaching on Western society for decades, and his book explores how this is changing the psyche of the individual. His concerns are warranted; as the dominant role of the value of privacy in North American life grows, so, too, does the disturbing trend in the rewiring of the average individual's personality toward narcissism. The psychological concept of narcissism roughly equates to *self-centeredness.* Psychologist Jean Twenge, at San Diego State University, performed a meta-analysis from data collected using a survey instrument called the Narcissistic Personality Inventory. Current scores now have no precedent in the proportional rise in the population—narcissism in students has increased 30 percent since 1982.[30] When we grow, develop, and live in a world where everything is geared toward the individual, how can we help but view the world with a more narcissistic lens? When we live with other people, share resources within the environment, and work through problems together, the outcome is an individual that is less self-centered and more psychologically flexible.

As we move forward in the twenty-first century, it appears that the loneliness trend line has only increased throughout the economically developed world. As journalist Johann Hari wrote in his 2018 book *Lost Connections*: "Loneliness hangs over our culture today like a thick smog."[31] An entire body of research is coming to the robust scientific conclusion that social connectivity is one of the most powerful drivers of individual health and wellness. In fact, loneliness can be downright deadly. Hari was fortunate to interview one of the preeminent scientists in the field of loneliness research, John Cacioppo, professor of psychology at the University of Chicago, before Cacioppo's death in 2018. Cacioppo had pioneered the field of social neuroscience for three decades.

His groundbreaking study entailed recruiting participants to help understand the effects of loneliness on human physiology. The subjects were equipped with a cardiovascular monitor (to measure heart rate) and given a beeper to record how lonely or connected one felt at any given moment. This would be immediately followed by spitting into a tube; the saliva would eventually be measured to assess cortisol (the stress hormone) levels. All this was to determine how loneliness affected stress levels out in the real world. What he discovered was that loneliness was highly correlated with cortisol levels; so much so, the experiment found, that acute loneliness was about as stressful as a physically violent encounter.[32]

But, as any good scientist knows, correlation is not causation, and so Cacioppo wanted to disentangle the causal direction of loneliness and depression. The research question was: Is it depression that makes one lonely, or as Hari noted in his interview with Cacioppo, "that if you become lonely, that might make you depressed?" In an ambitious study that recruited 135 people who self-identified as acutely lonely, and through an extensive battery of survey instruments, Cacioppo found that this loneliness status was also correlated with being anxious, pessimistic, afraid of others, and characterized by generally low self-esteem.

Yet, Cacioppo wanted to dig deeper, so he brought in David Spiegel, a psychiatrist trained in hypnotism. The lonely subjects were then divided into two experimental groups, where Spiegel hypnotized members of one group to recall moments in their lives that were profoundly lonely and isolating. Another group was prompted to recall times when they felt incredibly socially connected. That is, they were primed to feel hyperlonely or hyperconnected to others. They then retook the surveys. The hypothesis Cacioppo was testing was that if loneliness leads to depression, then loneliness would drive people to be more depressed. Feelings of social connection would reasonably have the opposite effect. The team's hard work paid off, as reported in an interview: "What John's experiment found was later regarded as a key turning point in the field. The people who had been triggered to feel lonely became radically more depressed, and the people who had been triggered to feel connected became radically less depressed." Cacioppo noted: "The stunning thing was that loneliness is not merely the result of depression. Indeed—it *leads* to depression."[33]

As Jacqueline Olds and Richard Schwartz summarize: "The degree of social connection has significant effects on longevity, on an individual's response to stress, on the robustness of immune function, and on the incidence and course of a variety of specific illnesses."[34] The effects of loneliness can even seep into your genes. For example, one molecular study examined more than two hundred genes that control immune response and showed that "subjective isolation"—which is your own perception of how isolated you are—influenced expression in these critical health genes. Loneliness literally changes your genes for the worse.[35]

A 2015 meta-analysis revealed that chronic social isolation increases the risk of mortality by a whopping 29 percent. In fact, one study uncovered the mechanism that is likely driving the stress response in lonely people. Using neuroimaging devices, research determined that isolated individuals could spot threats in half the time that socially connected people could.[36] Isolated people felt less secure and thus more vulnerable to threats and succumbed to hypervigilant states to compensate. The consequences are profound, as this creates a type of negative feedback loop, where loneliness begets depression and anxiety, behaviors that end up driving people away, in turn pouring more fuel on the loneliness fire.

Sadly, loneliness leads to insecurity, and insecurity has been theorized as a driver of suicide. David Lester, a British-American psychologist, "suicidologist," and emeritus professor of psychology at Stockton University, found that decreasing social integration on a societal level (divorce, declining marriage and birth rates) demonstrated a near-perfect correlation with homicide rates, which were also tethered to suicide rates.[37] Jean Twenge and Roy Baumeister, a social psychologist at the University of Queensland in Australia, hypothesized that acting out in antisocial ways by displaying aggressive behavior is directly related to social isolation. Specifically, they forward the idea that aggressive impulses are generally held in check by the social norms of our communities. When someone loses a sense of community, they have no "witnesses" to their moral actions and thus the antisocial impulses are likely to manifest in ways that are less restrained.[38]

Sebastian Junger described the counterintuitive statistical trend of improved mental health when people are bound together in purpose: "Rampage killings dropped significantly during World War II, then rose again in the 1980s and have been rising ever since.* It may be worth considering whether middle-class American life—for all its material good fortune—has lost some essential sense of unity that might otherwise discourage alienated men from turning apocalyptically violent."[39] Corroborating this trend, after the September 11, 2001, terrorist attack, there were no rampage shootings for the next two years. New York's suicide rate dropped by approximately 20 percent, the murder rate dropped nearly 40 percent, and the number of prescriptions filled

---

* When I hear of the latest mass shooting, and am told the nauseatingly familiar demographic statistic of "loner white male," I always ponder whether or not his life (and the lives of the people he irrevocably changed) would have turned out much different if he had been born into a hunter-gatherer band. How differently would his mind have developed if it had been stitched into the fabric of a real community, with genuine prosocial male role models, propped up by the symbols and language of a united worldview, and given meaning through social purpose? Would he then still have turned out to be an antisocial monster?

for antianxiety and antidepressant medication noticeably decreased. Even veterans who were being treated for post-traumatic stress disorder (PTSD) showed a reduction in symptoms after the attack. Another example can be seen from modern-day Israel. Israelis have compulsory military service. It is mandatory for all citizens regardless of biological sex as soon as they turn eighteen, and they must fulfill a thirty-two-month commission. Despite nearly everyone in their society serving in the military, and many having fought in Israel's major wars, they have one of the lowest rates (1 percent) of PTSD compared to other militaries. Perhaps it is the "we're in it together" frame of mind that is the source of such psychological resilience. Sociologist Charles Fritz (after studying the bizarre World War II data generated from an anxious British government with respect to its citizens' mental health) theorized that communities undergoing the worst oppression are the most socially resilient. Under these circumstances, people devote their energy to the community instead of themselves. He noted: "Class differences are temporarily erased, race is overlooked, and individuals are assessed simply by what they are willing to do for the group."[40]

Another study on the Belfast riots in 1969 found a similar trend. In a *Journal of Psychosomatic Research* article in 1979, "Civil Violence: The Psychological Aspects," the authors wrote: "It would be irresponsible to suggest violence as a means of improving mental health, but the Belfast findings suggest that people will feel better psychologically if they have more involvement with their community. . . . When people are actively engaged in a cause, their lives have more purpose . . . with a resulting improvement in mental health." It appears that the core feature of the survival mindset is that you transform from being a victim of your society into a resilient participant in its tribal survival, and there is something about this act that mitigates mental disorder and even strengthens mental health.

In our search for the smoking gun of mismatch's deleterious effects on our species, we have surveyed a number of different topics. We have mostly found data-driven and depressing clues from the negative effects of WEIRD societies' self-isolation. Fortunately, there is also growing scientific literature on the benefits of living a more socially oriented, prosocial life embedded with intentional proximity. We will dive into the applied literature in the second part of the book. Yet, skeptical readers may counter that we grew up in smaller networks, and so haven't we adapted to our more isolated environments? In the end, having been reared in the individualist cultures of our forefathers and mothers, aren't we especially situated to thrive in isolation? That may be the case, but if you wish to stick to your isolation guns, know what you're getting into. Here is a quick summary of the scientific evidence of what isolation does to your health and well-being:

- Loneliness is predictive of increased broad-based morbidity and mortality.[41]
- Loneliness predicts increased antisocial behavior.[42]
- Loneliness makes it harder for your adult brain to grow and change.[43]
- Loneliness increases baseline stress, leading to greater levels of anxiety and depression.[44]
- Loneliness changes your genes for the worse.[45]
- Loneliness predicts drug use and suicide.[46]
- Loneliness negatively influences your sleep.[47]
- Loneliness in women leads to more difficult births and predicts postpartum depression.[48]
- Loneliness in men leads to cognitive impairment, weaker immune systems, and special sensitivity to loss of social contacts in old age.[49]

The science is clear—loneliness is bad for humans. But why is it so devastating? This is not the case for many animals. *Presociality* is a term that applies to an animal that lacks any characteristics of eusociality beyond courtship and mating, and it is an evolutionary strategy that has evolved in many animals. Subsociality—when parents care for their young for short lengths but not for other adults—is common in winged insects and has evolved independently many times. Solitary animals, those that live mostly independently from groups but sometimes link up to eat or sleep together, are pervasive and even include several primates such as lemurs, lorises, and even a great ape—the orangutan.[50]

Evolution crafted within us an instict to be social. Our understanding of the evolution of our interdependence with each other is only recently coming into greater scientific resolution. A new term has been devised to describe this phenomenon—*the Social Suite.*

Nicholas Christakis, a Yale sociologist whose Human Nature Laboratory researches the evolutionary determinants of social networks, articulates in his book *Blueprint: The Evolutionary Origins of a Good Society* the power of the Social Suite to not only influence individual health and well-being but to drive the ultimate success or failure of societies and species. According to Christakis, a group's success depends on the degree to which it adheres to eight social laws. Individuals must be arranged in a group that attends to *individual identity* (without which tracking long-term patterns of reciprocal kindness is not possible), *love* (commonly oriented toward kin or reproductive partnerships), *friends* (long-term, non-reproductive unions), *social networks* (mathematically universal ways in which groups assemble themselves cross-culturally), *cooperation* (bias toward, and not against, cooperative behaviors), *ingroup bias* (a preference for one's groups of residence), *social learning* (sharing of information and norms), and *mild hierarchy* (hierarchy based more on prestige than dominance, where leaders engage in skill-dependent social teaching and learning). There are precise predictions about the optimal parameters for each of the eight social laws that can influence the effectiveness of the group. Later in the book we will explore those parameters in detail to see how we can leverage this information to better ourselves and our communities.

The Social Suite made it possible for our ancestors to carry with them a type of protective social shell that made them hyperadaptive in any environment. Thus, in an unforeseen reversal of Abraham Maslow's hierarchy of needs, we see that unlike most animals, human attendance to the higher order needs of transcendence—which can involve a sense of trust, loyalty, and duty to social groups and the sense of meaning we derive from group attentive behavior—can permit a more effective satisfaction of the basic needs of survival. In essence, it is by honoring the needs for friendship, cooperation, and socially directed learning that we as individuals can optimize our chances of finding food, shelter, and safety. In sum, those groups that find the proper balance among the eight social laws prove to be the most successful because they mirror the strategies our ancestors co-opted in their struggle to survive.

An Intervention—The Brigade for the Sea Turtles

It is fair to say that a first step to improving the quality of your life is understanding how you, your mind, and your body, were shaped by the evolutionary process. Not understanding this is like running with poor form. You can do it, especially if you have expensive Nikes that are so cushioned they permit you to overpronate. But over time it could result in small, chronic microtears that wear away your tendons, until one day you blow out a knee. So how do we fix humanity's running form? Is it even possible to do? Is it as simple as just being more like our ancestors?

With respect to the current state of play for humanity, Paul Atkins, David Sloan Wilson, and Steven Hayes observed: "Social interactions with strangers at larger scales, far beyond those of the tribe or village, create challenges for humans. These days, the value of the small group has been almost eclipsed either by a focus on the individual, or on vast groups at a notional or even global scale. . . . We are in effect, *in a state of evolutionary mismatch* [emphasis mine]." We evolved to be extraordinarily cooperative in small groups, but we are now faced with the challenge of a wholly new environment in which we're asked to cooperate, not just with those we know well but also with people who are practically strangers—that is, not just with "us" but also with "them." The institutions we have created are monstrously big and too unwieldy for us to see how they are influencing our well-being. Throughout the long and tortuous path our ancestors blazed, these institutions were never the *basic social units of* survival. The good news is, as a species, our science has accomplished the first step for us by identifying the mismatch. That is a battle, but it is not the war.

The terrifying truth is, we are caught between two worlds—the one that shaped us and the current moment that changes at speeds much greater than our ability to adapt in real time. Since the first human convinced another to adopt stone tools we became inherently reliant on technological innovation. Anarcho primitivism is not the answer. There is no rewinding the clock to get back to the "Paleo" basics. As a species, the option of committing to a Luddite strategy or an attempt to return to halcyon paleolithic camps is unfeasible nor is it even ideal. As the planet's last hunter-gatherers join most of the world in economic development, the "good old days" are gone, never to return. Pandora's box has been opened, and it will remain so until the last human breathes their last breath. Thus, we are presented with a dilemma. The core of the mismatch is that modern society has made us more physically isolated by decreasing our social support; all the while it has made us more mentally unstable by increasing social pressure, tricking us into thinking that low-grade online and institutional social interaction is good enough to live a healthy and fulfilling life. In this sense, the people who dwelled in the first tribes were not challenged as much as we are today. Their units were glued together in a common struggle for survival, not the weak ideological grounds many use as the foundation to their tribal social identities today.

EVERY YEAR, HUNDREDS OF VOLUNTEERS fly from all over the world to Costa Rica to perform an inter-species act of kindness. Many are young, in their early twenties, and most are night owls. But for love of the turtles, they become morning larks* for a period of time. They rise early and begin to patrol the beaches at 3:00 a.m. in search of stray sea turtles. With great effort, this brigade of determined conservationists may find a few hatchlings to reorient and send into the ocean, ever so slightly upping the odds of one in a thousand sea turtles making it to adulthood. We *Homo sapiens* are like these turtles, stranded in a deluge of fake light, albeit light of our own making. It confuses us and strands us on the seashore of evolutionary mismatch, luring us astray.

We have created an environment laden with massive institutions on

* The term chronobiologists use to label extreme morning people; the opposite term for those who are extremely oriented toward nighttime activity is *owl.*

scales almost incomprehensible to the human mind. Although safer and more secure than any preceding human environment ever crafted, it is proving to be uniquely hostile to human flourishing and well-being. Larger-scale social units we have relied upon in modernity are breaking down under the pressure of a system not in sync with our evolutionary past, when smaller groups of Paleolithic humans bound together for the purpose of survival. There are secrets of success to be gleaned in places and societies past and present. There is, after all, a good reason we survived this long. We move to the next chapter to better uncover these secrets.

In 2020, a drone flew over the Ostional National Wildlife Refuge in Costa Rica. It showed, in brilliant resolution, five thousand turtles floating in unison. In one part of the world, with a little help from a small number of humans willing to nudge them on the right path, sea turtles are now *thriving*. If we value ourselves and the people we love most in this world, perhaps it is time to form a brigade to seek out our fellow humans—humans who are hurting, humans who are confused and wandering in disorientation among the traps we have set for ourselves. We must learn to do for humans what scientists did for aquatic insects and conservationists did for sea turtles . . . help bring us back from the brink and into sync.

3

TRIBAL TRUSTS EVOLVE

Tribe Is the Answer to the Question

The figure of 150 seems to represent the maximum number of individuals with whom we can have a genuinely social relationship. . . . Putting it another way, it's the number of people you would not feel embarrassed about joining uninvited for a drink if you happened to bump into them in a bar.

–Robin Dunbar, 1996

This ability to use symbols to cross [tribal] lines is part of our DNA. It roughly coincides with the earliest dates for our species. All humans have the formal capacity to form more formal tribes.

–Alison Brooks, 2019

The Lykov Experiment

During the last ice age, pockets of the world were covered in monoliths of ice. Stretches of impossibly large icebergs expanded and penetrated the Earth's subarctic regions. One such glaciation completely covered northern Russia, and when it receded, it left, as a reminder of its previous domain, a taiga. A taiga is a forest. Not just any kind of forest—it is the forest of unforgiving cold. Nestled in the frozen bosom of the Arctic Circle, it is the only kind of forest that can grow between the tundra of the North and the temperate forests of the South. Alaska, Canada, Scandinavia, and Siberia all have taigas, but none match the scale and prominence of Russia's taiga, which stretches 3,600 miles from the Ural Mountains to the Pacific Ocean.

Imagine the most cold, remote, and inhospitable land for human habitation possible, and you may approach some measure of an idea of what this place is like. The soil beneath the taiga (geologists call it permafrost) is often a layer of permanently frozen soil. It can be crisscrossed with layers of bedrock that insulate and prevent the water's ability to drain, which leads to the formation of muskegs. Muskegs appear to be solid ground, but in fact they are wet and spongy patches of mossy, grassy soil that are breeding grounds for lichens and mushrooms. If the soil becomes thick enough, it can be rooted by trees. Spruce, pine, and fir lean into the long, harsh winters and survive the short, brief summers; and so, taigas are known for their epic thick forests of deciduous evergreen trees. All manner of cold-adapted animals live in the taiga. Rodents live close to the forest floor, and birds, such as eagles and owls, prey upon them. Moose—the largest type of deer in the world and native to the taiga—survive off the aquatic plants growing in its bogs and streams. But to the human eye, the most imposing predators are the bears, the lynx, and the Siberian tigers that hunt the moose and the native boar.

It was in this land, in 1936, that a "nuclear" family unwittingly began a seventy-year natural experiment. This was a Russian family led by the patriarch, Karp Lykov, his wife, Akulina, a nine-year-old son by the name of Savin, and a two-year-old daughter named Natalia. They were situated 150 miles from the nearest human settlement in the wilds of Siberia. There, in a place unfathomably isolated from civilization, the family existed and made no contact with other humans for forty years, until the summer of 1978 when a team of geologists encountered them while surveying for resources. The leader of the geologists, named Galina Pismenskaya, made first contact with the free-ranging nuclear family on "a fine day and put gifts in our packs for our prospective friends." But since humans are often more dangerous than animals in the wild, she packed a pistol just in case.[1]

After a tenuous, yet successful, first contact, several more visits with gifts followed. The reason the Lykovs left to this refugium was due to what Karp Lykov described as religious persecution. He was an *Old Believer*—a member of a fundamentalist Russian Orthodox sect that had remained so since its conception in the seventeenth century. Under the atheist Bolsheviks, the communists persecuted the Old Believers, and as a result Lykov's brother was killed. It was in this way that the Lykovs found their home in the wilderness, where they bore two more children (Dmitry in 1940 and Agafia in 1943). Remarkably, World War II passed by without their notice, and those two children never saw other human beings until the geologists chanced upon their cabin. At the point of first contact, this natural experiment had been running for forty-two years.

Together, in this chapter, we will explore the fundamental insights this one-of-a-kind experiment reveals about the human condition. There is a valuable lesson to be learned here, but before we can glean its meaning, we need to grasp the evolution of human societies.

IF TRIBE IS THE ANSWER to the question—then what is the question?

The ultimate question that all life is bound to ask is: How do I survive? Tribe is the uniquely human answer. This question can be broken down into three subquestions of human survival. In some way, every animal must overcome the same challenges in the game of survival and reproduction. These are the three critical subcomponents of the ultimate question:

- How do we get food and shelter?
- How do we avoid incest?
- How do we mitigate bad luck and know with whom to cooperate?

Each species has their own take, their own strategies, their own "answers" to the questions that accumulate over time to give them a shot at successfully playing the evolutionary game of life. The currency of the evolution game is "fitness." But what is fitness? Colloquially, the term is a marker of general health and overall well-being. In this context, we tend to think of people who are physically active, diet conscious, and well adjusted as overall "fit" individuals. The Darwinian definition of fitness, however, is distinct from its everyday application.

Specifically, fitness in this context means the ability of an organism to *survive to reproductive age*—and if they are sexually reproducing organisms, as we are, find a mate and produce offspring. The more offspring an organism produces during its life, the greater its fitness.* Humans are one of the rare species that can call many of even the most extreme and variable habitats their home. In the past ten thousand years our species population has increased a thousandfold.[2] As of this writing, there are now over 8 billion *Homo sapiens*. That's a lot of species-level fitness. Whatever humans have been doing, it has undeniably given us an impressive measure of evolutionary success.

Other species have different ways of accruing fitness points. A species with a sensitive olfactory system, such as dogs, may rely on scent to assess close genetic relatives to avoid inbreeding. A species that chooses reproductive quantity over quality, say ants, may try to overcome massive disruptions to their population, like genocidal attacks from rival ant species, by way of sheer numbers. And for those species privileged with enough brain power to remember previous interactions and associate them with other individuals in their group, like chimpanzees, quite sophisticated rules of conduct can manifest to help them figure out if it's worthwhile to either help or hinder their fellow chimp compatriots. Every species that ever existed had to have good answers to these three questions—or they went extinct. From a human perspective, the survival subquestions and answers we will explore in this chapter are:

- How do we get food and shelter? The human answer: we live in **camps**.
- How do we avoid incest? The human answer: we live in **bands**.
- How do we mitigate bad luck and know with whom to cooperate? The human answer: we live in **tribes**.

---

* In more precise, sciency terms, *fitness* is typically discussed in the context of genotypes; genotypic fitness is the average fitness of all individuals in a population that possess specific genes or a complex of genes. Thus, a genotype with the highest absolute fitness approaches a fitness of one, whereas a fitness of zero equates to no expression of the gene. Absolute fitness is the ratio between the number of individuals with a genotype before natural selection compared to after the selection occurs.

Since camps are nested in bands, and bands are nested in tribes, logically we are left with a stunning conclusion: "tribe" is the answer to all three questions wrapped into one.

If tribe is the answer, then it's worth exploring and defining. What, then, is a tribe? In this chapter, we're going to investigate natural history, population genetics, evolutionary theory, cultural anthropology, and more to figure that out. We'll survey the history of anthropology in an attempt to make sense of the types of societies humans use and have used throughout our existence. We'll lay the groundwork to understand precisely how humans were transformed from chimpanzee-like creatures swinging in trees to the thousands of ground-dwelling contemporary tribes that constitute our species today. It is a complex and sometimes bizarre story that includes incest-avoidance strategies, computational limits to social network processing in the human brain (called *channel capacity*), and good old-fashioned ethnography. Using the mathematical properties of groups, we will devastate the concept of the "nuclear family" and reinvigorate it with a new and fresh idea—the *nuclear camp*.

Let's uncover the secret recipe for the human success story. The tribal ingredients are: *a group, nested in groups, bound by symbols.* This is not just an academic exercise, as this precious knowledge will help us on our quest in Part II; we will learn to harness it not only to better our own lives and the lives of those we care about most, but also to improve the overall health of the larger societies within which we live.

It Takes a Village—Camp Is the Fundamental Group

Humans are obsessed with groups. We have been since the first moment one of our ancestors peered over a ridge and gaped at an unknown group of humans and puzzled as to why *they* were different. Dwelling in groups was a precondition for survival. No human alive today exists free of the mark of our natural history of social living. There are different ingredients to the recipe of social life that our ancestors used to build the world in which we exist.

The core foundational group that a tribe builds from is the *camp*. It is the *nuclear group*. That is, the basic unit of survival. The members of a camp share in the project of finding calories, keeping each other safe, and reproducing. Pair-bonded partners in reproduction, friends, and extended family divvy up roles among each other so that the camp succeeds. It is likely that the first fledgling human society stemmed from a camp-like structure; the early people adopting this special type of society, which persists even today, would have been something like the world's first hunter-gatherers.

INCEST TABOOS ARE CULTURAL UNIVERSALS. All human groups share an instinct to avoid genetically intermingling with too close of stock,* and it manifests in cultural norms that express themselves in every human culture that exists or has ever existed. That last claim sounds bold. How do we know this to be true? Because humans are not the only species to have incest-avoidance strategies. Most animals do, and brother-sister incest is not only formally prohibited in most humans, but is extremely rare or absent in mammals and birds.[3] On the population level, this decreased fitness (called inbreeding depression), can affect an entire population, and there is an important mathematical principle in population genetics that will guide our understanding of the evolution of early human groups—the *50/500 rule.*

What hinders a group's ability to survive, increasing the likelihood of extinction? The 50/500 rule is the mathematical answer to the question. It details the minimal number of individuals needed to ensure a group's survival. The term that is used in the fields of biology and ecology to explain this concept is the *minimum viable population* (MVP); in other words, the MVP is the lower bound on the population size of a species, such that it can survive in the wild. Scientists have quantified two things in particular that are bad news for small, fledgling species: inbreeding and bad luck. The 50 in the 50/500 rule is where fifty individuals are the minimum for a population size to counter the deleterious effects of inbreeding. We'll save the 500 in the 50/500 rule for later in the chapter.

* Thus, the evolutionary reason—the why—behind the gross, icky feelings we experience if we are to consider the incest proposition, is because it's going to handicap your offspring, and thus your chances of successfully reproducing offspring that successfully reproduce.

When considering the MVP and hunter-gatherers, there is plenty of data to mine. The Hadza, living in the same camp, range from twenty-one to thirty individuals (median = 21).[4] In !Kung, group sizes range between twenty and fifty people,[5] although this includes both large, almost-annual occupation campsites and ephemeral, short-lived camps. John Yellen, a research associate at the Smithsonian Institution's National Museum of Natural History, performed a critical study in 1977 on a set of short-term !Kung camps, which yielded an average of seventeen people per camp. Notably, this camp size value is very close to the Hadza value. It appears that using multiple channels of data camp sizes can be realistically estimated to be between twenty and thirty adults. And this is why the basic "nuclear" social unit is *not* the family, but instead the camp.

AT THIS POINT, IT MAY be helpful to know what kind of ancestral animal it was that began the brave social experiment of living in camps. Paleoanthropologists call this animal a *hominin*. This is the first of the prehistoric apes along the human lineage that were bipedal—meaning they could walk upright. I bring up these early hominins because they were progenitors to one of the lineages that led to the genus *Homo*—and eventually us. That moment occurred right around the time that they descended from the trees to habitually live on terra firma. Between 1.8 and 2 million years ago there was a hominin that possessed cranial capacities that varied between 510 and 755 cc (modern human brain sizes average 1,350 cc), and this is important because it pushes beyond the volume of even the brainiest of living apes.[6] The endocasts of *Homo habilis* present several humanlike features, including increased size of the frontal parietal lobes and a more humanlike gyral (brain folding) pattern.[7] The cognitive implications for changes to brain shape are almost entirely unknown, but the brain sizes themselves give us some room to speculate on *Homo habilis*'s social system and absolute group sizes. Scientifically, we can use the neocortex ratio to estimate the limits of an early human's mental capacity to process social information. Based on these estimates, *Homo habilis* likely had a breeding population that was a small number of camps that numbered between nine and sixteen adult individuals.[8] Thus, with the help of the archaeological record, we are beginning to paint a picture of the moments in natural history where early humans innovated new ways to dwell together.

Carl Zimmer, a science writer for *The New York Times*, while profiling the work that Charles Nunn and I did on the role of sleep in shaping human evolution, highlighted how important sleep was to this transition in an article titled "Down from the Trees, Humans Finally Got a Decent Night's Sleep."[9] We discovered that early humans were engaging in what Christakis would call a type of *social niche construction*. When our ancestors got to the ground, they went to work much in the same way that beavers modify their natural environment by crafting a dam or creating an impoundment to provide protection against predators and easy access to food during stressful seasons.

But instead of only crafting physical environments, our early ancestors also crafted social environments that afforded them a safer, more secure, mobile environment that allowed them to not only survive but thrive in this brave new world. The adoption of group living, after the discovery of fires, was one of the adaptations most critical to our species' survival.[10] Just as a snail carries its physical environment—a protective shell—on its back, human groups carry their social environments with them wherever they move.[11] Sleeping on the ground is perilous and of all the primates, only humans are habitual ground sleepers.[*] To help explain how early *Homo* overcame risks arising from the transition to terrestrial sleep, I have proposed "the social sleep hypothesis,"[†] which is the idea that early humans engaged in a type of socio-technological

---

* With rare exceptions being male chimpanzees, where predation is low, and male gorillas, who have few natural predators.

† When working with the Hadza, we discovered that synchronous sleep is incredibly rare. Using actigraphy (think superscience Fitbits), we showed that all Hadza adults only slept simultaneously a miniscule eighteen minutes over twenty days of observation. On average, eight people were awake at any given minute. This shocking level of asynchronous sleep was the engine of natural, sentinelized (a term I term in my scientific work) sleep that allowed the majority of individuals to go deep into their sleep stages while nested comfortably in the bosom of their evolutionarily crafted social niche. In an interesting juxtaposition to modern sleeping environments that are socially isolated, the key to great sleep from an evolutionary standpoint is other people. In essence, the capacity for flexibly timed, high-quality sleep improved because of, not despite, other people.

niche construction that enabled short and flexibly timed high-quality sleep.[12] In other words, camps—the ancient proto "neighborhood"—facilitate day-to-day survival, enhanced safety and security, and sleep sites, but still don't ensure that the population won't become inbred over time. How did our ancestors overcome this challenge?

SEVERAL BENEFITS WOULD HAVE RESULTED from our ancestors living in camps. In our sister species, chimpanzee mothers bear a tremendous burden of responsibility. One that, within the chimp society as a whole, goes entirely unsung. After a feverish conception period that involves approximately thirty copulations a day* for several months with every single male in the community, a female chimpanzee finally conceives. Now she gestates—a process that is immensely calorically expensive and takes 228 days, about 85 percent as long as human gestation. The entire birthing process from beginning to end is done independent of any aid from others in her community; with her hands serving as nature's forceps, she pulls the baby out of her birth canal. *Alone* in their "core areas," in which they live three-fourths of their entire lives, a chimpanzee mother holds her newborn infant. Due to their life history, chimpanzee growth and development is slow going. After a few days the infant begins to ride on the mother's back and will adopt this mode of transportation for the next three years. The infant nurses on demand until it is weaned.

Alyssa Crittenden is an anthropologist with nearly two decades of experience working alongside the Hadza. She is a dear friend and the principal collaborator on all the aforementioned sleep research with the Hadza. She, and her colleague Coren Apicella, note: "The relationship between infants and mothers is fairly straightforward—infants consume." This is even more acute for a chimp mom. She is locked in for a long, lonely ride where her life force for the next five years will focus entirely on keeping her infant alive. One human evolutionary solution to

---

* On average chimpanzee females have 1,200 copulations per birth. Chimpanzees are a very sexy ape, especially when compared to gorillas, which only average six copulations per birth.

mom's dilemma of having calorie-expensive kids was camp living. The key difference between the experience of motherhood for a chimpanzee and human mom is that humans practice a kind of cooperative breeding called *alloparenting*, which camp dwelling enabled. Alloparenting is parental care provided by a non-related individual to *your* offspring. By extending cooperation beyond the pair-bond and even the "nuclear" family, moms could leverage the increasing social power of the camp. In the wild, allomaternal care can vary in form and is co-opted by the needs of the species. This can manifest in alloparents helping with nest building, grooming, thermoregulatory assistance, food provisioning, transport of young, predator defense, and even nursing. Although alloparenting is not entirely unique to humans, like so many traits we share with other animals, humans tend to take things to their extreme.

A suite of characteristics unique to human patterns of reproduction and childrearing evolved during this critical period in human prehistory, patterns such as delayed maturation, prolonged dependence of children on their pair-bonded parents, and the provisioning required to keep these energy-greedy offspring from dying. Astoundingly, and markedly different than our ape cousins, women can give birth to new offspring while still having other nutritionally dependent children to care for.[13] Human children also stay children longer. They have a very long period of juvenile dependency, wherein their survival hinges upon them using those big human brains to learn critical life skills and practice the application of this information through play with large groups of other children. Although children in foraging societies do contribute a significant sum of calories to the camp, once they become proficient foragers, the sum accumulation *deficit* from birth to adulthood has been calculated to be around ten to thirteen million calories. Million-calorie debts cannot be paid by a single mother alone—they require a team of continuous helpers to pay them off.

Grandmothers that focus their attention on doting over their daughters' children enhance that daughter's fertility; this enhancement leads to shorter periods of time between births and ultimately more children to play the evolutionary game.[14] Yet, perhaps more importantly, moving away from direct reliance on kin and close family, humans are most remarkable in the striking feature of non-kin directly sharing food and effort in a network-distributed strategy for childcare. Not only does a camp share and distribute food through their central place-provisioning strategy, but unrelated allomothers provide a significant amount of direct child care by monitoring, nurturing, and caressing infants.[15] The human propensity to increase familial kin and friends between groups, fostering multigroup cooperation and coordination of survival strategies, may be one of the most distinguishing characters of human social structure compared to other animals.[16]

The suite of traits that were necessary to provide the fertile grounds for which our species would attain its unparalleled biological success began in the Paleolithic period with changes to our sleep and the way we shared the burdens of childrearing. This massive retooling of the way we utilize our social networks had profound effects in the evolution of our minds. Those groups that privileged the camp and had developed cognition that focused the individual's energy on the benefit of the camp were in a prime position to outcompete their rivals that had no such cognitive and social abilities. Thus, some element of the Tribe Drive was born out of the advantages given to the lineage of human beings that started protecting each other by sleeping in groups and taking care of each other's kids even when they were not their own. But living in a group is not all "sweetness and light." What were the challenges and costs of this new social compact?

The First Community—Band Is a Nested Group of Camps

Now we know that the foundational group, from which other super groups build their structure, is the camp. Assemble enough camps together, with a little social networking grout, and you've got the first group of nested groups—*the band.*

The band level is a constellation of a few lineages or small *kin groups* and serves as the population's secondary hedge against incest. Camps can function alone by finding food and ensuring each member's survival (and thus they are the basic fundamental social unit), but bands ensure that the genetic stock of the population does not become too inbred. For example, if you were to take two individuals randomly within a band and compare their genes, there would be a 25 percent chance they are closely related—significantly better than having only a single camp within which to find a mate and procure offspring.[17] Critically, the band and camp levels are all "face-to-face" in that upon meeting someone in your camp or band, you can instantly recall an index of memories that serve as biographical information for the individual; in most circumstances, by shared history of interactions you *know* the person who is your band-mate. In other words, each member of a band has a rich biographical story of each other member that is both known from experience and by way of the shared stories (otherwise known as gossip) of other band members' interactions with them. When things get dull, or perhaps too dramatized with conflict, members in a camp can trade spots with other people in other camps within the band to freshen things up. In terms of a loose modern analogue, if camps are neighborhoods, a band is a community of neighborhoods. The important takeaway is that people within a band network community do not physically live together all the time in the same camp, but remain intimately tied together in the shared project of survival.

WHY DO TELEPHONE NUMBERS HAVE seven digits? (Not including the area code, of course.)

Alexander Graham Bell, the engineer who is credited with inventing and patenting the first telephone, wanted to have as big a number as possible. Bell's logic was that the longer the number, the larger the capacity the phone would have; but Bell wasn't accounting for a hardwired cognitive limitation embedded in the human mind: "There seems to be some limitation built into us either by learning or by the design of our nervous systems, a limit that keeps our channel capacities in this general range," says Jonathan Cohen, a scientist who studies memory at Princeton University.[18] The end result of having a phone number eight or nine digits long would have been harmful to the success of the nascent technology because too many people using it would have dialed wrong numbers.

Cognitive psychologists call this phenomenon *the channel capacity*—which refers to the space allocated in our brain for certain types of information. The channel capacity for numbers, what the psychologist

George Miller describes as "The Magical Number Seven,"[19] is a robust scientific finding that comes up again and again in domains that require some type of proportional categorization. For example, if you are tasked with counting how many flashing dots you see on a screen, you would likely get the number correct up to about seven dots. This goes for other senses besides vision. If you were to drink thirty kinds of beer, and you had to rank them all by hoppyness, you would begin making mistakes after about seven categories. Another sense, hearing, would be capped by the magical number seven if you were played dozens of differently pitched sounds; the average person would likely ascribe differences in tonality to about six or seven such tones and then lose accuracy. There is a natural limit, when surpassed, where we become overwhelmed by the raw information. It turns out there is an incredibly important sense to the Tribe Drive that has cognitive limitations: our socially directed feelings.

In the 1990s, Dunbar discovered the human channel capacity for social groups. The cognitive limitation to the human brain's ability to maintain stable social relationships has been called many names: the *Rule of 150*, *Optimal Community Size*, or *Dunbar's number*. This social channel capacity is a law that governs the dynamics of how humans group together—something I call *concentric circles of sympathy*, which we will explore in greater detail throughout Part II of the book. Humans have the amazing capacity to live in structurally diverse societies, but there are successive cumulative layers of group sizes that appear to be social network "sweet spots" where humans perform optimally when navigating their social worlds.[20] The social sweet spots are characterized by distinct layers, or concentric circles, that represent different frequencies of interaction and tiers of emotional intimacy that bind members together.[21]

What is the evidence for the social channel capacity in humans? Dunbar's initial comparative studies revealed a tight, positive, linear correlation (that is, when the x-axis gets bigger, so does the y-axis) between primate brain size and average social group size.[22] Specifically, the theoretical backdrop for Dunbar's hypothesis is tied to the direct function of relative neocortex size, which presumably limits group size by capping your brain's social networking processing ability. Ingeniously, he extrapolated the results (with what is now called Dunbar's equation) from the primate data with *Homo sapiens* and concluded that an average member of our species can comfortably navigate 147.8 relationships before the energy involved in maintenance hits a failure state. Informally, Dunbar explained that this value is "the number of people you would not feel embarrassed about joining uninvited for a drink if you happened to bump into them in a bar." Perhaps an updated alternative thought experiment may be that this value is the number of people you would feel bad about booting from your Facebook friends list.*

Fascinatingly, when we observe natural community sizes in hunter-gatherer societies, African foragers' camps typically contain between 11 and 20 percent of the number of individuals that form their entire band, which corresponds perfectly to the estimates of social group size based on Dunbar's equations of neocortex size for 150-person networks.[23] Studies looking at twenty-one different hunter-gatherer societies, from the Ona of Tierra del Fuego to the Tauade of New Guinea, yielded an average village size of 148.4 people. It is likely not a coincidence that Dunbar's number is literally the mean value for the size of forager bands. To be clear, the band—characterized by its mean value of approximately 150 members—is approaching the fringe of the social self, beyond which our evolved cognitive mechanisms related to coalitionary alliance detection are pushed to their limits.

The reason for this remarkably consistent set of group sizes that function so much more effectively than other values is likely tied to the costs of caring. Plainly stated, caring is exhausting. And our capacity to care is highly constrained to what cognitive psychologists call a *sympathy group*. Think for a moment about the number of people who, upon hearing of their death, you would be truly devastated. If you were to craft a list of these names, studies show that it would likely be about twelve people long—this would be your sympathy group.[24] These people are embedded within the social network you spend the most time

* Incidentally, after challenging my students in class to cull their Facebook friends down to Dunbar's number, I went ahead and ran the experiment for myself. After only assessing "strong" ties, I ended up barely breaching one hundred friends. Without hesitation, I can say that the reduction to a more band-level Facebook was a major life improvement.

and energy in dedicating to cultivating relationships; you talk to them, think about them, worry about them, and identify with them more than any people you have interacted with—and doing so is not cognitively cheap.

The emotional energy required for good relationships is incredibly expensive. It also makes these relationships precious and, in many instances, irreplaceable.

Homo erectus IS WHERE THE resolution of the paleoanthropological record becomes finer. In 1984, Kamoya Kimeu, Kenya's famous fossil prospector, spotted a bone eroding beneath an acacia tree west of Lake Turkana along a dry stream. Nearly every bone of a boy who had died 1.8 million years ago was found. Named by the nearby town, Nariokotome boy* was reconstructed and presents the most detailed skeleton of an ancient hominin. The telling features of this early human indicated that it was decidedly a new variety, or *grade*, of hominin. Thus, with this find dated approximately 1.8 million years ago, our ancestral progenitors had revealed themselves as characters in the human story. Again, relying on Dunbar's equation to estimate the channel capacity of group size, early Pleistocene hominins group sizes reached an approximate channel capacity of one hundred individuals for *Homo erectus*.[25] The number of people in the band expanded, permitting the membership of a few more people per camp, which estimates suggest were about eleven to twenty adults. This means that using the record, we can trace the evolution of the necessary social networks required for a fully flourishing band that later becomes the basis for the first human tribes. But with more people comes more social cognition, and undoubtedly more

* Nariokotome was male (the pelvis gives this away) and its teeth showed an eruption pattern of an eleven-year-old. He was five feet, three inches upon his death, and if one extrapolates, he would have grown to nearly six feet in height. He had thin, narrow hips and relatively long arms and legs for his height. This indicates he had a body type that would have been ideal for heat loss and most likely was adapted for hot, open grasslands—much different than our australopithecine ancestors who liked dwelling in the woodlands. In other words, from the neck down this animal looked nearly like a modern human. The brain, however, would have produced an adult capacity of around 910 cc, significantly larger than the brains of *Homo habilis*, but not quite human.

problems. Our protoape ancestors, like our chimpanzee sister species, were default aggressive. How can such a "quick to anger" species live harmoniously with their new roommates? The biggest challenge faced by our ancestors during this 1.5-million-year period is what has now been identified as the *Tyrant Problem*.

Brian Hare, professor of evolutionary anthropology at Duke University, and Victoria Wobber, together with their advisor Richard Wrangham, developed an idea to help address this profound question. Their work aimed to test the self-domestication hypothesis, the idea that bonobo females had been actively selecting against aggression in males. The idea began with the observation that bonobos had diverged from a chimpanzee-like ancestor, and due to differences in ecology, where female resources were not sequestered in isolated patches but more evenly distributed, females could bond together, regulate, and *select* for less aggression in males in a type of *self-domestication* evolutionary process. In their 2013 book *The Genius of Dogs*, Hare and Vanessa Woods describe Hare's expedition to visit the legacy of the Soviet geneticist Dmitri Belyaev, who performed a massive, longitudinal domestication experiment that revealed empirically for the first time a *domestication syndrome*.[26] Hare flew to Siberia because he wanted to better understand the power of artificial selection on aggression and domestication, and he witnessed firsthand the experiment that transformed a wild fox into an adorable household pet.

Belyaev's model species for his experiments in creating a tamer psyche was the silver fox. The silver fox had been brought over from Canada in the 1920s, and given its unusual fur color, drove worldwide attraction. Belyaev hypothesized that aggression was a complex behavior that worked on several biological systems, including anatomical, neurological, and hormonal levels, but instead of attempting a complex experiment that targeted all those systems independently, he created a simple experimental paradigm: allow foxes to breed together if they are friendly to humans. Specifically, in 1959 he began the experiment by sifting through each generation of foxes and separating them based on whether or not they shied away from or showed aggressive behaviors (biting and growling) to human handlers when they were fed. A small proportion of the pups tolerated the handling; after being bred for just three generations, they were no longer showing aggression. After four generations the fox pups were approaching humans with wagging tails, and the sixth generation (marking the appearance of the "domesticated elite") were not only wagging their tails but whimpering, licking, and sniffing humans to attract their attention.

Despite only *one* behavior being actively selected, a suite or *syndrome* of traits emerged that was completely unpredicted: piebald spotting (patches commonly associated with cows, cats, dogs, and other domesticates), floppy ears, rolled tails, and narrower, feminized skulls.[27] These are all juvenilized traits that humans consider cute and adorable.* Hare, Wobber, and Wrangham built off the discovery of the mechanisms for self-domestication by inferring its implications for bonobo evolution.

Coming off the heels of these exciting discoveries, Wrangham was still contemplating how this all fit in with human aggression and the evolution of in-group and out-group psychology. A breakthrough in his thinking, which he describes in his 2019 book *The Goodness Paradox: The Strange Relationship Between Virtue and Violence in Human Evolution*, was connecting self-domestication with the different types of aggression exhibited by humans. In the silver foxes, and as he hypothesizes in humans as well, *reactive aggression* is what was being selected against. This type of aggression is defined as the immediate response to a threat that is based in fear, anger, or a combination of both. This type of sudden ignition, high-arousal aggression, is a conserved evolutionary trait and draws upon all the powers of the ancient limbic system as it co-opts the sympathetic nervous system in the *freeze-fight-flight response*. Its primary function is to eliminate threats, and it is responsible for releasing adrenaline, quickening the heartbeat, prompting metabolic upregulation of glucose, and dilating pupils—all while inhibiting non-essential biological processes. It is the "red alert" of threat response.

Proactive aggression, in contrast, is not a response to an immediate threat but is characterized by the presence of deliberate planning and

* Juvenilization is the process of making the adult form look more like juveniles. From an evolutionary standpoint, this process is one that hijacks the adult psychology into investing in offspring. We *feel* this effect as *cuteness*. Cuteness is nature's evolved mechanism to ensure we do not abandon children when they are crying and pooping on everything.

is conspicuously noted by the absence of emotion performed at the instance of the assault. Proactive aggression is premeditated, deliberate, and low arousal in nature. Put another way, reactive aggression is hot and proactive aggression is cold.[28]

So how did our ancestors course correct away from such a reactively violent society? How did they keep social order when tyrannical alphas could coerce others into getting their way? The answer is we leveraged our new capacity for proactive aggression by becoming slavishly adherent to social norms and enforcing those norms with capital punishment.

The paramount issue is social control. Humans have relied on two primary strategies. The one we are familiar with is what Ernest Gellner has dubbed "the tyranny of kings." Large agrarian groups needed strong, hierarchical, centralized systems of control that limited personal freedom and liberty. This is in contrast to a forager way of life, as "obedience and sovereignty are not found among mobile hunter-gatherers." A romantic perspective of living in small-scale societies is that it is utopic. There is an argument to be made on its behalf. Social support is embedded into the very fabric of these societies; people are buffered from depression, and generally, day-to-day life is quite peaceful. Everyone has inherent (but not necessarily equal) value and is given a special role to fulfil within the society that imparts meaning and purpose to their lives; all this without some monolithic political entity dictating and circumscribing their behavior. Sounds idyllic, right? But this is not the full picture.

Liberty is not absolute: "In the absence of domineering leaders, a social cage of tradition demands claustrophobic adherence to group norms . . . a *tyranny of the cousins* [emphasis mine]. The cultural rules are paramount. Individuals have limited personal freedom; they live or die by their willingness to conform."[29] The word *cousins* in this context means the group of adults whose collective decisions held absolute power. That power was backed up by lethal responses to norm violation in the way of banishment from the group or capital punishment. Thus, these early human bands created a mechanism to conquer the Tyrant Problem by *self-selecting against reactive aggression*. Over a million years or so, the domesticating changes to the human genome significantly altered the course of human evolution and provided the final launching pad for the Tribe Drive to come into its full expression.

Transcending the Channel Capacity—The First Tribes

By now you may have noticed how the same group size numbers keep popping up. Recall the 50/500 rule that explains minimum viable population. To reiterate, the 50 refers to the minimum number of individuals a population needs to counter the deleterious effects of inbreeding. The 500 in the 50/500 rule is the number of individuals that are needed to minimize the risk of bad luck. Geneticists term blind luck (either helpful or harmful) as *genetic drift*. When it's bad luck, it usually comes in the form of a population group disappearing for a number of reasons, ranging from simple chance (termed "stochasticity" by population geneticists) to a random and unpredictable factor (e.g., volcano eruption or disease outbreak). Taken together, the 50/500 rule is the set of values that give a group a decent shot at avoiding extinction.

In our discussion about group survival, you may have noticed among all these numbers an interesting problem. If 50 individuals is the minimum, and 500 individuals is the optimum, but Dunbar postulates the human brain maxes out at 150 relationships, how did the first human groups that expanded into larger and larger group sizes defy the Rule of 150? Framed another way, the problem is one of cooperation. In essence, how do we know who to cooperate with if we don't have a memory of interaction with a stranger? *The amazing and informative answer is we became a tribal species.*

Ours is the only species that uses symbols to signal coalitionary alliance. This is one of the reasons our species, relative to other animals, is unique. It is also why our species is uniquely tribal. Alison Brooks, a professor of paleoanthropology at George Washington University who works with the Ju|'hoansi (!Kung) foragers of northern Namibia, characterizes tribal level of organization as "beyond the face to face." Thus, tribe is an "imagined community" that is reified by symbolic tokens—these include visual, linguistic (e.g., accent, dialect, idiosyncratic colloquialisms), behavioral, gestural, ritual practice—and a body of material culture in dress, tools, and artifacts that unite its members in a shared origin story.[30] Essentially, tribe is *fictive kin*. This foreshadows our discussion to come in chapter five on the topic of trust signals. For dramatic examples, tribal trust signals of allegiance can literally be embedded in your bones. Several Australian indigenous groups knock out an upper incisor as soon as a boy or girl reaches maturation. Another example would be cranial deformation (otherwise known as head flattening) as a form of body alteration in which the skull is intentionally shaped during development; it was highly valued as an honest, costly signal of identity by tribes of both Neolithic peoples in Southwest Asia and East Germanic peoples of late Antiquity and in the Americas among the Inca and Maya.

The critical consequence of tribes relying on the human power to cognitively manipulate symbols is that you transcend the face-to-face shared experience and knowledge of the individual. This is key because the first tribes (averaging 1,500 individuals) emerged from bands (averaging 150 individuals).[31] What this social innovation meant for early humans was that if you have never met a person before and possessed no shared history of indexed interaction, you could still recognize a stranger as part of your coalitionary alliance if they emitted, and you recognized, the right tribal tokens. This innovation allows people to cleverly index weak social ties up to numbers in the thousands, thereby breaking the bonds of the Rule of 150.[32*]

Therefore, the operational definition of tribe is the following: *a tribe is a nested group of groups that uses symbols as tokens of identity signal-*

* For example, in 1974, Jeremy Boissevain, a Dutch anthropologist who worked with the Maltese population, published a pioneering study of personal networks and found that at the maximum range, one individual had a network of 1,750 persons "whom he had met or had dealings with in the recent or distant past. They formed the social universe of persons who could help him solve his problems." These were not best friends, nor were most of them even considered a friend, but more like prosocial acquaintances. Chaining together the extended network of friends of this socialite probably encompassed a significant fraction of the Maltese population. Intriguingly, this example illustrates that within the context of a single ethnic group, a single individual tops out at about the number of ties that equal close to the average number of tribal group size. Importantly, the consistent convergence of these numbers leads us to the important conclusion that humans are limited to the number of relationships they can process, and that ultimately this cap is likely a product of evolution of tribal societies.

ing group membership within an shared imagined order. Homo sapiens are the only primate that form meta-group social networks by maintaining consistent relationships among groups that permit intergroup flow of individuals,[33] and the way we tie each other together when the meta-group extends beyond the Rule of 150 is by gluing the groups together with symbolic information.

Humans are walking, talking sentient signs of coalitionary alliance. Once you see it, it is hard to unsee. After pruning the lexical brush that has grown indiscriminately around human grouping theory, we now have three crucial operationally defined concepts of human grouping patterns. Everything else—every other term we have discussed—is derivative of these three meaningful groups—camps, bands, and tribes. By breaking beyond the computational threshold of 150, humans solved one of the survival problems we discussed at the beginning of the chapter—how to mitigate the bad luck of randomness and chaos taking out your population. But with great numbers comes great computational limitations—which brings us to cooperation. When we are embedded in a society that has strangers, with whom do we cooperate?

TABLE 1: OPERATIONAL DEFINITIONS FOR THE THREE KEY UNITS OF HUMAN GROUPING PATTERNS

SPECIES	ORIGIN TIMELINE	LARGEST GROUPS	AVERAGE SIZE	NESTED PATTERN EMBEDDED INTO CIRCLE	FUNCTION
Homo habilis	2.0–1.8 mya	Camp	25–30	Pair-bonds (mate sets), family (nuclear set), friends (age sets), gender (role sets)	Basic "nuclear" unit of survival; shared projects of reproduction, calorie finding, and mutual protection.
Homo erectus	1.8 mya–0.3 kya	Band	120–180	Dynamic fission-fusion of camp sets within a given band	Multi-clan genetic and resource exchange network; hedge against incest.
Homo sapiens	0.3 kya–present	Tribe	1,000–1,800	Dynamic fission-fusion of bands of nested camps that coalesce around shared symbolic culture and identity	A symbolic corporate platform for large-group cooperation and action (e.g., warfare, raiding, territorial expansion, adoption of new technologies and practices); hedge against genetic stochasticity and genetic drift (bad luck); conflict reduction.

* * *

AMID THE COLD CALCULATIONS OF spreading our genes, the question of when to cooperate with and when to cheat somebody is a highly discussed topic in scientific literature. The topic that has united evolutionary biologists, economists, and political scientists is called *game theory*. One game, the Prisoner's Dilemma, has provided a set of parameters that allows us to test the mathematical models of cooperation that work best. The classic setup is this: two gang members (A and B) are arrested, and prosecutors lack evidence to convict them of a major crime but can get them locked up for a lesser charge, resulting in much less prison time. A and B can't communicate with each other when the prosecutors offer each a deal; the major sentence will be reduced if one cheats the other. Thus, the dilemma is whether or not to be loyal to (cooperate) or betray (defect) your partner.

Two of the biggest challenges to these mathematical strategies of cooperation center on when to start cooperating and in what cases to forgive defectors. It turns out, the solution to this dilemma can be found in a "green beard" gene. Richard Dawkins, in his book *The Selfish Gene*, coined the term that illustrated a thought experiment by W. D. Hamilton. Dawkins envisioned a green beard as a signal that leverages a kind of stripped-down kin selection to its advantage. The green beard effect is when an organism has both a gene that codes for growing a green beard *coupled* with a drive to cooperate with other green bearders—the logic being that when mixed in a varied population of non-green bearders and green bearders, the green bearders will bootstrap cooperation and outcompete all others without the green-bearded signal. Using more scientific phrasing: "the crucial requirement for altruism is genetic relatedness at the altruism locus [i.e., merely a multifaceted green beard gene] and not genealogical relationship over the whole genome."[34] Green beard genes are not theoretical but have been experimentally demonstrated to exist.[35]

In summary, we can cooperate with strangers using a basic tit-for-tat strategy that can theoretically lead to stable relationships, but that strategy alone is quite vulnerable to cheaters. Those who can bootstrap the initial startup by correctly identifying and perceiving green bearders will have slight advantages about what strangers to start a cooperation loop with and when to apply forgiveness. Perhaps a new adage is in order: beards of a color flock together.

Thus, with the tribal cheat sheet helping to align incentives for humans, our ancestors had game-theoried their way into cooperating, more often than not, within their identity groups. And so we have made it to the advent of modern humans. Brains of the humans living in this era—which begins approximately three hundred thousand years ago—are now supporting the social processing for camp sizes that are between 17 and 30 adults, band sizes of 150 individuals, and what we will detail in this section as a new class of human grouping—tribes, which number on average between 900 and 1,500 individuals.[36] Recall from the previous section that analyses have shown that modern human bands number approximately 150 people, and that this conveniently coincides with the channel capacity of human social processing. If Dunbar's number clocks out at one hundred fifty, then how do we cope with social network values that exceed it? It turns out, *symbols* are the heuristic "cheat code" that humans use to *stereotype* each other into groups; this results in significantly reduced cognitive bandwidth with which we can computationally cope beyond the critical Dunbar number of face-to-face relationships. The prerequisite to tribes was a novel, emergent use of humanity's new power—the mastery of symbols.

The Cursed Blessing—When Did Humanity Scale Trust Beyond Friends?

When did humans innovate tribal trust? We'll likely never precisely know. The modern take on the archaeological record is synthesizing an incredibly complex picture of the last half-million years. *Gradualism* is the term *du jour*. It appears that humans were accumulating cultural innovations at a steady pace, resulting in a very gradual and patchy accumulation of "modern" human behaviors.[37] Furthermore, they were interacting with their hominin neighbors (the Neanderthals and Denisovans) in complicated ways. Our goal here is to determine the best approximation of when humans started embedding themselves with an identity that spanned beyond face-to-face networks, into a larger, trust-bootstrapped "us" that also distinguished a "them." A tall order indeed, but our attempt may offer important insights to our quest to better understand the Tribe Drive. According to the archaeologist Alison Brooks, we are looking for evidence that past peoples created tribal tokens (material objects left behind in the record) that are characterized by behavioral signatures of a specific group. If it can be demonstrated that these tribal tokens were transported beyond the typical band geographic ranges, then the people transporting them were an "us" going to trade with a "them."

THE ADVENT OF HUMAN LANGUAGE, a powerful system of syllabic symbols, facilitated the self-domestication process occurring within groups. Chimpanzees lack language—at least, human language. Although human-enculturated chimpanzees that can express themselves in American Sign Language have working vocabularies of around five hundred words, chimpanzees have about a tenth of the linguistic capacity of humans and they cannot leverage grammar to form complex enough statements that allow for *displacement* (the ability to project into the future or recall the past). Displacement is necessary for all the gossipy record keeping humans love performing with their linguistic capacity.

Once human language evolved, we could keep moral tabs on each other. Another crucial outcome of the evolution of language was that while reactionary aggression was *reduced,* proactive aggression became a critical method by which actors within groups could premeditate a killing of a tyrannically aggressive male. This strength in numbers coalitionary strategy lowered the risk of injury to the group of conspirators. It could not have evolved without language because it requires the victim to be selected in advance; in other words, you have to *displace* into the future by way of complex linguistic manipulations to be able to be proactively aggressive. Modern human foragers also use language to sanction bad, tyrannical actors before escalation to coalitionary execution by cajoling, ostracizing, ridiculing, and even enacting separation of colocation either by leaving the individual or excommunicating him from the group.*

We can see it in the remains left behind by humans of this epoch. Outside of one's own group, another kind of evolution was occurring with the use of symbols. Paleo-archaeologists have uncovered evidence of symbolic behavior and social intensification that really ramps up by the Middle Stone Age (sixty to ninety thousand years ago). Examples range throughout East and West Africa from Namibia, Tanzania, Ethiopia, Morocco, and Algeria. By this point, the humans that were living in these societies had full possession of modern human language capacity.[38] The typical territory size for hunter-gatherers is not greater than 30 kilometers.[39] Therefore, a smoking gun–level of evidence comes from the earliest silcrete stone tools (seventy-seven to eighty thousand years ago) transported more than 150 kilometers from their site of origin across the Kalahari desert. To restate the deductive logic: if you have a rare resource that has behavioral signatures of a specific group embedded into it, and it is transported beyond the typical band geographic ranges, then the people transporting it were an "us" going to trade with a "them."

Going back in time (in reverse chronological order), ostrich eggshell bead technology probably originated in eastern Africa and spread via complex regional trade networks some 50,000 years ago.[40] Back in time another 120,000 years, one site had 43 percent of its artifacts made of obsidian (volcanic glass) that was transported 166 kilometers away from its point of origin. Even earlier at a site in Jebel Irhoud in modern-day Morocco, there is evidence of long-distance exchange. But the most compelling and earliest evidence goes to a Middle Stone Age research site in Olorgesailie, Kenya—dating back to 320,000 to 220,000 years ago—that found forty-five thousand small, non-local obsidian flakes associated with widespread use of pigment whose function was to mark and identify the source of the material.[41] Recent reanalysis of pigmented remains in South Africa's Northern Cape Province describes

* The other key strategy is to recruit the male kinsmen of the target to inhibit cycles of revenge killing. Much like Brutus being recruited for the assassination of Caesar, close kin—often brothers—are similarly recruited to execute the tyrant or freeloader.

pigmented materials that had definitive but less regular use starting 500,000 years ago, but widespread and regular use within limited areas by 300,000 years ago, that were transported over long distances. These researchers concluded that "brilliant ritualized display" was a hallmark of some human groups by half a million years ago.[42] To summarize a complicated picture of when the first Earthlings began to leverage the power of identity for collective action, we have a possible range of 50,000 to 500,000 years!

We can say with some level of confidence that somewhere in Africa, 300,000 years ago, the first person identified themselves with no small measure of pride as part of an "us." At least, this would be consistent with current reevaluations of the timing of complex human social organizations that argue for groups that were more sedentary, unequal, large, and politically stratified by the Late Pleistocene some 130,000 years ago. Critically, all these societal traits would have required large-scale cooperation—and therefore tribal identity—to be able to manage resources in a scaled way.[43] Regardless of the precise geological timing of that moment, when it happened, group-level identity and affiliation—undergirded by symbols—was born.

But this happened at a cost. I call this cost *the cursed blessing*. In fact, it was a recalibration of the scale of violence to a much greater, nastier form. Humanity had been blessed with greater benevolence within, but a storm was brewing in the Middle Paleolithic. The specter of pseudospeciation was cast onto our people, and it remains with us to this day. Fear of the other is humanity's greatest vulnerability to the hungry, dehumanizing ghost within. Its ultimate expression is humanity's greatest curse and the darkest component of the Tribe Drive—genocide. The complete and wholesale destruction of those who are not "us."

The question of *why people kill* has plagued philosophers for time immemorial. The sad and unnerving answer is that for certain human groups in our ancestral past, there were adaptive benefits to killing people they didn't identify with. The somewhat terrifying proximate mechanism (explained in greater detail in the next chapter) is that in certain cases, it even *feels good* to kill: "Evolution has made the killing of strangers pleasurable, because those that liked to kill tended to receive adaptive benefits. . . . Using cues such as the stranger's *weapons, dress, and dialect,* [emphasis mine] the warrior can . . . [regard the hostile] as non-human." Just like with warring chimpanzee communities, the grisly logic is sound; when each group relies on its own strength for survival, reducing the power of rival groups by killing their males brings inherent reward, and "the rewards do not have to be anticipated consciously. All that is needed is enjoyment of the kill. Sexual reproduction works in a parallel way."[44]

Polly Wiessner is a professor of anthropology at Arizona State University who has dedicated her professional career to working with the Enga of Papua New Guinea. She recorded the logic used by the Enga when they engaged in war: "Now I will talk about warfare. This is what our forefathers said: When a man was killed, the clan of the killers sang songs of bravery and victory. They would shout Auu! ('Hurray! Or 'Well done!') to announce the death of an enemy. Then their land would be like a high mountain and that is how it was down through the generations."[45] The enjoyment of the act of killing those who do not possess your tribal identity here is clearly communicated.*

The reason the curse came with the blessing is because humans have the strange cognitive property of being able to be manipulated into feeling more or less related to someone than we objectively are. When it's the former—scientists call this *pseudokinship*—we are overcome with empathy and perform amazing acts of altruism by donating to, advocating for, adopting, and loving humans that are not related to us; when it's the latter—scientists call this *pseudospeciation*—we can be convinced that these same non-related humans are akin to vermin, cockroaches, pathogens, and other various animals we consider nasty.[46] This is the playground of the propagandist ideologue that uses the power of the cursed bug in our hardware to drum up xenophobic hatred of the out-group.

The archaeological evidence demonstrates that intolerance and cooperation are not antithetical—in fact, they can go hand in hand. Unfortunately, the logic of proactive coalitionary aggression turned outside the group goes beyond safely eliminating a single tyrant and scales up to

* This may help explain why the vast majority of competitive video games are simulations of killing out-group. Also consider fantasy and science fiction narratives where orcs, goblins, and aliens are proto-dehumanized sentient lifeforms that must be exterminated before the hero's journey ends.

using strength of coordinated numbers to eliminate rival groups. Coalitionary aggression is incredibly ancient—it is also seen in wolves, chimpanzees, whales, some ants, spotted hyenas, and a few other animals—but it wasn't until the Tribe Drive emerged that it came into its ultimate form.

This expression needed ethnolinguistic groups that leveraged tribal signals to rally warriors into tribal warfare. The phenomenon is called cautionary proactive aggression and, once innovated, it made for a more nerve-racking human experience. Hunter-gatherers who are outmatched technologically and live adjacent to agropastoralists are relatively peaceful.[47] But when it is only hunter-gatherers, dwelling on a finite amount of land, rates of warfare increase substantially. Evidence for this comes from pre-colonial Australia, where there were six hundred ethnolinguistic tribes, and warfare was endemic throughout the continent.[48] The scale of warfare is at the raiding level, where small fireteams of six to ten males use coalitionary proactive aggression for nighttime attacks on other groups.

But warfare really took off after the agricultural revolution some ten thousand years ago. Statistically, hunter-gatherer median death by violence rates averaged at 165 per 100,000 per year. Incredibly, small-scale farmers averaged 565 per 100,000 per year. To put things into perspective, during World War II, the economically developed powers experienced a rate of 20 per 100,000 per year.[49] Once humans had mastered clumping resources, the patriarchy came into its ultimate prominence with a winner-take-all system where the most coordinated coalitionary aggressors could amass amazing, never-before-seen resources under the banner of a single chiefdom. Single male lineages ran the genetic table, with only a few surviving the bottleneck of continuous warfare and most male lines going extinct.[50] During this human era, life was "continual fear, and danger of violent death; and the life of man . . . poor, nasty, brutish, and short."[51] It's not like it was fun being female during this epoch of human history, but the evidence suggests females did not suffer the same odds of death by murder as males; being a male was nearly a death sentence for those who couldn't fight or lacked the skills to endear themselves politically to dominant male coalitionary alliances that revolved around patrilineal kin groups.

Ultimately, evolution's *blessing* was intensified cooperation within groups creating little dissension within the in-group bound by ritualized loyalty. But the blessing came with the *curse* of increased capacity for groups uniting in a conformist way in deciding who is an out-group enemy—and these considerations will often be extreme, resulting in dehumanizing pseudospeciation. Coalition cognition is exaggerated by unifying and discriminating effects. And so the *cursed blessing is revealed.* The stunning conclusion is that this "cursed blessing" is also the answer to a most profound moral question: When is it OK to be violent versus compassionate? It is the reason why the following moral intuitions, with all their context dependencies, run widespread in our species:

- **Within the Group:**
  - It is "OK" (fitness enhancing) to be compassionate when living in close proximity to other humans in a community that protects each other as sentinels to outside threats and when serving as alloparents in the project of cooperative breeding.
  - It is "OK" (fitness enhancing) to violently dispose of tyrants and freeloaders.
- **Outside the Group:**
  - It is "OK" (fitness enhancing) to be compassionate to green-bearded strangers—those who share signals of identity that bootstrap cooperation among large social networks beyond the human channel capacity.
  - It is "OK" (fitness enhancing) to dehumanize out-groups where competition is favored over cooperation with respect to a limited pool of resources, and to use violence against them.

The Tribe Drive was a genetic and cultural innovation that ranks comparatively in significance to innovations with tool use and language in terms of relative importance to our species' success. Those non-tribal, more wild variants of humans that were less inclined to share their intentions became extinct. As a result, we evolved moral intuitions. These are embedded in the Tribe Drive and they impact our religion, our politics, and effuse our day-to-day social connections. The factors involved in the evolution of the Tribe Dribe, especially the driving influences within the domestication syndrome in our species, resulted in a paradox: the fact that we are as peaceful within a group to each other as we are today came at the cost of the capacity to proactively, and without much emotional strain, decimate all humans within our group that do not conform to our social norms *and* other humans we do not identify as being from our own tribe.

The Lykov Extinction

The natural experiment of the Lykov family is about to come to a close. Since first contact, their story traveled far throughout Russia. They had a crude spinning wheel to fashion clothing but had no technology for replacing the metal tools they had originally carried to their homestead. The prized metal goods they had were kettles, but by the time the geologists came upon them, they had rusted over and been replaced by birch bark bowls, which made cooking a challenge as the Lykovs could not place them in the fire. The staple diet, by the time of their discovery, was potato patties mixed with hemp seed and ground rye supplemented by pine nuts, bilberries, and raspberries that grew in abundance. The family, though, was constantly in a famine state, and in the 1950s they experienced what Agafia recalled as "the hungry years," when they ate certain fallback foods, such as rowanberry leaf. When Dmitry became old enough, having no guns or bows, the male Lykovs taught him to hunt by digging traps or going on endurance pursuits where, like typical endurance hunter-gatherers, they would follow the prey until it collapsed by exhaustion. The anecdotes the family collated over the years were incredible feats of survival. One time, Dmitry was hunting barefoot in winter and was gone for several days; he survived by sleeping with a slain elk across his body during the night.

Hunger was their primary enemy, and in 1961 Akulina chose to die of starvation to save her children. When they were gifted salt again after human contact, the ones who had been alive long enough to have tasted it before said that living without it had been "true torture." After contact, although having brief visits to the southern settlements and even witnessing small towns, the family remained in their homestead.

In 1981, likely as a result of their limited diet, both Natalia and Savin died from kidney failure and Dmitry died of pneumonia. The family that had lived together in the wild for decades capitulated in only a few days, with three of the four children dying and leaving Agafia as the sole heir of their family's legacy. Twenty-seven years to the day after his wife, Akulina, died of starvation, in 1988 Karp Lykov—the man who had led the family into this natural longitudinal experiment—passed in his sleep. As of the last reports from 2019, Agafia, now over seventy years old, is the last remaining Lykov. She has no plans of leaving the taiga.

This Russian family showed incredible resilience and endurance. By all accounts, they were pleasant to be around and reports of their behavior recounted no descent into moral depravity that we know of. Yet, despite this, the family was doomed as soon as they opted to live in a social unit smaller than a nuclear camp. It's clear that a lone human would not permit even the most apt hunter-survivalist to successfully reproduce a lineage. Likewise, the belief that "love conquers all" would do little good for a male-female pair-bond that was isolated in the wild. If they did not die of starvation, and they created a nuclear family, then their offspring would be so inbred as to leave them little chance of not succumbing to natural disasters or a bad run of luck with genetic stochasticity. A nuclear family plopped into the wild alone and isolated would likely go extinct; attempts for nuclear families to "go it alone" have historically been met with tragedy by privation, starvation, incest, and extirpation.

In an ironic bug in the Tribe Drive software, the Lykovs justified their nuclear family's self-isolation because their religion (as we will see, a classic ideological trust signal) told them the end of the world was near. In a way, their instincts weren't too off the mark. Tribal coalitions had drummed up proactive aggression, producing a war the likes of which the world had never seen. By going it alone, the Lykovs didn't even know that millions from the Russian tribe had died in World War II. But the trade-off, in the end, was not worth it. Although they avoided a world war, their fate was sealed. If kidney failure, pneumonia, and starvation had not killed them, then they would have faced even greater long-term challenges of the eventual homogeneity of their genes. If they had left with a group of families, with approximately thirty sex-balanced adults and children of distinct genetic stock, they could have made a go of it. But they had no such healthy, fertile social network within which to run their experiment. And so, when the last Lykov dies in the Russian taiga wilderness, the experiment will come to its conclusion.

4

TRIBAL BEHAVIOR, TRUST IN ACTION

Eagles versus Rattlers

When we reach a certain level of familiarity with our close social others, to some extent our neurons don't know the difference between ourselves and other people. . . . Our brains do this . . . by including familiar others in our neural representations of self, in our brain's mapping of what is self and what is other, what is us and what is our friends.

–Sarah Rose Cavanagh, 2019

The insula (insular cortex) activates when we eat a cockroach or imagine doing so. And the insula and amygdala activate when we think of the neighboring tribe as loathsome cockroaches. . . . This is central to how our brains process "us and them."

–Robert Sapolsky, 2018

A Tribal Behavior Just Occurred. How Did It Happen?

On September 28, 1991, an estimated 1.6 million people gathered together in Moscow are locked in a chorus chant.[1] Metallica, arguably one of the greatest heavy metal bands of all time,* is playing one of their hit songs, "Creeping Death." The song is a jarring narrative depicting the apocryphal Old Testament book of Exodus, where a fog-like plague

* The band's instrumentals, fast tempos, and aggressive musicianship made them one of the founding thrash metal bands that revolutionized music in the United States in the early 1980s. Now inducted in the Rock & Roll Hall of Fame, the band has won nine Grammy Awards and as of 2018 has sold over 125 million albums worldwide.

afflicts ancient Egypt and takes the lives of all first-born sons (including the pharaoh's) who are not tribally protected by the lamb's blood aegis wiped onto the door jambs of the Jewish dwellings. The chorus chant is "DIE! DIE! DIE!" Over a million Russian and Eastern European voices echo this chant.* The authoritarian Soviet regime does not know what to do. Decades of repression, frustration, and collective resilience in the face of oppression are let loose in a giant roar, in one great anguished refrain.

The Soviet military begins walling the stage off, AK-47s in hand to serve as deterrents. Propellers beat harshly a few hundred meters above the metal music–loving tribe as a Soviet Mil Mi-24 gunship helicopter hovers overhead. Despite their orders, soldiers are losing themselves in the collective ecstasy. They begin to throw their guns to the muddied earth, launch their *pilotkas* (caps) into the air, and rip off their uniform overcoats and use them as tools to amplify their displays of oneness with the tribe by waving them in unison with the drum, bass, and soaring guitar solos. Never before, and perhaps never again, have so many human minds been linked in a musically driven mass synchronization of neuroendocrine states. In their physical prime, Metallica plays furiously—as would a cabal of warlocks and wizards crafting a high-level spell before its unleashing. They masterfully summon their incantation of controlled aggression and give to the tribe (in my humble estimation, having attended six of their live concerts) their all-time greatest performance.

From an artistic standpoint, it is a magical moment never to be recaptured. From a scientific standpoint, if you were to hook up an fMRI to the 1.6 million people participating in this song and dance collective, the graphs depicting the firing of neurons would be stunningly identical. In a very real sense, 1.6 million people were one—their neural patterns screaming "*We are legion*." The effect is so profound and significant for some that they request that the time and place be carved onto their tombstones, "I was at Monsters of Rock." Eighty-eight days after Metallica's performance, the Soviet hammer and sickle is lowered for the last time over the Kremlin, ending the Cold War—the Soviet Union

* With the number of people attending this concert, Madison Square Garden can be filled 175 times over.

is dissolved. Some credit the Soviet metal movement, and this concert, as playing a part in this world event.[2] The flag of the new Russian Federation flaps in the breeze.

A TRIBAL BEHAVIOR HAS JUST occurred. How did it happen?

If you are an endocrinologist, your answer would be, "Because circulating testosterone levels worked in a certain part of the brain to enhance aggression." If you're a psychologist, your answer would be, "Because the individual had been primed by identifying with a minimal group." And if you're an evolutionary biologist, you'd say, "Because over the course of millions of years, the ancestors of the actor in question that biased the direction of altruism to those within their tribe over other tribes were more fertile and left more copies of their genes." This is a type of "thinking in categories" strategy. It is the standard way we talk about and attempt to explain phenomena. It uses differing scientific disciplines of explanation to understand how behaviors occur. However, these things are intrinsically intertwined. As noted by Robert Sapolsky:* "When you explain a behavior with one of these disciplines, you are implicitly invoking all the disciplines—any given type of explanation is the end product of the influences that preceded it."[3] This technique is so valuable because each level of drives, by definition, has an interaction.

Behavior is super complicated. Categories can be arbitrary and the artificial lines we put between things sometimes cloud our ability to see the big interconnected picture. We spent last chapter thinking about the evolutionary *why* behind our tribalism, and with that foundation

---

* In fact, this entire chapter's theme is inspired by Sapolsky's *Behave*. In a tome that can be described as no less than a scientific epic, he masterfully tours all the levels to explain not just one type of behavior but *all* human behavior using this technique. As he notes: "If you say, 'The behavior occurred because of the release of neurochemical Y in the brain,' you are also saying, 'The behavior occurred because the heavy secretion of hormone X this morning increased the levels of neurochemical Y.' . . . You're also saying, 'The behavior occurred because the environment in which that person was raised made her brain more likely to release neurochemical Y in response to certain types of stimuli.' And you're also saying, '. . . because of the gene that codes for the particular version of neurochemical Y.' And if you've so much as whispered the word 'gene,' you're also saying, '. . . and because of the millennia of factors that shaped the evolution of that particular gene.'" And so on.

we can now explore the fascinating scientific terrain of *how* tribalism is expressed. This chapter will tackle many scientific specialties and weave a coherent narrative involving biological and cultural evolution, early life experience, prenatal environment, sensory cues, hormones, and brain chemistry in a series of comprehensible levels of drives. Instead of explaining all tribal behavior with a single discipline, we'll be thinking about myriad interdisciplinary levels. Our aim is to reverse engineer our way back in time, from the split second before the tribal behavior occurred back to the place where evolution first crafted the parameters for the behavior. To help in this task, let's draw upon a thought experiment that will guide us—arguably the greatest scientific experiment in tribalism ever executed—the 1954 Robbers Cave Experiment led by Muzafer Sherif.[4]

IN NEARLY EVERY CONTEMPORARY DISCUSSION on the origins of tribalism, on in-groups versus out-groups, no study is more cited than the Robbers Cave Experiment and the titanic, sandlot struggle between two artificially constructed tribes: Eagles versus Rattlers.[5]

Sherif, a social psychologist motivated by the events that had unfolded from World War II, was interested in exploring the drivers of group conflict. The legacy from his work is known as *Realistic Conflict Theory*, which proposes that group conflict, negative prejudices, and stereotypes are the result of competition between groups for desired resources. In his most famous study, Sherif tested the hypothesis in an ingenious field experiment that involved two groups of twelve-year-old boys at Robbers Cave State Park in Oklahoma. Sherif controlled for background, as the twenty-two boys were all recruited from white, middle-class families, all shared two-parent, Protestant homes, and all were previously unknown to each other. After being randomly assigned to one of two groups, the boys were kept separate from each other and were encouraged to bond as two individual groups through the pursuit of common goals. In the first phase, the groups were unaware of each other's existence. The boys went to work quickly on developing their group norms and forming their own unique identity and culture.

The Eagles and the Rattlers were the self-forged identities they created for their groups, and once they decided on their totems, they readily adopted the tribal tokens by stenciling the icons of snakes and eagles onto their shirts and flags. Sherif now engaged the next phase, where competition between groups came to a head. Competitive team tasks were assigned, including tug-of-war and baseball, and the winners received rare resources such as trophies, medals, and pocketknives. The Rattlers, having created a "toughness" and default aggressive cultural norm, boasted about their victories. They discussed violent ends to an Eagle that would dare deface their tribal icons. After a few losses, the Eagles burned the Rattlers' flag. The reprisal the next day was swift and furious. The Rattlers escalated by ransacking the Eagles' cabin. The researchers separated the boys after the aggression turned into physical violence.

After a brief cooling-off period, the researchers asked the boys to characterize their group and its members. Presaging the psychological work to emerge later in the twentieth century on minimal groups, the boys tended to characterize their in-group in very favorable terms, and the out-group unfavorably. For the first time, empirical evidence demonstrated that prejudice and discrimination does not need phenotypic or ideological traits (such as race or religion) to be leveraged against out-groups. Competition over scarce, limited resources had brought an extremely demographically homogenous group to the brink of hatred for each other. But, could they come back from the brink? And by what mechanism could they be restored? We will drop in, by way of a thought experiment, on the camp from time to time to understand basic biological principles and solutions to help overcome seemingly intractable tribal behavior.

Tribal Brains—One Second Before

The most proximal level to understand a behavior is the realm of the neuron. On this level we can come to understand how the brain dictates muscles producing a tribal behavior. Recall earlier our brain wave analogy that likened the brain to an ocean, with ionic swells of energy combining to produce a conscious experience. The second before a tribal behavior occurs, the molecules composing our neuroanatomy pulse with swells of electrochemical energy channeled through the neurons. If the axon hillock is pulsed with enough energy to break its threshold from -40 mV to 30 mV, then the wave crests upon the shores of consciousness and we feel, sense, or think a thing. The brain "is the conduit that mediates the influences of all the distal factors," and directs its energies toward its inner and outer social environments while expressing its Tribe Drive.[6]

The brain is not some homogenous indistinguishable tissue mass. It is chock-full of organization. It binds cell bodies of neurons with related functions together in particular regions; the axons are the projection cables that network the neuronal circuitry together. This results in a brain that has different parts that do different things, and these different parts guide the currents of electrical energy that ultimately expresses itself as a behavior. Our objects of interest are the neurons themselves and the brain regions and the circuitry that binds them together. We are not going to labor over Neurobiology 101,* so I will highlight only the key neuroanatomical players implicated in the Tribe Drive, but it is worth summarizing a metaphor, proposed by the neuroscientist Paul MacLean in the 1960s that has been a helpful (if not oversimplistic) conceptual model of the macro-organization of the brain.[7]

- **Layer One:** The "primitive" ancient brain endowed in all brainy animals. This layer regulates automatic functions key to survival, cold, hunger, injury, and stress responses.
- **Layer Two:** This is the emotional layer. Fright, flight, sadness, joy, desire for food, etc. A key aspect of layer two is the limbic system, which processes emotional cognition.
- **Layer Three:** Crowning both layers, on the top is the neocortex, where advanced, executive cognitive processes are dedicated to memory storage and abstractions using all of the above categories.

* Sapolsky presents a masterclass for those uninitiated in neurobiology. If you know nothing about the brain, but want to, look no further than *Behave*.

From here we can peruse the brain's different types of neurons that compose systems of functional clumps of brain tissue that ultimately guide tribal behavior.

The "Us" Brain

We arrive at the modern proxy of a Robbers Cave State Park summer camp. You get off the bus, are assigned your lodging, and after some time working together on crafting an identity with your camp mates, you are adorned with a freshly minted identity. You are now a proud member of the *Eagle tribe.*

You are hiking with your Eagles and have been on the march for a few hours. The terrain is getting tricky. Gnarled trees and extended roots protrude from the surface of the forest's leafy, matted floor. You are conversing with a comrade and, all of a sudden, she trips on a root, tumbles to the ground, and screams in pain. She is caught up in a knotted, thorny bramble. As you kneel down to help, you try and stay composed, but it's challenging as you literally feel her pain as your own.

Humans are compassionate and this capacity for empathy is contagious. On a neurobiological level, the underlying systems enabling the senses to perceive closeness with our in-group—when our conscious minds meld with the collective—is due to the breaking down of the physical and psychological divisions between us, caused by mirror neurons and the anterior cingulate cortex.[8] From the perspective of a neuron firing, seeing someone perform an action and actually performing the action are the same thing, and on this basis arguments have been made for the neural basis of empathy.[9]

The neurobiological crossroads of empathy is the *anterior cingulate cortex* (ACC). This is a frontal cortical structure that resides in layer three but also projects into layer two, and it is particularly sensitive to the perception of pain. It also monitors and responds to conflict and the expectation of conflicting behaviors. The ACC is really interested in the *meaning* of the pain. When our Eagle ally fell into the thorny brush, your ACC activated, and you felt her pain.[10] The more pain, the more the ACC is activated, and the corresponding activation predicts you doing something to alleviate their pain. When you knelt down to help your campmate, it was at the command of the ACC's action potentials firing.

You play a game of catch with two of your fellow Eagles. After you drop a few balls you notice your two compatriots start throwing to each other with greater frequency. You feel an inward (neurobiologists would call it an interoceptive) pain at being excluded from the game. You come up with an excuse to leave the field of play and go off to sulk by yourself.

Again, the ACC is implicated. Remember, it cares about the meaning of expected behavior, and it's just as concerned with the abstractions of emotionally laden social pain such as anxiety, embarrassment, and social exclusion. (In fact, a fascinating line of research indicates that anomalies in the ACC are associated with major depressive disorders. Poorly working ACCs can lead to the expression of horrible, world-darkening sadness.) When you were excluded from your in-group's activities, your ACC fired, and you felt inward, self-oriented, emotional pain.

You are playing baseball and your closest friend is up to bat. Down by three runs in the bottom of the ninth with the bases loaded on a 3–2 count, she crushes the ball over the fence, dramatically hitting the game-winning home run for your team. You feel a sense of ecstasy. It's as if you *hit the home run, though you're sitting in the dugout.*

The ACC erupts and you are flooded with positive emotions. Mirroring is a type of process that permits the individual to experience something as a social other would experience it. The scientific work demonstrating this principle has been mainly generated from neuroimaging scanners (for example, people hooked into an fMRI who begin tapping their own fingers and then stop to watch movies of others tapping their fingers) that show watching others perform an action produces the same effect in their neural regions as if they had performed the action themselves. One famous and controversial theory about how this works is that the mirror neurons help you understand what someone else is thinking by simulating the other's actions inside your brain.[11] Now that we have a better understanding of the brain on "us," it is time to consider the brain on "them."

The "Them" Brain

Humans are discriminatory about who they are compassionate toward, and this is a function of both social and physical proximity. Therefore, the ACC is involved in this process as well.

You have been summoned by the camp's directors to a nighttime fireside intertribal meeting, with both Eagles and Rattlers. Naturally, your Eagles congregate in a semicircle around one side of the fire, opposite the Rattlers. A bed of burning coals and simmering ash is spread in a path. A representative from each tribe is summoned to perform a fire walk. You witness your leader walk the glowing ember path and your heart races. You rub your aching feet as the leader conquers the fire walk. When you observe the Rattler leader perform the same act you feel a cold nothingness and an indifference to their experience.

The neural mechanism the brain uses to cognitively distance another person is disgust and, as we will see, the *insular cortex* (henceforth *insula*) is the prime mover of this sense-driven feeling.[12]

THE ACC IS HYPER–CONTEXT-SENSITIVE TO how we classify individuals we are observing. Jean Decety, a neuroscientist and empathy researcher at the Department of Psychology at the University of Chicago, remarked that "empathetic arousal [was] moderated early in information processing by a priori attitudes toward other people."[13] The closer the classification of the observed individual undergoing some stressor worthy of empathy, the greater the ACC activation. As a result, a stranger or a person from a group we don't like receives little to no ACC firing.[14] Finally, when the differences between you and the person in pain are great, there is an extra cognitive cost to processing whether or not to help them.

The profound and lasting implication of this processing cost is that the worst outcomes of tribalism are not the consequence of a moral crisis but in fact an energy crisis! The frontal cortex*—the part of your brain that is the utilitarian calculator of whether or not it is worth doing the hard thing now for a better outcome later—undergoes serious cognitive strain when assessing whether or not to help a stranger or out-group.[15] Close relationships do not get taxed this way, and so our empathy pool can be drawn on automatically and unconsciously with no cost to ourselves.[16] The *amygdala* lights up on brain scans when an animal is actively being aggressive.[17] In humans, if you stick an electrode in someone's amygdala and stimulate it, the product is rage. Crucially though, the amygdaloid function is also linked with both fear and anxiety. In times of social chaos, where people are unsure about their place and role within their group, the amygdala freaks out and causes fear and anxiety.

An incredibly important input into the amygdala is the sensory projection from the prefrontal cortex's insula. Humans have a significant feature (or bug) in our hardware that codes morality with gustatory disgust, similar to how we experience fear and anxiety. When people are hooked up to neuroimaging machines and asked to consider people doing bad things (i.e., social norm violations) the insula cascades this prefrontal cortex abstract of "moral violation" down the input into the amygdala. Tellingly, the same firing pattern emerges if you smell or taste something toxic or gross. This also happens when people observe others who have been stigmatized for social norm violations. This may in part explain much of the ills of social media and cancel culture. When someone gets "canceled," it's because of some perceived violation of norms, so we excommunicate the person from our group lest we experience a disgust response.

Norms are so powerfully linked with our brains, they can even manipulate the senses. One study demonstrated that visceral disgust made people more judgmental.[18] Evolution crafted a brain from old parts. It works with what it's given, and this results in embodied cognition that

* The frontal cortex has been implicated in much of our species' capacity for higher thought. It is listed as being critical for working memory, gratification postponement, regulation of emotions, reining in impulsivity, long-term planning, and strategically organizing knowledge.

can be literally felt in our nervous system. Moral violations trigger the same neural stuff that makes you reflexively spit out bad food, gag, or vomit. This reflex protects you from ingesting pathogens and toxins. We don't just *think* about moral violations—we *feel* them. The most powerful factors driving these cognitions are emotional and automatic. The implications are significant because this means that Us-Them distinctions, spurred on by the amygdala and insula, are created at lightning-fast determinations that precede conscious awareness.

The automaticity of out-grouping—by recruiting the amygdala and insula—is crucial to the Tribe Drive and is one of the best predictors of the dehumanizing behavior that is *pseudospeciation*. This is when we categorize the other as something gross, non-human, and worthy of extermination. Think about eating a cockroach? Insula and amygdala fire. Think about that other tribe you are warring with? Insula and amygdala fire. Different inputs, same output. When you hear that type of dehumanizing "she/he/they disgusts me" language addressed from someone toward the other, beware—it means the amygdala/insula complex is firing. When this occurs it produces several flavors of out-group disgust, signaling that the out-group is repellent, primitive, and disgusting.

Tribal Senses—Seconds to Minutes Before

Cues are incredibly powerful. This section focuses on the background environment seconds and minutes before a tribal behavior occurs. If you look at your cell phone today you will likely be cued by it to behave in a way you would not have otherwise. Psychologist Robert Cialdini details this science in his book *Pre-suasion: A Revolutionary Way to Influence and Persuade*. Corporations have spent hundreds of billions of dollars in both funding the science of persuasion and then using that science to manipulate your behavior.[19] It makes us click on websites and social media posts every day. How does our social environment persuade us to behave tribally?

Humans are ultrasocial and as a result we have the capacity to reach states of consciousness that psychologists have termed "the boundarylessness effect." This is when our in-group becomes indistinguishable from ourselves, so much so that the analogy offered by Cavanagh is that humans are beelike in their interdependence. Cavanagh repurposes with modern gloss E. O. Wilson's revolutionary discovery that *eusocial* species* are rare, unique, and exhibit the highest level of organization of sociality.[20] Cavanagh writes:

> Rather than flocking like birds or swarming like ants, human beings may synchronize through the processes of emotional contagion and social conformity. . . . When human beings interact face-to-face, they tend to mimic one another's posture and facial expressions, feel similar emotions, fall in step when walking together, and easily share mannerisms and patterns of prosody and even eye gaze patterns. Synchronizing with other people also seems to drive bonding with them . . . like honey bees in a hive.[21]

Cavanagh gives us the context of things "that nudge our consciousness into a shared frame of experience, such as dancing in sync to a common rhythm and chanting and singing together. Situational factors that blur the boundaries of our bodies—such as darkness, alcohol, extreme tiredness, shared repetitive movements, and certain drugs—also encourage a more communal level of consciousness." Thus, synchronous motor movements with your social self increases feelings of solidarity—with the corollary of being more willing to engage in aggressive behavior toward others on the behalf of those you have deemed your in-group. These are all cues, and for some, the feeling can be akin to ecstasy. Perhaps this is why Jonathan Haidt, professor of ethical leadership at New York University, has metaphorically noted that humans are "90 percent chimp and 10 percent bee."[22]

JIM COAN, A PROFESSOR OF psychology at the University of Virginia, is a social neuroscientist that has proposed the *Social Baseline Theory*.[23] The core of this theory is that in the same way our bodies strive for

* Eusocial species are ones that cooperatively breed, overlap in generations within a colony (or in human camps), and have division of labor. This definition binds humans and social insects like bees and ants together in a shared strategy that has been one of the most successful ever crafted by evolution.

homeostasis in thermoregulation (seeking out heat if you are cold) or glucose levels (finding something sweet to eat if you are hungry), so, too, does our brain seek a constant social baseline. In other words, we need other human beings and there is a draining cognitive tax to being alone. We spend more cognitive effort and absolute energy in isolation if we are not embedded in a social network. If Social Baseline Theory is right, then the question is not *which* cognitive processes are involved in social contexts—*all* of our cognitive processes are situated within a social context.[24] When we're embedded in a social network our bodies and minds run more efficiently.

Coan's remarkable experimental innovations were critical to uncovering how this all happens. Most studies rip the individual from their social context and study their responses to stimuli in isolation. Coan hypothesized that this was a bias, and given the ultrasociality embedded in our species, he decided to introduce other people besides the participant into the experimental paradigm, with the hopes that doing so would reduce the level of artificial isolation bias. Rebecca Saxe, a neuroscientist at MIT, discovered that if you put people in social isolation, the areas in their brains that would normally show craving for food start craving social connection. You can literally see this craving effect in the brain.[25] The reason being that if you are socially isolated, you are vulnerable; vulnerable individuals in the wild better be careful because danger could come at any second, and you have only yourself to take it on.

Mark Levine, a social psychologist at Lancaster University, and his colleagues performed two experiments that demonstrate helping behavior bias. In the first study, intergroup rivalries between soccer fans were used to examine the role of identity in emergency helping. They discovered that an injured stranger wearing an in-group team shirt is more likely to be helped than when wearing a rival team shirt or an unbranded sports shirt. In a follow-up study, it was shown that helping is extended to those who were previously identified as in-group members but not to those who lacked signs of their own group membership. In sum, these studies show that when emergency strikes, those strangers with shared in-group identity have an increased likelihood of being helped.[26]

Moving on to pain perception, being around an in-group literally transforms our perception of pain. Coan first demonstrated this concept in work where participants were in a threatened state with a probability of getting electroshocked on the ankle. Subjects who held the hand of someone they knew well experienced reduced threat processing compared to subjects whose hands were not held at all.[27] The effect seemed apparent even when handholding was not involved and close familiar social connections such as family members, relatives, and siblings were simply present in the room. This effect has been corroborated with more recent work that demonstrated that handholding with a partner reduced neural circuitry related to pain interpretation and reduced self-reports of pain intensity.[28]

During a titanic struggle where you beat the Eagles in a rugby game, you ended up with a concussion. Now in the infirmary, your fellow Rattlers huddle around your bedside as you recover. When you woke up alone, you were sullen and downcast because you wouldn't be able to help them during this week's activities. But when your teammates came in, they quickly started cracking jokes, and as though being taken over by a contagion, you were laughing along with them. It felt wonderful, and while they were there you totally forgot about your injured state.

Humans, then, may possess beelike superpowers we are not consciously aware of, but as with all such powers, what is the catch? What is the human kryptonite to the beelike collective consciousness we can experience? Ultrasociality has brought us the paradoxical combination of selflessness and selfishness, and it is manipulated by our immediate social environments. Experientially, the key driver for this phenomenon is *familiarity*. Remarkably, when we reach a certain threshold of familiarity with others, we stop differentiating between us as separate selves. It's part of the cursed blessing, the yin and the yang embedded in our humanity. On the one side there is compassion, love, and selflessness; on the other side is bigotry and hatred. Once we understand this, we can begin to intervene and uphold those tribal signals that benefit humankind.

Tribal Hormones—Hours to Days Before

We leave neurology and now and enter the kingdom of hormones. "Hormones don't determine, command, cause, or invent behaviors. Instead, they make us more sensitive to the social triggers of emotionally laden behaviors and exaggerate our preexisting tendencies in those domains."[29] Two important denizens of this kingdom are *testosterone* and *oxytocin*. Testosterone is famously linked with aggression. More of it plainly means more aggression—right? Not necessarily. Oxytocin has a mythic and romanticized narrative surrounding it as the love hormone. More of it means more "kumbaya" togetherness—right? Again, not so. As we'll see, our old friend "context" is critical to understanding how these hormones prime the Tribe Drive.

IN GENERAL, TESTOSTERONE CAN REDUCE fear and anxiety while increasing optimism and confidence.[30] The "winner effect" of testosterone is seen in rats that win physical challenges, in humans that win board games and athletic competitions, and even in brokers on a good stock market bull run. The trade-off is that it also can lead to overconfidence, resulting in a personality effusive with egocentric cockiness. As an added kicker, testosterone secreting into your system is pleasurable to experience, and it also boosts impulsivity and risk-taking, nudging choices toward the option to "do the easier thing when it's the dumb-ass thing to do."[31] Its relationship with aggression is nuanced. A 1977 study dosed middle-hierarchy male talapoin monkeys with testosterone. It didn't create new patterns of aggression but exaggerated persisting ones.[32] We don't have to be physically in a challenge to experience testosterone's effects either, as simply *observing* your favorite team* win (pick any sport) will raise it, while losing will lower it.[33] This gives us a powerful clue that this hormone is about the psychology of identification, status, and dominance.

Suffering a string of competitive defeats at the hands of the Rattlers, your Eagle tribe is in disarray. Not only do you feel the loss of group

* Your doctor tells you that you've got low testosterone and you need a supplement? Go the homeopathic route and move to a city with a winning sports team.

status throughout the entire camp, but it has caused a shuffle among the strongest and smartest girls in Eagle tribe. You are fighting with some of your camp mates, but you are spending more time with others and gifting them precious camp resources from your own stash to solidify your bonds. If you play your cards right, you may actually be calling the shots in the Eagles when the time comes to stand up to the Rattlers.

Far from the nasty reputation testosterone has for increasing aggression, the latest scientific evidence about the function of the hormone points to a much different role. Testosterone isn't the aggression hormone—it is the *status maintenance hormone.* And if maintaining status requires one to be more generous, kind, and prosocial, testosterone will prime those behaviors. In one study this was demonstrated by elevating the participants' sense of pride based on honesty. When subjects decided how many resources they would keep versus publicly contribute to common pools, testosterone actually prompted the players to reduce their lying in the game and increase their generosity.[34] In the right context, where good reputations increase social standing, testosterone is prosocial.

One quick but fascinating note about estrogen: women's ovulation influences coalitionary cognition. Carlos Navarrete at Michigan State University demonstrated that ovulating white women express more negative views directed toward Black men.[35] In a follow-up study in 2008 where subjects were presented with different shades of brown to black and asked which most accurately matched U.S. president Barack Obama's skin color, women who viewed him as darker were less likely to vote for him while ovulating.[36] Amazingly, the propensity of Us versus Them can be driven by such under-the-skin forces as our hormonal cycles!

Oxytocin and vasopressin may be the most important of all tribal neuromodulators. They are a neuropeptide cocktail that is produced by the hypothalamus.* Touted by evidence of its role in social bonding in work done with prairie voles, it has famously been dubbed nature's love

* An ancient, evolutionarily conserved region in the limbic system of the brain responsible for regulating eating, drinking, temperature and energy maintenance, memorization, and stress control.

potion.[37] And it can facilitate both love within and between species. For example, this love potion is released between humans and dogs when they share prolonged gazes.[38] It is all about the business of bonding units of life together . . . but there is a catch.

The peptides also have a dark side. Importantly, these neuromechanisms appear to be central to the psychological underpinnings of attachment, the development of in-group solidarity, and the formation of out-group hostility.[39] The context in which it gets nasty is when strangers are involved. One economic game study showed that if the other player is anonymous and in a different room, then oxytocin decreases cooperation and even boosts behaviors like gloating (in the case of good luck) and envy (in the case of bad luck).[40] Shockingly though, when faced with the prisoner's dilemma, the same oxytocin increased the odds of backstabbing the other player.[41] This could mean that one of the key tribal functions of these peptides is to enhance our social ability to make us better at identifying who is an Us so that we can direct more prosocial behaviors toward the "Us-es" at the expense of the "Thems." Instead of the love potion, we could be calling vasopressin and oxytocin the tribe potion.

One final, yet important note on *stress* before we leave the kingdom of hormones. It is of no coincidence that some of the most important behaviors we'll ever make will be performed under stress—from giving a lecture to being chased by a lion. These stresses produce glucocorticoids in the classic "freeze, fight, flight" behavior that enables you to mobilize life-saving energy stores at a moment's notice. Glucocorticoids manipulate who we direct empathy toward in times of stress. The crux is that cortisol *reduces* who we count as in-group.[42] Stress narrows our "Us" by making it easier to reject "Them" from our sympathy group. Importantly, stress shrinks our moral circles of concern.

Tribal Learning—Days to Months Before

We now know a few of the factors that are involved in our tribally driven behaviors: the neuronal axon hillock's threshold for initiating the action potential; the familiarity of the people in the setting; the excitability of the neuron based on the flood of hormones that primed it—all nudging, cajoling, and pushing us to do the tribal thing. But what additional factors come before the behavior? Surprisingly, there are substantial changes in the brain based on preceding experiences. Scientists call this domain *neuroplasticity.*

We will call this *learning.*

In time, a tribe can experience a famine, a devastating disease, or war with another group, all of which are sufficient to radically alter brain structure. It's time now to return to our Eagles versus Rattlers camp experiment to understand how some of these learned behaviors can be modified to create unity from two diverse groups.

You are a Rattler through and through. For your loyalty, determination, and Rattler grit, you are one of, if not the most, influential people in the group. Maybe that's why you were chosen by the camp directors among a few other die-hard Rattlers and Eagles to have your tribal membership switched in an effort to diffuse the animosity. You hate this change, as now you must bunk alongside the kids that have been your sworn enemies for weeks. The first days are miserable. You miss your Rattler friends and the Eagles do things so differently that it's disorienting. One practice the Eagles do is morning drills and exercises. At first, you found the early morning wakeups annoying; but after several days, you begin to find a rhythm to the routine. As you perform these daily rituals, where you move together through several types of physical challenges, you notice the hate toward the Eagles ebbing, and when you are in sync you even feel a sliver of unity with them.

Humans have heavily relied upon reenactment. Functionally, reenactment has been described as practicing for future events that may draw upon similar skills. For example, in hunter-gatherers, specific dances that reify group cohesion are intended to help lead the group to later successful hunts. These reenactments multiplied over time lead to the creation of many other cultural norms.[43] For example, William H. McNeill, a World War II veteran and professor of history at the University of Chicago, wrote of the euphoric feelings experienced even under conditions of duress while he was a soldier and practicing combat drills—what he termed a "collective ritual"—with his fellow soldiers: "A sense of pervasive well-being is what I recall; more specifically, a strange sense of personal enlargement; a sort of swelling out, becoming bigger than life." McNeill hypothesizes that "boundary loss"—the blurring of the sense of divisions between the collective and the self—occurs by syncing physical and mental inputs and outputs through "muscular bonding," leading to a greater group cohesiveness, kinship, and oneness. If there are any prosocial outputs of the military, the one reported most from soldiers is the feelings of kinship for their fellow comrades.

Neurogenesis is the creation of newly minted neurons by the brain.[44] Neurobiologically, on average, an adult replaces about 3 percent of their neurons each month. Importantly, the science shows that certain types of activities are particularly good at spurring on neurogenesis; among the prime movers of new cells being crafted are exercise, environmental enrichment, and brain injury with an inhibitory effect caused by the classic stress response.[45] Special types of exercise that combine learning new motor movements (think martial arts, yoga, dancing, Olympic lifting) with physical repetition top the list, because not only do they stimulate the body's core motor processing, they do so in conjunction with increasing growth factors in the brain and body.[46] Combine muscular output with group coordinating, and you've got the perfect recipe for enhanced group-level bonding.

Psychologist Henri Tajfel discovered that it takes very little coaxing to separate "Us" and "Them," which he dubbed the *minimal group paradigm* in 1970.[47] Tajfel discovered that the creation of categories—even something as simple as wearing the same T-shirt—results in people instinctually (and mostly unconsciously) identifying with one category at the expense of the other. The scary result is that this unconscious drive explains much of human behavior that drives ethnocentrism and xenophobia.[48] This can be explained by a phenomenon known as *mutual fate control*—meaning that even though one person may not be able to directly influence the behavior of another, the fact that they share a group identity means their fates are nonetheless bound together.[49] If group members feel they share a fate, then this togetherness facilitates expectations of mutual reciprocity and aid—ultimately leading to greater evaluation of the in-group. When we are tasked to reward our in-group, we expect to be more favorably treated by its members, and this is independent of whether or not you think your in-group is superior among other groups.[50] This is why we bias our groups. It's core to the Tribe Drive.

The good news is that because minimal groups are arbitrary, the boundaries of a minimal group can shift and change along with the experiences of the individuals within them.[51] Experience can change our identity, which can change our brains.

Tribal Development—From Utero to Before

All animals develop. Humans are no exception. Are there conditions in fetal development, childhood, and adolescence that prime the individual to express a tribal behavior?

In an attempt to uncover the developmental origins of group-level bias, in 2011, researchers led by Yarrow Dunham at the University of California, Merced, published a foundational study titled "Consequences of 'Minimal' Group Affiliations in Children."[52] A cohort of five-year-old children was distributed into minimal group categories based on randomly assigned red and blue shirts over the course of their time in a summer camp. As part of the experiment, the children were shown pictures and told stories about members of their in-group (having the same color T-shirt) or out-group (having a different color T-shirt). The result was a most compelling demonstration of implicit bias toward another group—the children demonstrated several strongly encoded unconscious beliefs about their in-group relative to their out-group.

First, they believed that their T-shirted allies were more generous and hence in any future exchange of resources, they biased members of their group. Second, when shown movies of an equal number of children in each group performing "good" and "bad" actions in precisely the same proportion, the children's free recall was highly biased against recall of "bad" actions for their in-group and skewed heavily toward recalling the "bad" actions of the out-group. This manifested into a third discovery: that overwhelmingly, the children preferred their own in-group as playmates. In the real world, kids start demonstrating sophisticated empathy around the age of seven, and by ten to twelve years old, that empathy has the capacity to be abstracted categorically (for example, empathy toward the poor). The trade-off is that when this capacity develops, they also begin negatively stereotyping other groups of people.

But how early do these traits manifest?

Consider studies that have shown young infants' visual preferences for new faces they have never seen before to be linked to their *familiar face* categories. For example, infants of African descent will stare for a much longer time at darkly pigmented faces than at lightly pigmented faces if they reside in African communities where such skin pigmentation is most common. Contrast this with Israeli-born Ethiopian infants who look equally to Black and white faces.[53] Additionally, language is also a factor. Five- to six-month-old infants gaze longer at people speaking their native language compared to people they hear speaking in a foreign language. It appears that the accents are the tip-off for the babies.

In a series of studies, ten-month-old infants in the U.S. and France were shown videos of a native French speaker and a native English speaker. The people appeared to be speaking directly to the children. The same infants were then shown events in which the two speakers appeared together without speaking. They held up two identical toys and silently, and in synchrony, offered the toys to the infant. At the very moment that the toys disappeared from view on the screen, two real toys appeared, giving the infants the illusion that the toys were generated by the individuals on the screen. American infants preferred English speakers' toys to French speakers', and vice versa, despite the toys being identical and having never been paired on screen with the language.[54] Young babies, it seems, develop the precursors to in-group preference throughout their infancy.

The reality of adolescence is that your frontal cortex does not reach its full number of synapses until your mid-twenties.[55] This has incredibly important implications. Delayed frontocortical maturation explains a lot about adolescent behavior. The key point being that evolution could have jump-started and ended our frontal cortex development earlier, but it didn't. It lets us flounder around in adultish bodies with an underdeveloped frontal cortex for nearly a decade. We are neurologically adolescent until about age twenty-five. Where is the evolutionary advantage in being stunted?

The neocortex "makes you do the harder thing when it's the right thing to do." It's the most recently evolved brain region and accounts for much of the neuroanatomical difference between a chimpanzee and a human. One important subregion is the prefrontal cortex (PFC). This is "the decider" when you're presented with conflicting options: to go outside to enjoy the nice weather and play hooky from work, or stay inside and finish writing a book. One would think it would be important to have the whole PFC fully operational as soon as possible. From the perspective of a fully developed adult brain, adolescence is a frustrating challenge to overcome; but from the perspective of evolution, it's a valuable rite of passage that sets up (the surviving) adults for success. Adolescence is the time in one's life where an individual is most likely to run away from home, innovate artistically, marry someone from outside their own group, commit a crime, commit themselves to activism, commit themselves to a religious cult, kill, be killed, or participate in genocide. In other words, this is the period of one's life when seeking novelty and taking risks nets you your most valuable future asset as an adult—your peer group. As a result, at this moment in your life you are most vulnerable to the siren's call of the Tribe Drive, and that is owed all to your underdeveloped PFC.

These studies on infants and language, the T-shirt studies, and the delay of PFC development until the mid-twenties in tandem significantly advanced our understanding of the developmental origins of group bias. It is clear that the behavior emerges early in human development (presumably before overt bias can be taught); it is likely instinctually embedded in the makeup of the Tribe Drive, following us, sometimes perniciously, throughout our growth and development.

Tribal Genes and Tribal Cultures—From the Moment of Conception to Millennia Before

Culture serves as a data storage system. It encodes the rights, rituals, customs, and accumulated intergenerational knowledge that aids human populations to overcome life-or-death environmental challenges. Let's consider courtship. Cultures took millennia to develop the rules that help the players navigate the complexities of human mating to its main end—the meeting of sperm and egg. The joining is a biological big bang of expansion. Genes interact with culture and culture interacts with genes—a process known as biocultural evolution. Cultures vary incredibly across the globe, but there are also a stunning number of cultural universals.[56*] Yet, in what ways do our genes and culture produce different types of brains that produce different types of culture (and vice versa) that drive the ultimate expression of a tribal behavior?

The classic paradigm of cross-cultural analyses in psychology relies on comparisons of collectivist and individualist cultures. In practice, this usually means generating samples from a survey in both East Asia and the United States, because each represents the extreme on the collectivist-individualist continuum. Collectivism is about conformity, interdependence, emphasis of group (not individual) needs guiding behavior, with the end goal being social harmony. In contrast, individualism is about uniqueness, personal achievement, autonomy to decide one's actions despite group desires, with the end goal being the actualization of the needs and rights of the individual within society.

One example from this extensive literature was a study that measured subjects' competitive drive. When requested to draw a diagram of their social groups, where circles were drawn to denote themselves among their friends, a pattern emerged where Americans typically placed the circle representing themselves as the largest, most centered circle on the page, whereas East Asians had more equal sizes to their bubbles with less center-focused placement of themselves.[57] For example, the Japanese have a special term for in-group–out-group designation: *uchi soto. Uchi* (内) literally means home, while soto (外) refers to outside. The core concept revolves around the idea of dividing people into two groups, an in-group and an out-group. Your family and close friends

* The anthropologist Dan Brown gives a lengthy list, partially constructed here: concern with magic, males and females having different natures, baby talk, gods, induction of altered states, marriage, body adornment, murder, prohibition of certain types of murder, kinship terminology, numbers, cooking, private sex, names, dance, play, distinctions between right and wrong behavior, nepotism, prohibitions of certain kinds of sex, empathy, reciprocity, rituals, afterlife mythology, music, color terms, gossip, binary sex terms, language, humor, and—important to the Tribe Drive—symbolism and in-group favoritism.

are considered *uchi* (in-group), as well as your coworkers and superiors at work—because in Japanese culture your workmates are considered extended family.

You have been waiting with anticipation all summer for the time when the sister camp in Korea visits for a week as a camp abroad international event. They arrived with lots of fanfare, and immediately you are struck by the behavioral differences between your group and theirs. You volunteered to lead some activities in the group. That's when you noticed that no matter what the camp counselors said, none of the Korean kids volunteered to be a group leader, whereas you leapt at the chance. When you played a group game, they worked harmoniously, whereas your groups were much more chaotic, with individuals in competitions wanting the spotlight.

What commonality binds the Tuvans of Mongolia, Gujjars in India, Maasai in East Africa, Bedouins in Arabia, and the Sami of northern Scandinavia with two of America's most famous nineteenth-century feuding families—the Appalachia-dwelling Hatfields and McCoys? The answer, which is eerily predictive of their tribal behavior, is subsistence strategies. Although the feud stems from chance encounters in the Civil War, the best-documented recorded instance of violence in the feud occurred thirteen years later over disputed ownership of a hog. In 1878, Floyd Hatfield, a cousin of "Devil Anse," had a hog, but Randolph McCoy claimed it was his, saying that the notches on the pig's ears were McCoy, not Hatfield, marks. In June 1880, star witness Bill Staton was killed by McCoy brothers Sam and Paris, who were later acquitted on the grounds of self-defense, which only spurred on the sense of injustice felt by the Hatfields. All these rather minor incidences preceded the Hatfields' and McCoys' infamous thirty-year feud, which resulted in dozens of deaths and peaked in a frenzied clan confrontation at a New Year massacre raid and the Battle of Grapevine Creek.

There are several well-quantified correlates of pastoralists that predict elements of their culture, ranging from militarism, authoritarianism, polygamy, down to the kind of afterlife you'll experience and even the number of gods* you worship.[58] Anthropologists note that most stem from the typically tough environment and the inability for centralized governments to enforce rule of law in widely dispersed areas. Unlike hunter-gatherer populations that have edible plants and animals to hunt or agriculturalists that have crops to harvest, the vulnerability of being a pastoralist—whose primary assets are herded animals (such as llamas, yaks, horses, reindeer, goats, sheep, cows, or camels)—is that your animals can be stolen by rustlers with a penchant for raiding. The key takeaway is that pastoralism cultivates *cultures of honor*. When your culture hasn't outsourced violence to the state, if you don't take matters into your own hands when you've been raided, then you are likely to be raided again until not only your herd is gone, but so, too, your family.

Tamler Sommers in *Why Honor Matters* makes the distinction between honor cultures and what he calls the dignity-based culture that, starting with the eighteenth-century Enlightenment movement, replaced traditional honor-based cultures. Dignity does not have to be earned. In dignity cultures, everyone has equal human dignity simply by being born. But for Sommers, there appear to be inadequacies in certain aspects of dignity cultures, which he argues are incapable of motivating people to actually struggle against injustice. In a pro-honor argument, he posits that honor encourages self-reliance and independent action, where dignity relies on a state apparatus to protect our rights—a protection that it very often fails to provide.

Thus, reputation is everything, because a weak reputation fosters attacks on your clan. Reputations are to be guarded with your life. And they often are. Self-worth and reputation were the literal, not figurative, center

* Monotheism is a rare religion. Those who proclaim this as evidence of divine providence need to pay attention to the following: although monotheism is rare, it overwhelmingly comes from desert pastoralists. Ancient Hebrews and Muslims alike were pastoralists. When surviving in a monolithic environment like a desert, this makes sense: "I am the Lord your God" and "There is but one god and his name is Allah" are statements that are congruent with the sun- and sky-dominated ecology of these pastoralists. Compare this to the religions endemic to peoples surviving in rain forests, which are overflowing with life in all its variations. Often these cultures have an equal number of deities to worship as species of animals within their environment.

of a person's livelihood. The Hatfields and the McCoys were simply following an ancient edict of pastoralists—"death before dishonor."

It also sheds light on the culture endemic to the southern United States. Long have the values of hospitality, social etiquette, and chivalry from men directed toward women been associated with the southern honor cultures. The South has also emphasized clan- and family-level identity and has historically protected this legacy by being characterized by a long cultural memory. It's no coincidence that those who settled these states came from some of the most ferocious honor cultures in the world—the lowlands of Scotland, northern counties of England, and the Ulster region of Northern Ireland.* Parenting in honor cultures is typically hierarchical and authoritarian and children are expected to staunchly and aggressively respond to honor violations.[60] One famous study measured the stress response of both Northerners and Southerners to a verbal slight from a stranger. The Northerners shrugged it off with ease, whereas the cortisol spike exhibited by Southerners was drastic.[61]

But it's not all aggression and violence. There are strict cultural codes of conduct in terms of hospitality, courtesy, and civility. Often elevated above all other codes is the hospitality to weary travelers. This may help explain the targeted rates of violence in the South. Unlike the northern states, where violence is often a feature of large urban regions in attempts to gain and steal resources, southern violence is disproportionately rural and among people who know each other. The evidence is striking: as long as you don't flirt with your friend's wife or cheat your neighbor in a poker game, the likelihood of you suffering from violence is actually *lower* in honor cultures.[62]

INDIVIDUALIST VERSUS COLLECTIVIST, HONOR VERSUS dignity—these are important dimensions of the culture construct. Yet there is an ultimate "primal template of culture" that has revolutionized our understanding

* My mother's maiden name is Monroe, whose lineage can be traced back to the Scottish Lowlands deriving from clan Monroe. Perhaps this is why I have always found it especially difficult to shrug off verbal slights?

of the mechanism by which cultures drive behavior—*tight* versus *loose.* In essence, tight cultures have strong social norms and little tolerance for deviance, whereas loose cultures have weak social norms and more tolerance for deviance. In the words of Michele Gelfand, the cultural psychologist whose lab has pioneered this research, the former are rule makers, whereas the latter are rule breakers.[63]

Places like the United States have relatively loose cultures, where casual norm violations, like jaywalking, are commonplace. In contrast, a place like Singapore, a tight culture, has pristine pavements due to the severe punishment for offences such as littering gum on the street. The factors that influence the continuum are the same for all cultures and are most predictably answered by one question: How much threat do we face? Tight versus loose is driven by population density and diversity and subsequent resource vulnerability, disease exposure, frequency of natural disasters, and outside threats from other humans. Think of the tight-loose cultural mechanism as the lever that amplifies or diminishes our coalitionary cognition.

When threats are high, it is time for cultures to elevate tribal signaling—thereby shoring up coalitionary alliances—by getting people to act in lock-step. To do this, a culture amps up the punishment for norm violations. From this perspective, a "cult" may be nothing more than a community that has amplified social norm violation tightness to fervent adoption. A cult is just a tribe whose members intensely signal their membership. In fact, in the lead-up of the 2016 election, the desired tightness of people in the United States predicted support for Donald Trump better than any measure. For example, tightness predicted a Trump vote with forty-four times greater accuracy than even authoritarianism. It is no surprise, then, that Trump had the greatest support in states whose citizens felt threatened from internal and external threats—and whether someone was tight or loose was governed by this primal cultural reflex.

THERE IS HOPE ON THE horizon. Some recent experiments have shown that we can defang dehumanizing behaviors. This requires employing *superordinate goals* among groups. Recall Sherif and his study at Robbers Cave. The campers had gotten out of control and the Eagles and the Rattlers were at each other's throats. It was here that the researchers enacted the final phase: reconciliation by way of superordinate goals. The boys were brought together on multiple tasks, including pooling their money together to watch a movie, fixing a camp water tank, and working together to pull out a truck stuck in the mud. The divisions that had rocked these two warring tribes began to melt away. After the shared superordinate goals had been accomplished, the boys had restored each other's humanity. A group of boys singing together in unison was heard echoing from the one bus that drove them away from Robbers Cave.

Out of the entire summer, your final week of camp has been your favorite. Eagles, Rattlers, and the rest were all dissolved and you got mixed up in different cabins. You and the others collectively were rewarded tokens for working together to improve the camp—wash equipment, perform yard work to improve the landscaping, and do arts and crafts to hang up in public spaces. With these shared resources you and your friends pooled together camp tokens to watch movies, eat pizza, and play video games. It's been so much fun. You are really sad that you have to leave, but by the time you get on the bus to depart, you and your friends, Rattlers and Eagles alike, together sing your sadness away. In this instance, leaving the summer camp together, you feel like one unified tribe!

5

TRUST SIGNALS

When Tribe Trumps Truth

The truth won't make you free, it will make you extinct.
—Donald Hoffman, 2019

You are a member of a coalition only if someone (such as you) interprets you as being one, and you are not if no one does. We project coalitions onto everything, even where they have no place. . . . We are identity-crazed.
—John Tooby, 2017

Almost everything you believe is someone else's myth/meme/story. You can't avoid believing myths, but you should try to understand why you believe them.
—@punk6529, 2021

The Sombrero Code

I awake with a slight headache. With no small measure of displeasure, I note the early hour; it is just after 5:00 a.m. We must hurry, rise to brush our teeth and zip up our go bags. My roommate, Dr. Eric Shattuck of the University of Texas at San Antonio, acknowledges me with an exasperated sigh. I return his sigh and then nod. We need not verbalize that perhaps we imbibed too many Banderas the night before. A Bandera is an alcoholic beverage that takes its name from the Spanish word for "flag," so called because the triple shots of lime, tequila, and tomato juice resemble the Mexican flag when lined up in the order they are to be shot. Nothing is said, but we understand each other as Eric is an old grad school friend and this wasn't our first rodeo. We head downstairs to the pre-appointed rendezvous outside our hotel in downtown Guadalajara, Mexico. Our driver is there, waiting for us along with the rest of our team, Igor Martin Ramos Herrera, a professor at the Center for Health Sciences at Guadalajara University, and one of my senior graduate students, Leela McKinnon. Eric and I cram our backpacks into our SUV. I begin mentally preparing myself for the thirteen-hour drive into the Sierras of Jalisco by tipping my hat over my eyes and reclining as much as possible in an attempt to achieve a meditative state of Zen while it is still dark outside.

An hour out of Guadalajara and beginning up the long leg of tarmac into the mountains, my Zen is broken as Eric begins talking with René Crocker Sagastume, our guide to the Wixáritari tribe. René is a professor of nutrition and public health at the University of Guadalajara. He has made this drive hundreds of times, as he has worked alongside the Wixáritari people for a quarter of a century. He tells us of a time when "on this very stretch of road," one of the then-warring cartels had pulled his car over, beaten him, and taken all his belongings and he was forced to walk back to Guadalajara. This was why we are making the trek so early this morning, so as to avoid a similar fate. His story strikes me, not only because of the impending reality of where I am, but also because it is a stark example of the remote refugia of the Wixáritari people. Because of their isolation, they had thrived as one of the only indigenous peoples not conquered by the Spanish colonizers. But that wasn't the only reason they had thrived while others perished. Their tribal identity is so strong it has imbued them with resilience unmatched by most.

René is a man in his late fifties. His eyes gleam with intelligence and his mind is filled with knowledge both academic and arcane. Over the next week, I will come to appreciate his depth of life and intellectual experience, with one foot in the modern world, and his other foot firmly anchored in indigenous culture and history. He self-classifies as Mestizo—meaning he has both Spanish-European and indigenous descent. Yet, he also identifies as Wixáritari. The evidence he proffers is embedded in his hat, to which I am instantly drawn. The sombrero has a bright, beautiful, and brilliantly textured and colored band around its base. The pattern of the band is intoxicating. Its bright geometric patterns of topaz, orange, and blood red draw my attention for hours on the drive. I lose myself for a time in them. René has spent decades helping the Wixáritari attempt a difficult move, a transplantation in a healthy and sustainable way, into the economically developed regions of Mexico—all this, without losing their identity. Because of his care and concern, and this delicate balancing act, they have accepted him into their tribe as one of them. What I am about to realize on this journey to San Andrés to work with the Wixáritari is the answer to the burning question I have been contemplating for the past several years. It is right in front of me. The question: What is a tribal signal? I didn't fully realize it at the time, but the answer was tied neatly around René's sombrero.

The Tribal Translator—Cracking the Code

If two people are walking on the street and you are called upon to classify each as either a banker or a professor, do you think you would be able to do so accurately? The answer to the question is yes—we could, with a startling degree of accuracy, deduce which human was a banker and which was a professor. The reason why you can easily intuit this answer is because you have instructions embedded in your DNA for your brain to receive, decipher, and transmit tribal signals by other human beings. This chapter explains how, in milliseconds and mostly unconsciously, we do this extraordinary mental feat.

What then is a tribal signal?

A tribal signal is a symbolic token representing membership to a coalitionary alliance. In other words, it is a signal that helps define a *team* and the distribution of members through space and time within that team. Tribal signals announce your team allegiance. Additionally, they can also announce the constellation of other teams with whom you may have alliances. You perceive signals of team membership and you signal your own team membership to others by way of broadcasting the probability that you can be trusted. Our primary goal in this chapter is to tune in to this tribal trust data being emitted everywhere, so that you can rip it away from its subliminal, unconscious status to peak awareness.

Tribal signals are everywhere, but we are, for the most part, not conscious of them. They effuse around us, and the computational powers of our brains are programmed to spend incredible amounts of energy churning out predictive models of human behavior on the basis of these cues. Just as body language is an important component of human communication, so, too, is our *tribal language*. If humanity is to survive the twenty-first century, we need to become conversationally fluent in the tribal languages instinctually spoken by different coalitions within our species. Unless we tune in, we will remain consciously ignorant of their power. And if we remain ignorant, we will remain at their mercy. This effort makes this chapter one of the book's anchors.

WHAT IS A SIGN? WHAT is a token? In 1863, at the peak and frenzy of the Battle of Gettysburg—the bloodiest battle of the American Civil War, claiming fifty-one thousand lives—a confederate general, Lewis Armistead, was shot and mortally wounded. Bleeding out upon the field of battle, he raised his hand to the heavens and formed a sign. The symbol signed with his hand was ancient and secret. It was a token of shared experience and identity. Its origins can be traced back to medieval stonemasons that constructed the cathedrals and castles of Scotland and England over seven centuries before the Civil War.

Hiram Bingham, a Union officer, in that very moment was also in the fray of life-or-death conflict. He recognized the sign raised by Armistead. Bingham and Armistead had never met in person, but they were bound together by the amazing power of a symbol embedded with immense sums of information about the identity of the symbolizer. In an instant, amid the chaos of the picket charges, the "Us versus Them" of Union versus Confederate evaporated and another "Us versus Them" was elevated in its stead—Freemason versus non-Freemason. Bingham leapt to Armistead's aid and protected him as he dragged his bleeding body to a Union field hospital. Although Armistead would succumb to his wounds days later, Bingham ensured that his personal possessions, which included a pocketbook, a watch, his spurs, and a chain with a masonic emblem, were returned to his family.[1]

Again, what is a sign and what is a token?

A sign is a medium for symbols. And symbols convey information. When a visible or tangible representation of symbols is recognized by others, we call it a token. Thus, a human can be a sign, and tokens are how we embed that human with a matrix of trust probabilities we can simply call tribal data.

EVERY HUMAN BEING IS A walking, talking tribal beacon. There is a nested hierarchy. There are *primary signals* that the brain will automatically assess and track when encountering a new individual. These signals fall into two categories: *age* and *sex*. The brain elevates these above other signals because we all need to know if we are dealing with a potential *mate* or a potential *ally*. In evolutionary terms, on average, most ancestral groups had a mixed balance of age and sex. This is demographically typical for all tribes. Primary signals are fixed, static, and computed automatically upon first meeting someone.

In contrast, *secondary signals* fall into three categories. The first two, *language* and *ideology* are signals that are not fixed because they traffic in the symbolic. Steven Pinker in *The Language Instinct* notes that language "is not a manifestation of a general capacity to use symbols: a three-year-old . . . is a grammatical genius, but is quite incompetent at the visual arts, religious iconography, traffic signs, and other staples of the semiotics curriculum. . . . The complexity of language, from the scientists' point of view, is part of our biological birthright."[2] *Ideology* is the web of ideas that the group's society constructs by way of language to build consensus on reality. Combined with language, humans craft the symbols of group culture and consensus reality central to coalitional identity formation. The third category of secondary signal is *phenotype*. This is a set of observable characteristics of a person's genes expressed within a particular environment. Simply put, it is our appearance, and what we look like is a combination of our genes expressing themselves as we develop. All three combined, the categories form the bulk of what constitutes the tribal signature of an *ethnolinguistic group*—anthropologically speaking, this is the most stable unit of social organization ever measured.[3] It is safe to say that most, if not all, tribes throughout prehistory were ethnolinguistic groups.

Tertiary signals are the raw materials—both physical (like a Mason's ring) and memetic (like a Masonic ritual initiation ceremony) that codify the data through linguistic and ideological filters—that are amplified by your tribal signaling beacons. Many of these are essentially tribal memetic *tokens of loyalty* you use to identify yourself to the world and that others use to interpret your coalitionary alliances. Tokens of group membership predictively cluster with each other. You use them to forge your identity, and identity is how you model yourself to the world. In contrast, *stereotype* is how the social world models you. As tertiary tribal signals, all tokens fall within one of the symbolic *secondary signals* that generally compose ethnolinguistic groups.

The tokens are infinite in their variation, and as cultures evolve into and out of different tribal identities, so, too, do the tokens change. Because they are the stuff of symbols, they are limited only by the human imagination. Yet, typically, tokens fall within the tertiary categories of *dialect, worldview, politico-economic, population* stereotypes (e.g., race); here I provide a non-exhaustive list of some broad categories.

- Music
- Religion
- Art
- Philosophy
- Sexuality
- Class
- Skin color
- Body modification
- Profession
- Consumer behavior
- Clothing / style
- Gait
- Body morphology
- Manner
- Idioms
- Food (halal, kosher, vegan, etc.)

Take a moment to run through the following thought experiment. You meet someone for the first time. What are the characteristics and traits you pick up consciously? What in particular? Perhaps they have pink hair. Is their tie bold? Do you note body modification, like a tattoo or piercing in their lip? Are they sporting the jersey of a rival professional sports team? A T-shirt emblazoned with the cast of the hit series *Friends*? If you converse, what additional data does your mind instantaneously process? Do you speak the same language? Does their accent give away information on regional origins? How pigmented is their skin? Are there any facial features that give you a clue as to their ancestral population? Evolution endows each *Homo sapien* on the planet with an extremely sensitive, probabilistic tribal affiliation detection system.[4]* This system has predictive value on behavior.

Robert Sapolsky writes: "We feel positive associations with people who share the most meaningless traits with us."[5] Our "arbitrary" tribal signals have been yoked with identity markers that have real fitness-enhancing consequences. Stereotyping the in-group versus out-group mentality means the brain is always scanning for non-random clustering of greater predictive clues that allow positive interactions with others. It's hungry for cooperation; it just needs an excuse to keep playing nicely in the prisoner's dilemma game that is life.

If the Tribe Drive is embedded in our DNA, then how do we get from a gene to a tribal signal? Since genes provide the blueprint to the neuroanatomical features of the brain that produce coalitionary cognition, each species has a unique set of signals, and some signals are more "honest" or costly than others. Costly signals are useful because they usually are honest signals of the individual's genome. *Costly signaling theory*, originally proposed by the biologist Amotz Zahavi, arose to explain the many

* This evolved system in our brains is driven by a processor that is used to forecast the future. This processor is based on Bayes' theorem—named after Reverend Thomas Bayes, who, in the eighteenth century, first used conditional probability to provide an algorithm that uses evidence to calculate limits on an unknown parameter. In other words, Bayes' theorem describes the probability of an event, based on what you think you know already about what the event is, combined with any data about the event that you are actively collecting in the current moment. In essence, the output is not a binary, yes-or-no answer to whether future events will unfold, but a distribution of likely answers of future events that can continuously be updated. Applying this to human tribal detection systems, stereotyping is the process of assessing an individual's traits and predicting on the basis of the pattern or cluster of traits that the individual belongs to a specific coalitionary alliance. Speculatively, this Bayesian system that is responsible for the human ability to stereotype likely evolved in its modern form around the advent of true tribal-level societies around three hundred thousand years ago.

animal traits that are hard-to-fake indicators of animal fitness.[6] Just a few examples of these head-scratching traits include (i) male lions being characterized by massive, dark manes despite the fact that they cook the lions' brain with savanna heat; (ii) impala strotting behavior (a sort of hopping that certain gazelles do when they sight a predator) that wastes energy and diminishes a head start to escape; and (iii) the male peacock's tail that makes it more conspicuous and more difficult to evade predators.* The main reason costly signals are useful is because they are a good way to test if the individual is lying about what they are attempting to signal.

In the words of the evolutionary psychologist Geoffrey Miller, "Costly signaling theory offers a solution to this problem of lying: if a signal is so costly that only high-health, high-status, high-condition animals can afford to produce it, the signal can remain evolutionarily reliable."[7] Therefore, honest signals are essentially strong signals that serve as a proxy for expensive, high-amplitude, and energy-costly fitness indicators.† There is a mountain of evidence, beyond the scope of this chapter, that we use *beauty signals* (for example, with female waist-to-hip ratio and male perception of attractiveness, or male shoulder-to-waist ratio and female perception of attractiveness) to advertise mate quality.[8] Like a panhandler sifting through a stream for precious metals, the "receiver psychology" is looking for gold where there is often only bits of silver, copper, or more likely lead, and a massive amount of irrelevant junk. Good panhandlers end up with lots of fitness gold, and bad panhandlers with a bunch of useless metal.

---

* It is fun to speculate about Zahavi's handicaps being expressed in humans. For example, the propensity for males to purchase Veblen goods as a form of conspicuous consumption as a wealth display. Or the adolescent male tendency to overconsume alcoholic beverages. The principle could also explain the male sex bias seen in motorcycle ownership—what better way to display high-quality genes than to ride a machine that doctors in emergency rooms have dubbed a "coffin on wheels." Weak-gened buyers beware.

† Science has uncovered mountains of evidence that the *beauty drive* feeds off numerous fitness-relevant signals. Body morphology, scent, voice, symmetry, skin texture, eyes, hair, and a host of other traits that are embedded with multiple, complex, interacting qualities that serve as advertisements of the quality of one's potential mate. Colloquially, people in Western cultures often use a single word to substitute for the complex desire algorithm that governs our mate choice preferences: *hot*. As in: "Damn, that *Homo sapien* is hot!" We are masters of distillation of complex phenomenon into easily communicated terminology, and in this instance a single word suffices.

The same logic we discussed with those hard-to-fake indicators of animal fitness can also help us understand the logic of how certain human tokens are weighted more than others to better predict tribal coalitions. For example, a retweet of support for your favorite sports team is less costly than a twenty-dollar team T-shirt, which is less costly than a face tattoo of the mascot of your team. The same can be said for the context in which the token is being displayed, as it may be a costlier signal to wear a burka in rural Alabama than cosmopolitan Toronto. Both the presence or absence and cost of tribal signals within specific contexts reveals an incredible amount of coalitionary alliance information and enhances a predictive model of behavioral interaction.

The human body can adorn and emit tribal signals that simultaneously represent membership in many different coalitionary alliances. The juxtaposition among signals can be fascinating to deconstruct. There are *weak* and *strong* tribes. Tribes can be measured by the strength of their capacity to hold purchase on one's identity. For example, a baseball game can elevate the identity holders to the forefront, but as soon as they leave the stadium it smoothly moves back into the subliminal. In contrast, a practicing orthodox Jew may still wear a traditional Hasidic garb to a baseball game, taking their strong tribe with them everywhere.

One caught my eye while watching the Toronto Raptors win their first franchise NBA title, defeating the favored Golden State Warriors. It was an epic confrontation of Canada versus the U.S.! The tribal energy (led by the #WetheNorth slogan) that invigorated Toronto and the entire nation was palpable. The news media featured Nav Bhatia (known by Torontonians as the Raptors Superfan because he has attended every Raptors home game since 1995). Fleeing the persecution of Sikhs in India, Bhatia moved to Canada in 1984, where he became a top-selling car salesclerk and eventually the owner of the Hyundai dealership where he worked. At games, Bhatia sports the Sikh dastaar turban and his characteristic Nike-sponsored Toronto Raptors jersey. He sports a trimmed beard with prominent mustache. The turban is central to Sikh identity, and there is arguably no better Western icon than the Nike symbol. It represents the capitalist, pro-Western, pro-democratic institutions that facilitate its international prominence. His wrist is adorned by an iron *kara*, which is a community-binding symbol of God having no beginning and no end; it also doubles as a kind of brass knuckles that could be co-opted in a physical altercation if need arises. Bhatia's facial hair is laden with signaling; the *kesh* is one of the strongest pillars of Sikh religion, in which one's hair is grown out of respect for the perfection of God's creation. In the Sikh religion in its most fundamentalist forms, it forbids cutting and/or trimming hairs from the body. This is *khalas saroop* or "pure form." The idea is that upon your death you should go back to God in the form he gave to you. Here, Bhatia strikes a middle ground that many Western-dwelling Sikhs now practice. The most ardent practitioners of the *kesh* do not groom their hair their entire lives, whereas Bhatia's beard is kempt. To my eye, I see the clever and harmonious elevation of tribal tokens to prosocial ends. Canada seems to agree, as the Royal Bank of Canada named Bhatia the recipient of the Top 25 Canadian Immigrants Award. Bhatia is a super-tribal diplomat, as he founded the Superfan Foundation in 2018 to bring diverse people in Canada and internationally together through sport. The science described in this chapter indicates how his integration of multitribal, multilevel signals approach works, endearing him to both coalitions he represents.

There is an interesting correlation between politically conservative college students and those of a more liberal persuasion. Conservatives are more likely to have sports décor, laundry baskets, and postage stamps in their dorm rooms, while the more liberal students tend toward maps as decoration, art supplies, and many books.[9] Why the difference? The answer is that humans are social learning machines that tend to copy others we value or consider prestigious. These cultural clusters make stereotypes and the predictions they create about others heuristically valuable even though at times they may be wrong. In complex societies they have a way of being contradictory and systematically misapplied. A great example of this contradiction is that of the competing stereotypes that migrants are both hardworking and industrious *and* lazy.[10] This obvious inconsistent paradox, according to evolutionary anthropologist Cristina Moya, derives from two causes:

> First, the same social learning mechanisms that can give rise to clustered distributions of traits can also be vulnerable to acquiring inaccurate or outdated information from others. . . . We may have

> genetically evolved heuristics for privileging certain *cues* [emphasis mine], like language use, as indicators of cultural group membership. In most environments during the course of human evolution these heuristics likely produced accurate stereotypes on average. However, today we may systematically misapply them in contexts where language is not associated with cultural clusters. Furthermore, we live in more complex, dense, and large-scale societies, with more cross-cutting group identities today. This means that social group boundaries no longer simply reflect neighboring regional groups.[11]

The bad news is that our judgment of others' tribal signals can be wrong, but the really good news is that because of our brain's forecasting ability we constantly update past experiences with current data; that means inaccurate stereotypes can, over the long run, approach greater accuracy.*

EVERYONE LOVES A GOOD SECRET. Our tribal identities can, at times, serve like a secret code, and like a Masonic hand sign, convey information to a stranger about the identity of the signaler. In fact, secrecy plays such an important role in the human experience that human biocultural evolution has manifested a special class of meta-groups that can both strengthen the backbone within a tribe and transcend the boundaries of one tribe to another. This special meta-group is called a *sodality*. Sodalities are worthwhile examples to explore because they demonstrate many of the signaling factors that go into the Tribe

* What we are discussing here, in terms of our brains using stereotypes to predict the future, may strike some readers as dangerous ground. They may think that it might, in some way, make us prone to one of the alleged great sins in the humanities today—essentialism. Essentialism is the view that categories of people, such as women and men, or heterosexuals and homosexuals, or members of ethnolinguistic groups, have intrinsically different and characteristic natures or dispositions. Even though we can behave as essentialists (which can oftentimes lead to antisocial behavior) in any given moment, our brains are decidedly not essentialist. Even though stereotypes, as average tendencies for a category, are often accurate, this does not mean that an individual corresponds to the average. Upon first meeting, you have no idea where an individual falls on a probability distribution calculated by your brain, or if you even made an initially correct judgment of their particular tribal signals.

Drive algorithm. A sodality is a ritualistic, typically (but not always) male-gendered guild* and has the unique capacity to span across several levels of social organization. A key aspect of a sodality is that as a group, it uses secrecy and ritual to confer members with hard-to-fake tokens of achievement and prestige. Secrecy is paramount to the function of trust building in a sodality. Sodalities use an economy of secrets as a trust-building mechanism among its cohort. There is a "paradox of secrecy" in the fact that secrecy is procedural; in other words, it is the telling of secrets in the right context and to the right people that lends the secrets their power to bond and build the community among its membership. As a person moves up the hierarchical rungs of the sodality, everyone eventually becomes a secret recipient, holder, and giver; the paradox is that the secret economy exists by the virtue of the telling of secrets.[12]

Perhaps it is this instinct to need to keep and learn secrets that can account for our fascination with secret societies—groups formed by people who keep their inner workings shrouded from outsider interrogation. Secret societies are in several ways the most extreme expression of coalition cognition. Membership is limited and initiation rites can be complex and difficult to perform. That's because secret societies, like all effective groups, are looking for hard-to-fake *honest signals* of coalitionary alliance. For example, a Master Mason must go through intensive memorization of oaths in complex rituals to attain such a title, and the sheer energy that goes into such performance is difficult to fake.

When we think of secret societies, we tend to think of conspiratorial organizations popularized by the media, such as the Illuminati

* There have been numerous female sodalities throughout history. The Amazons are a famous, if not apocryphal, para-military sodality from antiquity. The Eleusinian Mysteries were a unisex, pan-tribal sodality that persisted for over one thousand years that was available to all Greek-speaking peoples. Female secret societies, such as the Daughters of Temperance and Daughters of Rechab numbered in the thousands and dotted the East Coast of the United States in the nineteenth century. The continent of Africa is still home to hundreds of female sodalities, often distinguishing between married and unmarried women. *Divine Secrets of the Ya-Ya Sisterhood* by Rebecca Wells sold over 2.5 million copies, demonstrating a public appetite among women to partake in the ancient, timeless art of sodality crafting.

or the Islamic secret society known as the Assassins. But sodalities are part of the human experience even in small-scale societies. Both genders typically have some kind of secret ritual performance upon reaching sexual maturity that indicate they are "initiates" into the male or female adult sodalities of their societies. Illustrating how important the function of secrecy is to their sodality, some of these societies even guard their secrets with threat of violence. Lucas Bridges had direct experience with this when, in 1874, he became the first European born on Tierra del Fuego, where he lived among the Yamana and Selk'nam hunter-gatherers. He grew up speaking Yamana and upon becoming a man, joined a men's society, where he was instructed that if he were to ever reveal their secrets to uninitiated men or any woman he would be killed.[13] What is instructive about this example is that killing the unauthorized secret-sharer would be the responsibility of his brother or father—thereby demonstrating that the power of the sodality could actually be more important than direct kinship.

Another example showed the direct outcome of such a violation. The anthropologist William Lloyd Warner, when he was in the field with the Yolngu of Australia in the 1920s, witnessed the result of the discovery of a sodality's secret:[14]

> Some years ago the Liagomir clan was holding a totemic ceremony, using its carpet snake totemic emblems (painted wooden trumpets). Two women stole up to the ceremonial ground and watched the men blowing the trumpet, went back to the women's camp, and told them what they had seen. When the men came back to camp and heard of their behavior, Yanindja, the leader, said, "When will we kill them?" Everyone replied, "Immediately." The two women were instantly put to death by members of their own clan with the help of [men from another group].

ALTHOUGH THE GOALS OF SUCH groups can be nefarious (think no further than the infamous Cosa Nostra mafia) as with the latter example, so, too, can sodalities be organizations with prosocial tendencies, such as the American Legion or the Shriners.* In fact, the Mason's motto is "Make good men better."

That is because the sodality can be viewed as a type of social tool to enhance the social insurance of its participants. This is what gives the sodality its guild-like properties. A guild is defined as an association of people for mutual aid or the pursuit of a common goal. The insurance policy of being a guild member is self-evident. If you need help, your guild mates—even if you've never met them before—are there to provide backup. Where each sodality is unique is their self-defined common goal. One way of thinking about it is that a sodality is friendship scaled, in that the functions of a friend and the functions of a sodality are nearly identical—take care of people you trust when they most need it. It's the ultimate human counter to the Banker's Paradox.

Sodalities tend to follow basic categories of function: paramilitary groups (e.g., the Knights Templar, Assassins) use force and violence to forward their agendas. Many sodalities exist in the context of confronting what is perceived as an outside threat. For example, the Maasai, Crow, and Cheyenne military war camps are sodalities with a shared purpose of tribal defense. They established the defensive and protective parameters from within which their tribes could exist and thrive. Religious sodalities (e.g., Opus Dei) can organize around either traditional or unorthodox theologies to mediate group behavior. Criminal groups (e.g., Cosa Nostra mafia) use illegal activities to benefit their in-group.

Yet, the core of a sodality's function is that participants have an added layer of *social insurance* if times get tough. The origin of the Masons, for example, goes back at least to the fourteenth-century master

---

* To draw upon an example from fantasy, George R. R. Martin's *Game of Thrones* depicts a prominent sodality within his narrative: the Night's Watch. It is pan-tribal, as members are drawn from all the seven (often warring) kingdoms of Westeros. They are a military order that holds and guards the Wall (a northern boundary of the seven kingdoms) from "wildlings" (cold-adapted tribes) and other more ancient and mythical foes (sorry, no spoilers). Once the initiates arrive at the watch, they undergo a sacred ritualistic ceremony and swear an oath: "Night gathers, and now my watch begins. It shall not end until my death . . . I am the sword in the darkness. I am the watcher on the walls. I am the shield that guards the realms of men. I pledge my life and honor to the Night's Watch, for this night and all the nights to come."

craftsmen who ran guilds in Europe and formed local congregations, called lodges, to take care of sick or injured guild members and extend such support to their families. The lodges became meeting places, where masons conferenced about their craft, trained apprentices, and socialized. Eventually, the groups developed secret passwords, handshakes, symbols, and oaths to protect their trade secrets and identify fellow guild members. In summary, it can be argued that the primary function of a sodality is truly tribal in nature—to culturally construct green beards that are invisible to those who are uninitiated and to help protect and safeguard members from harm. Sodalities can be healthy, prosocial instruments that help bond a community.

In summary, each of us has been endowed by evolution with a powerful detection system—a tribal signal receiver. This receiver parses the incredibly complex bits of information cascading into our orbit, among the thousands of multilevel, multicoalitionary alliances from the hundreds of human beings we encounter on a day-to-day basis.

How Tribal Truth Trumps Reality—The Tribe Bias Revealed

We perceive the world through our five senses. They have been exquisitely crafted by natural selection as foraging instruments with the primary duty of hunting for fitness. Those organisms with less-obsessive fitness signal perception systems ended in extinction. The implications of this are startling. Our perception did not evolve to understand reality. Its primary purpose evolved to fulfil *the prime directive: find food, have sex, and don't die in the attempt.* The latest scientific data on our perceptual equipment is revealing a totally counterintuitive fact. If the choice is between truth or fitness—fitness trumps truth.

LET US SUPPOSE THAT OBJECTIVE reality exists . . . that is, there is a *there,* there. For most of us, this should be an intuitive starting point. This position is held by countless philosophers (although not all*) and

* The epistemological position that holds that knowledge of anything outside one's own mind, including the external world and other minds, cannot be known and does not exist is called solipsism. For what it's worth, I don't hold this position.

is the starting assumption for most scientists exploring the frontiers of knowledge. The standard model is that our senses are reliable guides because they tell us the truth about reality. They are reliable because, generation after generation, they evolved this way. Offspring of those who saw reality more accurately survived and reproduced more than those whose perceptions were less precise.

Yet, our species has a poor track record of perception-based claims of knowing objective reality. Aristotle, Parmenides, and Pythagoras informed the ancient Greek world that the Earth is round despite the testimony of our sense of sight that the world is flat.[15*] Humans have a tough time understanding the evolution of species because of the immense time it takes compared to the brevity of our lifespan. Three centuries after Galileo Galilei championed heliocentrism, Charles Darwin circumnavigated the globe and uncovered evidence that species are not fixed, static, and unchanging but constantly in flux and evolving. Half a century after that, Albert Einstein discovered light has a speed limit and mass is interchangeable with energy. We certainly did not *see* that one coming.

Why did people think the Earth was flat? Why did they think the sun revolved around the Earth? Why did we ever assume that species are unchangeable? Why was it a surprise to twentieth-century humans that time objectively changes relative to the speed you are traveling? Finally, why do we think that our senses perceive reality as it truly exists? The answer for this series of questions, quite simply, is that *it looks and feels that way*. I believe we are on the verge of a series of scientific discoveries that will match the sheer magnitude in scope of each of the

---

* The authors of the Old Testament share their perception of the Earth's location in the universe in Joshua (10:12–13): "'Sun, stand still over Gibeon, and you, moon, over the Valley of Aijalon.' So the sun stood still. . . . The sun stopped in the middle of the sky and delayed going down about a full day." The clear implication here (which was used by theologians literally interpreting scripture up until the aftermath of Nicolaus Copernicus's book *On the Revolutions of the Heavenly Spheres* published in 1543) was direct scriptural evidence that supported a geocentric model of the Earth as the center of the universe. Part of that aftermath entailed Galileo looking through a telescope and observing that (like the Earth's moon) there were moons orbiting Jupiter, and phases changing Venus. For a proper understanding of the genre of the Book of Genesis and the entire Pentateuch, see Thomas L. Thompson's seminal work, *The Historicity of the Patriarchal Narratives: The Quest for the Historical Abraham*.

preceding discoveries that shook the foundations of how our species views its place in the cosmos.

I also believe the Tribe Drive is an important part of these discoveries and, once understood, will help us come to terms with what to do about our tenuous grip on objective reality.

Donald Hoffman, a cognitive psychologist at the University of California, Irvine, proposed a startling theorem in his book *The Case Against Reality*.[16] His theorem helps us contextualize the "true" value of tribal signals. His contribution, the Fitness-Beats-Truth (FBT) theorem, states that "evolution by natural selection does not favor true perceptions—it routinely drives them to extinction. Instead, it codes for survival and reproduction." And that coding has occurred through the only "interface" we've had—our five senses. That is, until the advent of science and the technological innovation of machines to augment our five senses (e.g., microscopes, hearing aids, MRI machines, compasses, etc.). Instead, natural selection has favored perceptions that hide the truth and guide useful action. Hoffman's guiding metaphor is that of a desktop icon: your mouse moves on a screen, clicks on this file, and allows the user to accomplish their task mercifully unhindered by the massively complex code and circuitry that underpins the entire computer. Using evolutionary game theory, Hoffman conjectured (and his colleague Chetan Prakash mathematically proved) the answer to the question: Does natural selection favor veridical (i.e., accurate) perceptions? The mathematics of natural selection* describe a counterintuitive theorem that *the probability that we see reality as it is equals zero*.[17] In other words, natural selection parses reality into species-typical fitness-maximizing data formats. *The truth is in the fitness.*

Life is successful because, among the vast majority of potential stimuli it encounters, it filters only what is needed for survival. It is this filtering that has led to the evolution of the incredible diversity of perception sys-

---

* In the quest to test the hypothesis that natural selection favors veridical perceptions, Hoffman and Prakash found the answer was a resounding no. They developed several game theory simulations to mathematically investigate their Fitness-Beats-Truth theorem, which states: Fitness drives Truth to extinction with probability at least (N-3)/N-1), where N is the sensory strategies that are capable of N perception types in an objective reality being characterized by N states. Thus, fitness is only tuned in to what are called "fitness-relevant payoffs."

tems exhibited by all animals. For example, pit vipers detect temperature differences of 0.01 °C and perceive infrared radiation; water scorpions detect minute changes in hydrostatic pressure to sense their depth; dolphins and bats (although they are separated from their shared ancestor by sixty million years) hone in on their social and environmental maps by way of precise echolocation; bacteria use magnetic waves generated by an ever-churning mass of molten iron as the guiding force to orient themselves; bumblebees detect and experience the polarization of sunlight; Chernobyl fungi thrive off eating the radiation that drove humans from the city; African freshwater fish summon an electrical field that can allow them to feel intruders that register the slightest of perturbations within it; a fecund female silkworm sprays a fraction of a billionth of a gram of sex pheromone and draws suitor males within a several-mile radius. What is it like to be a male silkworm that detects this scent? Imagine, then, an alien biologist asking readers of his latest book on *Homo sapiens*—what must it be like to be a human being sensing their life?

To summarize, you are a *Homo sapiens*. Your ancestors, the lineage that resulted in your existence, faithfully obeyed the algorithm's creed: perceive fitness-relevant objects, above all else—even objective truth about the material world—and you will be rewarded with replication. In other words: the FBT theorem postulates that genes that survive the process of natural selection do not code for truthful perception. Surprisingly, a complete understanding of both physical and metaphysical ontological truth and fitness are seemingly mutually incompatible, and not a complete picture of veridical truth. And as we will explore, for a significant span of time throughout our species' evolutionary history, the tribe was the critical social interface that mediated life, death, and sex. It is no wonder that when a human is forced to pick between truth and tribe, truth ends up in second place.

WHAT IF I WERE TO tell you that your sense of identity literally drives your perception of what is true? At first glance, the statement seems stupefying. What does identity have to do with truth? Reality shouldn't care about what coalition, team, or family you share membership with. These things objectively are orthogonal to each other, right? Terrifyingly, our perception of what is truth and our identities are intimately interlinked. Just as belief in science can be tribal, so, too, can disbelief in science be rational. Let me state that once again.

A disbelief in science can be rational. And a belief in mythical, tribal falsehoods can be just as rational.

THE FITNESS-BEATS-TRUTH THEOREM IS HOW the maxim plays out in the human mind. The core of this phenomenon is what psychologists call *identity protective cognition*. If you mix FBT with identity protective cognition, you can then grasp another important component of the Tribe Drive and the crux of coalition cognition.

In what instance can a disbelief in science be rational? When fitness is at stake. The move is rational on a fitness level—whether or not it aligns with veridical truth. This is an astounding revelation that every human should reflect upon. Any time you are confronted with a statement from someone else that you vehemently and emotionally disagree with, ask yourself if your identity is at stake and whether or not that is the source of the emotional reaction to the statement. If it is, you are likely being swayed by what some call the "my side" bias and I call *the Tribe Bias*. Put in another way: a belief in your worldview can be tribal, and disbelief from others in your belief can be on their part rational—independent of whether or not your view is correct or truthful.

In addition, the more extraordinary the beliefs become, the costlier the tribal signal and therefore the more powerful the tribal token is as a signal of membership. Thus, when tribes need unity in times of stress, they tend to impose greater and more severe litmus tests to assess allegiance. Dan Kahan, professor of law and psychology at Yale University, has demonstrated through a series of fascinating experiments how beliefs become tribal creeds—and thus powerful tokens of allegiance.[18] Summarizing Kahan's work, Pinker notes:

> People affirm or deny . . . beliefs to express not what they *know* but who they *are*. . . . People's tendency to treat their beliefs as oaths of

> allegiance rather than disinterested appraisals is, in one sense, rational.... A person's opinions on climate change or evolution are astronomically unlikely to make a difference to the world at large. *But they make an enormous difference to the respect the person commands in his or her social circle* [emphasis mine]. To express the wrong opinion on a politicized issue can make one an oddball at best—someone who "doesn't get it"—and a traitor at worst.[19]

There is a scientific consensus that humans in fact do play a significant role in the current trend of the warming of the Earth. If the world warms 4° C more, the results will be devastating if left unchecked, no matter how many people benefited from an in-group bump in social status among their climate-denying peers. This is what Kahan calls *Tragedy of the Belief Commons*, which concludes that tribal esteem is rational to seek on the individual level even though it is irrational for the health of the individual's society.[20] *Because,* not in spite of, the world's specialists agreeing on the concept, "global warming is a hoax" provides a powerful symbol of cultural allegiance.

And so, identity-protective cognition is rational (on the small-group, face-to-face level) because it evolved to confer fitness advantages to those whose minds were characterized by the Tribe Bias. To a rural Trump voter that relies every day on their community in the shared project of survival, why risk believing in global warming? What do those "liberal secular scientists" do for them? Do they babysit the kids? Do they see them at PTA meetings? Do they take them to the hospital when they are sick? The token of the denier is not a comment on the veridical nature of global warming—it is instead a proclamation that says: *no matter how much real evidence exists, it does not matter, I am still with you. We are one.*

In other words, the more extraordinary or outlandish the idea you display as in-group performance, the better, more convincing the token. Solidarity leading to enhanced fitness is (unfortunately) the currency of survival—not objective truth. Our brains, crafted by evolution and subject to the Tribe Drive, *rationally* influence us to happily discard the truth of reality if it ever conflicts with the truth of our tribe.

Language as a Tribal Signal—The Tower of Babel

If you want to threaten the Judeo-Christian God, according to the Bible, just become ultrasocial super cooperators. It is written in the Book of Genesis (11:1–9) that after the time of the worldwide flood, the devastating population bottleneck produced a single tribe. They spoke one language and began rebuilding civilization from the alleged post-flood apocalypse. According to the story, they were so united in the project that their building technology created a wonder of the world—a tower of such height that it was starting to encroach on God's turf, for heaven was his domain. Multiethnic populations united as one tribe, with one language, were so effective at living together that they created in the ancient world a little measure of utopia. Oddly, this appeared to make the ancient Jewish God jealous, although the story really does not indicate why this should be a threat to an all-powerful deity. After all, heaven and the fulfillment of life is his theological turf.

> And the whole earth was of one language, and of one speech.
>
> And it came to pass, as they journeyed from the east, that they found a plain in the land of Shinar; and they dwelt there.
>
> And they said one to another, Go to, let us make brick, and burn them thoroughly. And they had brick for stone, and slime had they for mortar.
>
> And they said, Go to, let us build us a city and a tower, whose top may reach unto heaven; and let us make us a name, lest we be scattered abroad upon the face of the whole earth.
>
> And the LORD came down to see the city and the tower, which the children of men builded.
>
> And the LORD said, Behold, the people are one, and they have all one language; and this they begin to do: and *now nothing will be restrained from them, which they have imagined to do.* [emphasis mine]
>
> Go to, let us go down, and there confound their language, that they may not understand one another's speech.
>
> So the LORD scattered them abroad from thence upon the face of all the earth: and they left off to build the city.

> Therefore is the name of it called Babel; because the LORD did there confound the language of all the earth: and from thence did the LORD scatter them abroad upon the face of all the earth.

This trope of humanity contesting with the gods is found in much older Babylonian, Egyptian, and Greek religious traditions. But the key troubling factor here for the Jewish tribal God is the threat that one unified language appears to present to him. Language is the tie that binds the members of a group into a data-sharing network that can capitalize on the trial and error of its constituents and the innovations of a few of the group's geniuses. Language permits teams to coordinate their efforts into prenegotiated end states. It is also one of the most powerful tribal signals humans can emit. This appears to be a spiteful act by the God of the Hebrew scriptures. The human tribe was to be robbed of the gift of communication. It is difficult to consider a crueler fate to befall a nascent community struggling to survive and prosper.*

Stemming from the *secondary signals of language,* there are *tertiary signals* such as dialect. One of the most powerful ways individuals can project their community identity signals into the social world is by way of *ethnolect*. Ethnolect is the use of variety in language features that mark association with a specific ethnic group. Dialects and subdialects fall within this category and are contrasted with institutionally sponsored *standard dialects*. Such institutional support may include government recognition or designation; presentation as being the "correct" form of a language in schools; and published grammars, dictionaries, and textbooks that set forth a normative spoken and written form. Ethnolects are powerful signalers that demarcate whether a group is increasing affiliation or distancing in association with other groups. This is a perfect way to signal coalitionary alliance when moving between multiple tribes. In some instances, when two tribes are in conflict, it can mean the difference between life or death.†

---

* For an insightful inspection of the origins of the Pentateuch, see biblical scholar Russell Gmirkin's seminal work, *Berossus and Genesis, Mantheo and Exodus: Hellenistic Histories and the Date of the Pentatuch.*

† One telling example from the Bible about two Israelite tribes demonstrates this at least in principle. The word *shibboleth* in ancient Hebrew meant "ear of grain" (but

The cross-cultural importance of language in demarcating social boundaries suggests that ethnolinguistic boundaries are of evolutionary relevance. Given the speed with which languages and accents evolve, and the apparent difficulty with which we learn a non-native accent as adults, languages were likely predictors of native group membership throughout our prehistory. Additionally, humans can be incredibly sensitive to socially meaningful minor variations in speech patterns and use this data to discriminate against an out-group.[21] In India, 19,500 mother tongues or dialects are spoken, with 121 major languages spoken by more than ten thousand people. Linguistic encoding appears to be more automatic and robust than phenotypic differences—that is, how people physically appear—and is less flexible than even variables like skin color that are used to encode "racial" categories. In a study titled *Blinded By The Accent! The Minor Role of Looks in Ethnic Categorization*, the authors found that when ethnic cues of appearance and accents were combined in a cross-categorization task, there was a clear bias toward accents as the most salient signals.[22] This study suggests language is a more powerful predictor of tribal coalition than race.

Is MUSIC A LANGUAGE? CERTAINLY, it can be said spoken languages are musical. Pitch is used in the more tonal languages from East Asia, such as in Mandarin Chinese. Mandarin has four pinyin that refer to tonal pitches used in word pronunciation. (The words are spelled the same but pronounced in slightly different pitches, which completely changes

---

some translations also determine it translates as "stream"). One tribe, the Gileadites, pronounced it with a *sh* sound, but another tribe, the Ephraimites, pronounced it with a soft *s* sound as they did not possess the *sh* phoneme in their dialect. As the story goes, the Gileadites defeat the Ephraimites in a bloody battle and in the aftermath they set up a blockade across the Jordan River to catch the fleeing Ephraimites who were trying to get back to their tribal homeland. The sentries asked each person who wanted to cross the river to say the word *shibboleth*. The Ephraimites, who pronounced the word with an *s* were unmasked as the enemy and slaughtered. "Say now Shibboleth: and he said Sibboleth: for he could not frame to pronounce it right. Then they took him, and slew him at the passages of Jordan, and there fell at that time of the Ephraimites forty and two thousand." Language, as acoustical signals, is readily marked in association with coalitions because of the ease with which the signals are interpreted and the quickness with which the predictions can be computed.

the meaning.) Another example is from southern Africa. Languages from southern Africa use percussive consonants that produce what we in the West hear as very strident tones. In some rarer and fascinating examples, there are whistled languages from the Canary Islands, and even drumming and xylophone languages from Africa, which are musical surrogates that communicate in similar ways to spoken language. Therefore, it should be no surprise that music serves as an important acoustical signal that is universally expressed in every human population ever observed—and co-opted by the Tribe Drive to serve its ends.

Remarkably, both music and language share a significant overlap with the types of neural structures that facilitate uniquely human forms of communication. Acoustical signals ascend the pathway from the brain stem to auditory cortex and on to more specialized cortical regions. Pitch perception, harmonic structural processing, syntax, semantics, timbre, and rhythm all share in a *discrete combinatorial system*. But language and music have basic logic and grammar (the instinct hardwired into our neural systems) that can produce a breathtaking amount of novelty in the ways they are expressed. It explains the incredible breadth of how one instinct—the drive to speak—created all the varied languages of the world.[23]

Race as a Tribal Signal—The Fallacy of Phenotype

At the turn of the twenty-first century, a group of psychologists began what may be one of the most societally significant experiments ever performed. Robert Kurzban and his mentors and colleagues, John Tooby and Leda Cosmides, are all pioneers in the field of evolutionary psychology. In 2001, in a publication in the prestigious *Proceedings of the National Academy of Sciences* titled "Can Race Be Erased? Coalitional Computation and Social Organization," these scientists set out to test the hypothesis that encoding by race is a reversible by-product of cognitive machinery that evolved to detect *coalitional alliances*. The results they yielded could very well save the human species. *They discovered the mechanism—irrespective of what race you identify with—of how to change a person from out-group to in-group.* Moreover, they revealed the shocking amount of time it took for the transformation to occur.

Categorizing others is a precondition for treating them differently. You cannot be racist if you do not note someone's race, but we do, and unfortunately it is not *learned* but *instinctual*. There is nothing inherently important about race as a category. It is like any of the many other categories that we use to stereotype. There were simply too many fitness-enhancing benefits to early humans being able to instantaneously and automatically stereotype—create a predictive model of behavior—other individuals based on their phenotype. Those who did not have the requisite cognitive machinery to compute these models did not survive and reproduce—which has led to stereotyping being a universal behavior in modern humans. The critical follow-up question is: If the automatic determination of skin pigmentation, or any other signals we emit, are immutable cognitive processes, then is it possible to override downstream? If so, how long and under what conditions could someone first determined to be potentially in the out-group be accepted as a true, inveterate, new tribal member of our in-group, working to be an ally to our cause?

As it turns out—the remarkable and almost unbelievable answer is *four minutes*.

Our ancestors lived in social worlds that, although impressively complicated for the time, would pale in comparison to the complexities inherent in a twenty-first-century megacity. You recall that age and sex were two factors that enabled an individual to generate probabilistic models of inference about the members of their tribe. Did you notice that I had left *race* out as a variable? This is a critical point—because until the nineteenth century, *race* did not exist as a relevant social construct. Our modern categories of race are the result of the (profoundly erroneous) attempt of nineteenth-century European scientists trying to classify members of their species within the broader context of the natural world. In contrast, ancestral hunter-gatherers, traveling on foot, rarely made residential moves greater than forty kilometers. They would have never encountered populations genetically distant enough to qualify as a different "race." As noted previously, Paleolithic ancestors had only two concerns: (1) age group—because they were your immediate collaborators in survival and (2) sex—because you needed to know who could be a potential mate. Modern skin color–based racism was likely not a feature of the middle Paleolithic.

But what about race today? Remarkable experiments by Kurzban and colleagues demonstrated that "despite a lifetime's experience of race as a predictor of social alliance, less than four minutes of exposure to an alternate social world was enough to deflate the tendency to categorize by race." That is, any observable feature—no matter how arbitrary—can possess social significance and be part of the cognitive representation generated to detect patterns of alliance.

It has never been about race—at least not biologically.* What humans instinctually crave upon first contact with another human being is a predictive model of their coalitionary alliances (race being a single signal among many). Specifically, the Kurzban study showed that racial cues are not fundamental during the process of social categorization. In the experiment, the subjects observed photos of eight individuals. Each photo chained together sentences that formed a heated discussion. Each of the eight people of different races in the photo were represented identically, except for (one of two) colors of their T-shirt. The scientists told the subjects that they would observe individuals of two basketball teams that were rivals and had a history of fighting, and that these individuals uttered the sentences in the context of a group discussion. After a short (strategic) distraction, respondents were asked to recall who said what. This happened in two different conditions. First, all individuals displayed in the photos wore T-shirts of the same color. In contrast, in the second condition, players wore T-shirts of different colors (hinting at team membership). This experiment crucially demonstrated that the extent of racial categorization is strongly reduced when even arbitrary coalitional cues are present. In other words, T-shirt color became the most salient cue—more so than skin color—to the subjects' determination of coalition.

Yet, the same cannot be said for coalition versus biological sex or age despite strong amplification of coalition cues, as subjects continued to

* Race is not "real," biologically speaking. It was, primarily, an invention of one coalitionary alliance to justify the domination of other coalitionary alliances. The strategy is not new. It is as old as intertribal violence itself. Dehumanization is the primary tool every in-group that has ever existed has used to justify acts of violence against other out-groups. After centuries of the link between skin color and coalitionary alliance being reified within our societies, the stereotypes need updating.

strongly categorize on the basis of sex and age. Kurzban and colleagues summarized their results:

> The sensitivity of race to coalitional manipulation lends credence to the hypothesis that, to the human mind, race is simply one historically contingent subtype of coalition. Our subjects had experienced a lifetime in which ethnicity (including race) was an ecologically valid predictor of people's social alliances and coalitional affiliations. Yet *less than 4 min of exposure to an alternative social world* [emphasis mine] in which race was irrelevant to the prevailing system of alliance caused a dramatic decrease in the extent to which they categorized others by race. This implies that coalition, and hence race, is a volatile, dynamically updated cognitive variable, easily overwritten by new circumstances. If the same processes govern categorization outside the laboratory, then the prospects for reducing or even eliminating the widespread tendency to categorize persons by race may be very good indeed.

It is astonishing that for most of our Paleolithic journey, the most important (fitness-critical) categories for us to measure when interacting socially was not race, but age and sex. All groups during this period (unless savaged by disaster) had relatively equal sex-balanced groups and demographic variability. Class and race, two variables we consider as important today, were powerless to influence egalitarian forager societies. So age and sex are *primary signals.* They are immutable. You cannot help but notice these at first glance, and they will instinctually influence the way you interact with another person. All other signals, including the ill-formed concept of "race," are secondary signals used to run through a cognitive algorithm to assess coalitionary alliance. The only thing about race that makes it an exception is that the social construct correlates with a hereditary trait: skin color. In contemporary terms, it would be like having to wear the same baseball jersey (announcing your fealty to your local sports team) daily without ever having the option of taking it off.

It is worthwhile to further illustrate how this works.

Imagine you are a fly on the wall for a modern-day screening of the movie *Guess Who's Coming to Dinner.*[*] In this scenario we will run some updates on the variables at play. You are at the home of an upper-class family in San Francisco and a thirty-seven-year-old African American, who we will call Jamar Drayton, returns home for a surprise visit to his parents. Accompanying him to dinner will be his fiancée, whom he met ten days prior on a vacation to Indiana. She is a twenty-three-year-old African American woman named Joanna Prentice. Jamar's parents are elated to find out their son is getting married until Joanna walks in to greet them wearing a MAGA (Make America Great Again) hat and sporting a World Wrestling Entertainment T-shirt and an Indy 500–themed tattoo sleeve on her arm.[†] Her liberal, coastal elite hosts are stunned at their son's companion. She is politically incorrect, prone to locker room talk, pontificates on the "hoax of global warming," and makes several *Duck Dynasty* references that do not enthrall her fiancé's parents. This tribal signaling might endear her to many families among the working middle class throughout the Midwest and southern United States; yet, her authentic self ends up triggering too many trip wires of an ideologically *rival coalition.* Race (i.e., her skin pigmentation) means little to the Drayton family in this context, and Jamar and Joanna's love may be doomed.

The implication of Kurzban's research is profound, and once you see it, you cannot *unsee* it. Skin color is equivalent to a coalitionary alliance token (tracked as a default), but subsequent tokens can amplify or negate the weight of tribal allegiance. Throw in a few more tribal tokens, such as dress, dialect, manner, gait, religious affiliation, and purchasing habits, and humans will track it greater than race, and they do so in a remarkably short period of time—as we have noted with Joanna. In just under four minutes, Jamar's parents have made a judgment. As Kurzban aptly noted: "A few minutes in an experiment in which race was not pre-

---

* *Guess Who's Coming to Dinner* is a 1967 American comedy-drama that was one of the few films of the time to depict an interracial marriage in a positive light. In the film, Joanna Drayton, a twenty-three-year-old white woman, returns from her Hawaiian vacation to her parents' home in San Francisco with Dr. John Prentice, a thirty-seven-year-old Black widower.

† The Indianapolis 500, formally known as the Indianapolis 500-Mile Race, is an annual automobile race held at Indianapolis Motor Speedway in Speedway, Indiana, United States. The event is billed as The Greatest Spectacle in Racing and is part of the Triple Crown of Motorsport, one of the most prestigious motorsports events in the world. The race is a treasure of Hoosier culture.

dictive of coalition could not detectably impact the accumulated effect of years of exposure to a world in which race mattered." We can see from this example that skin pigment is simply one signal among many of the tribal tokens that people use to probabilistically predict behavior. Our forecasting brains can quickly be co-opted to transcend race, class, and religion, but only if we downplay the importance of certain tokens as predictive of coalitionary alliances.

Data like this should be sounded from the highest rooftop—for it is proof of concept that an individual's race should only be as interesting as someone's hair color and no more. This is the possible future world that Dr. Martin Luther King Jr. envisaged, where his children would "not be judged by the color of their skin, but by the content of their character." Dr. King would be greatly encouraged to know that this is also the world that science describes for our species. The tribe concept is in fact a *creed*, not a "race," and as such, coalitions are the domain of the symbolic and abstract. Mercifully, they are not fixed, and thus possess inherent flexibility in reorientation. Skin color is a tertiary token commonly correlated with coalitionary alliance, but it is not solely predictive.[24*] *The science shows that* Homo sapiens *are not actually born racist—but we are born coalitionist.* This is a godsend, because coalitions are, by nature, malleable. We can shape them in prosocial ways. We are not destined to be a racist species. With science's help, it is a disease we can cure.

Ideology as a Tribal Signal—The Ultimate Loyalty Test

In 2007, a small group called One Mind Ministries residing in the Baltimore area and numbering no more than a dozen people was accused of murdering an eighteen-month-old infant named Javon. To eradicate "a

* An article by Arun Gupta, "Why Young Men of Color Are Joining White Supremacist Groups," reported that the norm of multiculturalism has transformed far-right political movements so that they now attract members from African American, Latino, and Asian demographics. A lawyer who represents one such group, called the Proud Boys, said in a statement: "The only requirements for membership are that a person must be biologically male and believe that the West is the best." In essence, it is a kind of nationalism that leverages the superordinate goal of inherent Western superiority to transcend skin color. Again, coalitions are not equal to skin pigmentation.

spirit of rebellion" that had overtaken the boy because he ceased to say "amen" before meals, the group—including the boy's mother—deprived him of food and water. The leader of the cult ordered her members not to feed the child, and this was followed by a physical beating before being isolated in his room. Charges filed by detectives stated that the child was denied necessities that led to the boy's death. After the boy died, the cult leader directed everyone to pray around Javon's body, and she told everyone that "God was going to raise him from the dead." Of course, the resurrection never took place, but while waiting on the Lord, they decided to stuff the corpse into a green suitcase and eventually placed it in a shed. The body was discovered more than a year later.

During the trial the mother offered her cooperation to the prosecution against her codefendants under the condition that the charges be dropped if her son rose from the dead within the year. Prosecutors allowed the plea under conditions that the reanimation be "Jesus-like." Needless to say, she has since been convicted. The mother was evaluated by a psychiatrist who found her to be "not criminally insane." They found that her beliefs were indistinguishable from other religious beliefs. Reported by Marcus Baram of ABC News, the mother's attorneys said in a statement: "She wasn't delusional, because she was following a religion."[25] The shocking tale of the religious group appears bizarre on many levels; the most obvious paradox is that a mother's love for her offspring is usually among the most powerful examples of kin-selected altruism in nature.

In terms of evolutionary fitness, at face value, there is precious little that should theoretically be able to alter a mother's behavior to prize something else besides the survival of her offspring. Yet, here is a force more compelling than the safety and preservation of her own child. What is it? You may be tempted to answer: *religion*. But that only gets you half credit—as there is something deeper, more ancient, and more primal at work. This force was embedded into our ancestors and it predates the innovation of religion. Are the One Mind Ministries simply an isolated incident? Not at all: Christian Scientists denying their children lifesaving treatment, exorcisms, anti-vaxxers—are all modern examples. But if we pull the veil of time back slightly, we can see human sacrifice, female genital mutilation, and cannibalism added to the list.

It is difficult to see the One Mind Ministries example as anything but aberrant—an outlier and rounding error of collective group behavior in our species. In actuality, the One Mind Ministries behavior was characterized by its internal logic: *Obey group consensus—the tribal truth—about how the world works. Yield to the Tribe Drive.* This logic is an adaptive rule of thumb that got our ancestors out of a lot of scrapes, but when adhered to too strictly, can have bizarre and maladaptive consequences.

Tribal signals are the abstract and symbolic elements emitted from our social worlds that are filtered through our senses. Language (and music) funnel through our acoustic sense. Phenotypes that are embedded in our bones and skin and shaped by our material culture are sieved through our visual sense. And, perhaps most important, *ideology* is filtered through the cognitive subroutines that calculate logic and reason, which have been warped by an evolved bias for group consensus. Because ideology is based on the replication of linguistic memes and is a discrete combinatorial process, it can invent an infinite number of shades, textures, and colors to any worldview; it is the perfect vetting system to serve as a mental passcode to coalition membership.

On the topic of memes, the evolutionary culturologist Joe Velikovsky describes something perhaps even more important than the meme itself—how memes interrelate with each other through spacetime in the form of *memeplexes*.[26] Memeplexes are complex memes. If a meme is a unit of culture, then a memeplex is a network of units that are part of a larger whole. The memeplex can be a transmedia story (like Marvel's Avengers), a religion, or a Kuhnian scientific paradigm. For our purposes though, a memeplex is interesting because it is the most valuable tribal currency. Tribes live and die based on the quality and quantity of memeplexes that they host and transmit. When tribes are transmitting memeplexes at a fantastic rate, it can be said they are thriving, and when tribes lose relevant memeplexes to transmit, or their memeplexes become less selected (or popular), they are "archived" and die. One of the more pernicious memeplexes out there is *ideology.*

Nature is replete with instances of infanticide where mothers need to perform cold calculus to jettison offspring so that they and their future offspring can enhance their fitness. For example, ovicide (the destruction of eggs) as a practice has been observed in many species throughout the animal kingdom, including insects, fish, amphibians, and birds.[27] Startlingly, human history provides us with examples of children abandoning or killing their parents at the behest of the darker, more sinister expression of the *Tribe Drive.*

IT APPEARS THAT THERE IS a critical distinction to be made with the human adaptation that is the Tribe Drive and the phenomenon of *political tribalism. Tribalism* and *political tribalism* are used so often by mainstream media pundits that it is difficult to distinguish between the two. The main distinction is that modern political tribalism is a tool of statecraft. There is growing evidence that tribalism is the systematic weaponization of the human Tribe Drive and will be one of the twenty-first century's greatest military tools in global competition. It may be the greatest threat humanity faces.

To use a twentieth-century example, we can go to communist China. China from 1966 to 1976 underwent perhaps the grandest experiment in ideologically driven political tribalism the world has ever known. The experiment, which historians call the *Great Proletarian Cultural Revolution*, resulted in the persecution of thirty-six million people—including public humiliation, imprisonment, hard labor, and torture—and a mass genocide of three million out-group deaths.[28] This out-group was uniquely defined by only one dimension—ideology. In the 1987 autobiographical book *Born Red*,[29] author Gao Yuan recounts the tale of growing up in Yizhen (the name of the city was changed) as a young boy during the Cultural Revolution: "Virtually the entire populace of Yizhen, old and young, men and women, workers and peasants, housewives and peddlers, had taken sides. Once harmonious families were breaking up. Siblings refused to talk with one another, or communicated with curses and fists. Once loving husbands and wives were filing for divorce. Best friends had become enemies."

The students had, at the behest of the top-down dictate of Mao

Zedong, overthrown the teachers by "making cultural revolution" in which teams of students would criticize teachers by way of public pronouncements through leaflets and posters. This eventually led to trials by students of teachers on the basis of them being "capitalist roadsters" and sympathetic to American imperialism. Kangaroo courts led to devastating public shaming, and in the most severe cases, the disappearance or death of the teacher. A young girl named Yuling, a friend of Gao, was compelled to give up her parents:

> Yuling said her father's students had searched their house, found his diaries, and accused him of being a CIA spy. Yuling's mother was in trouble also. The charge? Teaching bourgeois music. "I'm really worried," Yuling said, crying. Many children of teachers under criticism had publicly disowned their parents and helped to expose their parents' crimes. It was only practical . . . finally she decided to forestall any trouble. She put up a poster announcing that she was severing relations with her parents politically and economically.

They were all victims of a drive they could not resist.

ONE OF THE MOST POWERFUL ideologies is religion. This should come as no surprise—*religion is a tribal signal.* Its primary function is tribal—that of bonding coalitions together and eliminating competing coalitions. As humans began living in densely populated urban strongholds, an important evolutionary innovation was that of a moralizing god. Animism, totemism, and shamanism were the default forms of religion that small-scale societies and foragers intuited. Yet, after agriculture took hold to permit complex societies, these first forms of religion failed at scaling. The *moralizing gods hypothesis* offers a solution to the scaling problem: it proposes that belief systems culturally evolved to facilitate cooperation among strangers in large-scale societies.

Harvey Whitehouse, of Oxford University, and his colleagues published the first test of the moralizing gods hypothesis titled "Complex Societies Precede Moralizing Gods Throughout World History."[30] Using fifty-one measures of social complexity and four measures of supernatural enforcement of morality, they systematically coded records from 414 societies that span the past ten thousand years from thirty regions around the world. Their analyses confirmed the association between moralizing gods and social complexity and revealed that moralizing gods follow—rather than precede—large increases in social complexity with million-plus populations. Moralizing gods are not a prerequisite for social complexity, but appear to be an innovative answer to the problem of how to sustain and expand multitribal groups into a supertribe. Like political tribalism, it's another tool of the centralist state to enforce cooperation from the top down.

The groundwork for moralizing gods appears to be *ritual*. Ritual, as a compact and transmissible behavioral code of tribal signaling, standardizes religious traditions across multiple populations. It appears that ritual practices were more important than the particular content of religious belief, and those rituals bind the society together like a glue binds the pages of a book. It does not really matter what the words in the book say, as long as the glue is binding.

The Axial Age, coined by the German philosopher Karl Jaspers, literally means "pivotal age" and characterizes the period from about the eighth to the third century BCE as one of great moral evolution occurring simultaneously in multiple civilizations.[31] Jaspers noted the nexus of thinkers and philosophers (in China, India, Persia, Judea, and Greece) that were simultaneously and independently laying the foundations of how to live together in complex societies. Carl Sagan and Ann Druyan note some of the benefits that transcend the religion, itself: "Over the course of human history, some religions, it is true, have become much more than this—at their best transcending intimidation, hierarchy, and bureaucracy, while providing comfort for the powerless. A few, rare, religious teachers have acted as a conscience for our species, have inspired millions by the example of their lives, have helped us to break out of baboonish lockstep."[32] As noted in the beginning of this chapter, the Wixáritari people were one of the only indigenous groups to outlast Spanish colonization. While there is increasing migration to surrounding cities, many Wixáritari maintain at least some of their traditional religious practices.[33]

One of the most sacred values of the animist Wixárika religion is to attain wholeness and harmony by way of religious rituals. They maintain family god houses and community temples. Unlike in the West, there is little distinction between religious and secular life. The two prominent features of Wixárika religious views and worldview are importance of the Wirikuta—the location from where they believe their ancestors came—and the cultural complex formed by deer, maize, and peyote, which are central to one of the most powerful tribal rituals I have ever encountered—*the peyote hunt*. A pilgrim group is led by the *mara'akame* (peyote shaman) to reify their unity and identity. The peyote cactus (*Lophophora williamsii*) contains a psychoactive alkaloid and hallucinogenic mescaline and is consumed for ritual and healing and is an integral aspect of Wixárika identity. The ritual consumption augments a strong sense of unity with both the natural and social worlds they inhabit.

The pilgrimage is a six-hundred-kilometer journey where participants experience sleep deprivation, fasting, and physical exhaustion: "Many times in the course of the peyote hunt, the peyote pilgrims affirmed that while their life is poor materially, full of travail and danger, they were not to be pitied, for their symbols make them rich."[34] This is an honest signal, too expensive to fake. The fact that the ritual enacts their myths with unquestioned repetition that must be executed to specific detail is a strong litmus test of tribal allegiance. Finally, Wixáritari have paradoxes embedded in their mythology, such as the trinity of their symbols—deer, maize, and peyote—being fused into a single entity. Anthropologist Claude Lévi-Strauss made the argument that the purpose of mythology is to create paradoxes (for example, the Holy Ghost is a paradoxical trinity where God the Father, God the Son, and God the Holy Spirit are one and the same). It is a mystery that must be believed nevertheless. Paradoxes rationally serve as tribal loyalty tests because the more logic-defying the premise, the stronger the tribal signal.

Not unrelated, sharing concepts that may not be literally true, yet inherently believed, can facilitate group-level cooperation by the sharing of *sacred values*. A sacred value is one that an individual observes as absolute and inviolable—in effect conceiving of breaking a sacred value is a social taboo. Sacred values differ from material or instrumental values in that they incorporate moral beliefs, such as the welfare of family, commitment to country, or identification with a particular religion that is thought to be absolute and inviolable.[35] Sharing of these stories demonstrates honest reliable signals that an individual values the group and its goals. Finally, and perhaps most importantly, the sharing of these sacred values is essential to the formation and maintenance of group identity. Ultimately, because these concepts—which may not be literally true—can still signal group membership and maintain purpose and meaning in that membership, their power rests in the fact that they are tokens of coalition alliance.

The Wixáritari, in their tradition, have always had one eternal identity, which has remained resilient "even when at the mercy of powers outside their ability to control or comprehend." I believe the Wixárika religious practices were crucial tribal signals that buttressed their identities, making them resilient in the face of Spanish encroachment and allowing them to survive centuries after the fall of the Aztecs and conversion of countless other indigenous Mexican groups. Barbara Myerhoff explains: "The treasure of Huichol culture is aesthetic and spiritual. These people envy no one, at least that is their ideal. They have in abundance a culture's greatest gift—an utter conviction of the meaningfulness of their lives." As we turn to the second part of *Our Tribal Future*, I hope that its application in practice can bring a similar sense of conviction to the meaningfulness of your life.

This Is the Way

A tribe is not a race, or even a population . . . a tribe is a creed; it is a team that has agreed upon a set of symbols—including sacred values—that identify membership. A creed is a mechanism that glues together disparate small camps and bands of cohabitating humans into a singular identity and shared purpose. Those who know the codes have in their possession a social passport.

The Mandalorian is Jon Favreau's live-action series that reinvigorates the *Star Wars* universe with a depiction of a character named Din

Djarin, a Mandalorian bounty hunter. The show follows his exploits beyond the outer reaches of the New Republic that dominates the galaxy after the events of *Return of the Jedi*. The epic tale highlights the Mandalorian people who originate from the planet Mandalore with a strong warrior tradition. In the opening of the series, the question *What is a Mandalorian?* is centrally placed. Are the Mandalorians a race? Are they a single species or population within a species? (Caution: spoiler alert ahead.)

Generally associated with moments of great personal sacrifice, one Mandalorian reminds another Mandalorian of their *creed*: "This is the way." Meaning, this is *our way*. This action defines Us from Them and gives our lives meaning and purpose. Throughout the entire season, you never get a glance at Mando's face, for he wears the traditional head armor of the Mandalorians, which resembles a cross between a stormtrooper's helmet and an ancient Greek Corinthian helmet with a T-shaped visor embedded with slits. Sabine Wren, a revolutionary leader of the Mandalorian people, indicates what is an important artifact of the tribe: "The armor is part of our identity. It makes us who we are."

For the Mandalorians, having your helmet removed from you in a battle risks you losing your identity and membership in the tribe. It is a powerful scene at the season's end when the answer to *What is a Mandalorian?* is revealed, in a flashback to Din's childhood after having had his helmet removed for the first time. Our hero is in fact an orphan of the Clone Wars who was saved by a unit of the Mandalorians. Despite being a member of a different species, he was raised as their own and taught their *creed*. Thus, their creed is the onus to membership and the ultimate tribal token. A Mandalorian is a pan-species coalitionary alliance that exists as a single tribe under one constellation of agreed-upon symbols—the Mandalorian creed. The creed tells team members: *this is the way*. Put in more academic terms, a memeplex of tribal symbols project group phenotype—created by *our* culture and embedded in *our* actions—that give life meaning and purpose.

My fieldwork guide to the Wixáritari, René Sagastume, was both a Mestizo and a member of the Wixáritari tribe. He implicitly knew this wisdom. His helmet was the sombrero he wore with the band of mesmerizing geometric patterns, hand-woven by a member of his tribe. His life embodied the Wixáritari creed balancing the old and the new in ways that honored the tribe but also ensured its survival; if you know a tribe's creed, then you know the soul of the members within it. The implications of this run deep, because if you have no creed and no tribe to share it with—do you have a soul?

PART II

THE PRACTICE OF TRIBALISM

6

TRIBAL BENEFITS

How Others Can Help You Achieve Your Dream

Loneliness is tied to being more likely to die at any time of any cause at any phase of life.

—Sarah Rose Cavanagh, 2019

When the snow falls and the white winds blow, the lone wolf dies, but the pack survives.

—George R. R. Martin, A Game of Thrones

The Roseto Mystery

In the late 1950s, Stewart Wolf discovered an astonishing outlier.

This discovery rocked the medical establishment and, by the twenty-first century, would fundamentally change the way scientists perceive the foundation of health and well-being. At the time, in the mid-twentieth century, Wolf taught in the medical school at the University of Oklahoma, and he spent his summers on a Pennsylvania farm, not too far from a town called Roseto (Italian for "garden rose"). He had lived close to the town most of his life, but it was insular enough that he had never really noticed it until he was invited to give a talk at the local medical society. It was during this visit, when speaking to a local physician, that his career would take a fateful turn.

What was this astonishing outlier? The physician noted that heart disease was nearly non-existent in Roseto. Wolf was shocked by this anomaly and decided to spend an entire summer there with a research team to rigorously measure the entire population's cardiovascular health. The results were revelatory, as summarized by the journalist Malcolm Gladwell in his book *Outliers:*

> In Roseto, virtually no one under fifty-five had died of a heart attack or showed any signs of heart disease. For men over sixty-five, the death rate from heart disease in Roseto was roughly half that of the United States as a whole. The death rate from all causes in Roseto, in fact, was 30 to 35 percent lower than expected.[1]

But it wasn't just the apparent strength of their heart muscle that proved to be outliers. Wolf teamed up with a sociologist, John Bruhn, to take a deep dive into Roseto's society. After knocking on doors and interviewing nearly every adult over the age of twenty-one, they discovered that as a whole, nobody in the community was on welfare, they had no incidents of suicide, no alcoholism, no serious drug addiction, and barely any crime.

The leading cause of death in Roseto was *old age.*

The research team was dumbfounded. The heart disease outlier alone was remarkable; before the advent of cholesterol- and hypertension-lowering drugs, the rest of the United States was embroiled in an epidemic of heart attacks. Wolf felt that if they could discover the secret to the town's fountain of wellness, they could transplant that knowledge for the betterment of the rest of the developed world. They feverishly went to work to unlock the mystery of Roseto.

Their first hypothesis was that it had to be dietary. This was quickly ruled out, as physiologists showed Rosetans were getting a massive 41 percent of their calories from fat embedded in pizza, cooking lard (instead of healthier olive oil), and pork livestock. To add insult to injury, the Rosetans typically didn't exercise, were heavy smokers, and many were clinically obese. The second hypothesis was that it was genetic. They tracked down Rosetans that lived in other parts of the United States to determine if they were imbued with some kind of molecular protection from the ill effects of modern life. Deepening the mystery, Rosetans outside the town were dying from all the same things as every other American. The investigators, now getting desperate, called their third hypothesis "is it something in the drinking water?" In other words, was it a regional factor that buffered people's health and well-being? They went to the neighboring towns of similar size and scale, Nazareth and Bangor, but the medical records revealed another dead end, with men over sixty-five suffering from three times the rate of heart disease of Rosetans.

The secret to the Roseto mystery, Wolf and his team surmised, could not be diet, exercise, genes, or location, and therefore had to reside somewhere in the *way* Rosetans, as a group, were living their lives.

What's in It for You?

By necessity, our paleolithic ancestors were small, interdependent groups that were joined together in the shared project of survival. They had a tribal lattice that guided everyday actions. By default, they were born into a deeply connected society. Upon arrival into the world, they automatically inherited a social network that would remain integral to them throughout their lives. This contrasts starkly with our vulnerable mismatched modern world, where our single-person homes and single-serving friends* dwell relatively independent from the rest of the institutions that comprise economically developed societies.

The primary aim of this chapter is to demonstrate that even the smallest steps toward greater social connectivity can enhance your quality of life. The issue is, society doesn't make this easy. If fact, our culture has disincentivized connectivity in dramatic ways. The United States, and now much of North America, props up the primacy of the individual and independence from reliance on others as a sacred value. It offers wealth and prestige to those who abandon family clans and local community to seek success in far-off financial centers. To many, intentionally giving up a pay raise or new job opportunity for the sake of friends or family seems wildly heterodox and too costly to consider. But there are drastic, hidden costs to social isolation that the new psychological sciences are

* This "single-serving" reference comes from a favorite movie of mine: *Fight Club.* Edward Norton's character, narrating in the film, speaks to someone he *thinks* he just met on a plane and says: "Everywhere I travel, tiny life. Single-serving sugar, single-serving cream, single pat of butter. The microwave Cordon Bleu hobby kit. Shampoo-conditioner combos, sample-packaged mouthwash, tiny bars of soap. The people I meet on each flight? They're single-serving friends."

only now uncovering. Thus, in this chapter we will list the ways in which living a life of absolute independence from true social connectivity can be inherently costly. Once we comprehend the trade-offs, we can, for the first time, make truly informed decisions about how connected we need to be, and whether we perhaps should move toward establishing more social connections with family and friends. We will then consider the benefits of being intentional about how we cultivate relationships with others around us.

You may not have the time, energy, or resources to build a tribe or a community, but you are empowered with the ability to intentionally cultivate social networks on the individual, family, and camp levels within your constellation of existing tribes. You can do this by decreasing distances within your group—social, physical, emotional, and spiritual distances that hinder group cohesion and well-being. This is what I call *intentional proximity*. Intentional proximity can be measured on one side of the continuum in small increments, such as a weekly ritual (brunch and mimosas or a poker game with friends). You can also be deliberate about where to live, by arranging to dwell in the same city as your childhood best friend or a beloved family member. And, when taken to its most intentional form—dwelling together within a community. Intentional proximity, expressed as *intentional community*, is a kind of communal living that can range anywhere from being collocated on the same neighborhood street, to building a cohousing community, or cohabitating within a larger group.[2] This topic, and the strategies that one can use to pursue greater intentional proximity, are the essence of the following chapters on *tribal practice* and what I call *campcrafting*.

You may be thinking, all this talk about intentional proximity is nice, but we don't live in caves anymore; only a handful of humans left on the planet approach dwelling in the ecological conditions that our ancient ancestors did. For those of us living in the economically developed, postindustrial world, we have smart homes, AI-governed light, climate-controlled environments, Teslas, and drones to deliver our Amazon packages that took us only seconds to order online, bringing us any goods imaginable. We have access to Italian espresso and French wine even in small suburban and rural strip malls, and an infinite amount of entertainment to consume via streaming platforms for shows and video games. And we even have smartphones, which are able to track us with billion-dollar algorithms that introduce specifically designed products and advertisements to ensure that not a moment of our lives is spent with our attention idly disconnected from our digital landscapes. With such abundance, what could possibly motivate us to undertake the hard work needed to craft intentional proximity to a group of humans? At this point, ask yourself the following questions:

- Have you ever felt depressed?
- Do you ever feel financially insecure?
- Have you ever felt like you were in physical danger?
- Have you ever failed at a romantic relationship even though you put your all into it?
- Do you ever feel like you have friends you cannot trust?
- Have you ever felt like your life lacks meaning and purpose?

If you answered in the affirmative for any of the above, then you certainly have something to gain by exploring and examining the ways in which you can engage in intentional proximity. Can intentional proximity help us live better lives? Let's find out. What's in it for you?

Reason One: Social Connection Improves Your Health

Is isolation really that bad for you? In chapter two we covered some of the disturbing trends and dangers associated with social isolation. While society has been focused on battling one virus in particular (COVID-19), there are other less-sung vectors running rampant through society; one such disease is the epidemic of loneliness. It is such a problematic trend that both Britain and Japan have named "ministers of loneliness." Former British prime minister Theresa May inaugurated the government role because more than nine million Britons reported often or always feeling lonely. The same pattern is showing itself throughout Western civilization. A study of adults living in high-income countries showed elevated levels of social isolation in Australia (25 percent), France (31 percent), and the United States (21 percent). One would think that loneliness is the province of the elderly, yet over a third of young Americans reported "serious loneliness" in a recent Harvard survey, including 51 percent of mothers with young children and 61 percent of young adults aged eighteen to twenty-five.[3] So we're a lonely society. So what?

The longest and still ongoing study in the history of science that started in 1939 with 268 men* at Harvard and 456 men who grew up in the inner-city neighborhoods of Boston. The article "This 75-Year Harvard Study Found the 1 Secret to Leading a Fulfilling Life" by Melanie Curtin revealed that the key rests in the duration and quality of male friendships.[4] As of this writing, the original cohort is now in their nineties. Researchers have looked at every aspect of their health, using blood samples, brain scans (once they became available), and self-reported surveys, as well as interview data. A single, striking result shone through: the men with the best social relationships are not only the ones in the best mental health but are the longest lived. Robert Waldinger, director of the Harvard Study of Adult Development, boiled it down: "The clearest message that we get from this seventy-five-year study is this: good relationships keep us happier and healthier. Period."

The scientific study of loneliness is extensive and points to a firm conclusion—*isolation is deadly*. We touched on this in chapter two, but let's dig deeper. The aim is to grasp the full biological consequences of not experiencing enough meaningful social interaction and how isolation, specifically, affects the brain, sensory systems, hormones, ability to learn, and all of your development—even down to your genes!

- *The socially starved brain.* The socially isolated brain craves connection like it does food.[5] In one study, after being experimentally manipulated by being socially isolated for ten hours, participants reported increased social craving, loneliness, and discomfort as well as decreases in happiness.† Notably, later in life a lonely brain also has a higher cortical amyloid burden—a marker of preclinical

---

* Unfortunately, there is no study with women that has been ongoing for such a duration of time. At the time of the formation of the research, Harvard was a male-only institution.

† The underlying neural mechanisms quantified with fMRI was dopaminergic midbrain responses in the substantia nigra pars compacta and ventral tegmental area (SN/VTA).

Alzheimer's disease. Thus, our brains are literally trying to motivate behavior that brings us closer to other human beings like it drives us to eat, sleep, and find shelter; if we don't get this baseline, later in life we suffer cognitive impairment.[6]

- *Mental and physical resilience cued by our immediate social network.* When in an environment where we can sense the presence of people we care about, we are more resilient to a number of challenges—including pain perception. In one experiment, holding the hand of a partner reduced self-reports of pain intensity after being electroshocked.[7] During childbirth, touch-related support substantially reduces labor pain.[8] A mother's touch reduces perceived pain by children coping with disease.[9] If we are in social isolation, our autonomic system regulates temperature less effectively.[10] In other words, when our environment presents direct challenges to our bodies, the presence of trusted group members enhances our resilience.
- *The stress hormones of loneliness.* Being psychologically stressed—that is, neurotic, paranoid, prone to hostility, and generally anxious—is a state that is primarily governed by glucocorticoid stress hormones.[11] The brain visibly documents the causal role of both perceived and *non-perceived* loneliness on stress.* A study measured perceived lack of companionship on the neuroendocrine system in adults and demonstrated that lonely individuals displayed significantly greater fibrinogen† and elevated cortisol response over the first thirty minutes upon waking.[12] In one group study of non-psychotic psychiatric inpatients, the individuals with little social support had greater levels of cortisol in urine than in patients with greater social support.[13] Short-term acute stress is adaptive, but long-term chronic stress hormones have devastating

* Additionally, studies show that loneliness, measured by way of PSI (perceived social isolation), has been associated with glucocorticoid resistance, and thus diminished neutrophil-to-lymphocyte ratios in white blood cells (denoting a compromised immune system) because of social isolation. Another study with herpes patients found that lonelier subjects had a much greater negative impact on their health and recovery (as indicated by having higher antibody titers indicating ongoing infection) than socially connected patients.

† A soluble protein present in blood plasma that is an indicator for inflammation and tissue damage.

effects—leading to more aggression, less empathy, and reduced prosociality.[14]

- *Alone we lose the capacity to learn and adapt.* As we recall, neuroplasticity is all about learning. Retreating from our social groups impairs our capacity to learn and even deal with trauma.[15*] This was demonstrated in a study where the dopamine system (which is critical in learning by rewarding behaviors) was activated less in lonely people when presented pleasant pictures of novel objects. The lonely participants were effectively less incentivized to learn new things about their environment. Another study discovered that lonely individuals regard pleasant interpersonal interactions to be less pleasant, whereas individuals with social support saw the same interactions as positive, thereby providing a positive feedback loop to prosocial behaviors in group contexts.[16] In mice, a study found that isolated subjects received less brain growth from exercise.[17] In sum, lonely people not only have less capacity to learn, but when they do learn, their brains physically grow and the new growth is likely to enhance the ability of the individual to see the world in a more positive way.
- *Isolated children have stunted development.* By depriving a child of a loving mother and peer-to-peer interaction, you are setting them up for devastating long-term effects. A study showed that more-socialized children have enhanced perspective-taking capacity.[18] Studies have shown less-socialized individuals tend to have lower IQs, poor cognitive skills, problems forming attachments, severe anxiety and depression, and a higher likelihood of being institutionalized. Tellingly, their brains are typically smaller, except for an enlarged amygdala.[19]
- *Loneliness negatively affects gene expression.* Gene expression is the turning off or on of protein synthesis, and it can have significant

* Emotion tags experience with relevance, and so from an evolutionary perspective PTSD can be seen as adaptive learning. PTSD is one example of the power of social support to augment an individual's capacity for resilience in the face of trauma. One study followed victims of motor vehicle trauma that were clinically diagnosed with PTSD. The most powerful predictor of the severity of trauma experienced by individuals was a factor the researchers termed *negative network orientation*, whereby victims retreat from social interaction after the event and view the utilization of social support as inappropriate, useless, or dangerous.

> effects on human wellness. This can affect tissue function and condition, as exemplified by one study that discovered subjective social isolation is associated with net reductions in the expression of genes bearing anti-inflammatory glucocorticoid response while over-expressing genes linked with pro-inflammatory transcription factors.[20] This finding also stands for decreased lymphocyte sensitivity, resulting in less capable immune systems to fight off infection.[21] Lonely people, therefore, are more likely to suffer from one of the most predictive factors of poor health that are mediated by epigenetics—inflammation and decreased strength to immune function.

All of these elements combined directly impact your health. By every factor—from our neural architecture to the kind of cultures we create and participate in—loneliness is one of the most powerful predictors of well-being. Lonely people exhibit (i) worse cognition, (ii) less resilience, (iii) less prosocial behaviors, (iv) less ability to learn and adapt, (v) more anxiety and depression, (vi) poorer immune strength, (vii) and live in a state of greater social risk. Ultimately, living the life of a loner is inherently more dangerous.

Reason Two: Social Connection Strengthens Your Romantic Relationships

Have you ever had a critical problem sneak right up underneath your nose while you were focused on something else? In the military, there is a term for this: *danger close target*. Snipers often find themselves behind a high-powered scope observing potential targets at far distances. However, the most dangerous threats to their team inevitably are "danger close." These are the seemingly mundane phenomena that can unexpectedly evolve into critical challenges. Snipers are trained not to be so overly focused on the far-off targets that they risk compromising the mission by losing sight of the danger close targets.

Supported by the University of Chicago since 1972, the General Social Survey (GSS) is a powerful tool for societal analysis and has provided politicians, policy makers, and scholars with a clear and unbiased perspective on what Americans think and feel about many issues, including confidence in institutions.[22] According to the GSS, one institution that has aged well is that of marriage. Yet, this has been at the expense of essentially all other critical relationships in our lives. For example, when asked a similar question with respect to how much we confide in our friends, advisers, neighbors, coworkers, family members, parents, siblings, and children, the likelihood has diminished in every category except spouses.[23]

This has led to feelings from married couples that their spouse is their best friend.[24] *Dyadic withdrawal* (or more colloquially, "cocooning")—is the pair-bonded version of social isolation. Specifically, it is when two people fall in love and disappear from everyone else. And given all the torrid effects to the individual level of social isolation outlined above, it should be of no surprise that couples can fall into the same trap. In this instance the strength of a couple can actually become more brittle as they become over-reliant on each other. The key question is: "When do social ties compete with one another, and when do they strengthen one another?" Both processes are clearly at work in the complex relationship between the vitality of marriage and the vitality of other social bonds.[25] The likelihood of divorce is actually increased by cocooning because the intense closeness, over time, places irreconcilable burdens on marriage, thus increasing its fragility.

On the level of biological sex, the effects of cocooning can be devastating for both men and women, but manifest in different ways. With respect to Millennial men in their twenties and thirties, fathers have doubled down on their commitments to their nuclear families. Yet, the activity budget came directly out of their social networks, and the male-male friendship was placed on the backburner. One study on the effects of childcare arrangements on marriage demonstrated that almost *every* father had lost contact with his male friends, and nearly as many experienced sadness due to those losses.[26] "These fathers stepped back for the best of reasons. They were busily productive in their careers and had romantic dreams of becoming wonderful husbands and fathers. Critical relationships they had forged throughout their adolescence found themselves atrophying, which has more detrimental effects than most of us expect."[27] The driver of this predicament is in the difference between how friends are cultivated and maintained between the sexes. As noted by Geoffrey Greif, professor at the University of Maryland, "Men tend . . . to have shoulder-to-shoulder friendships and women face-to-face friendships." That is, male friendships are more activity based and female friendships are more disclosure based.

There is good evolutionary theory to underpin the male/female difference in friendship styles. As noted by Robert Seyfarth and Dorothy Cheney of the University of Pennsylvania: "Friendships are adaptive. Male allies have superior competitive ability and improved reproductive success; females with the strongest, most enduring friendships experience less stress, higher infant survival, and live longer."[28] Many of the friendships that males create are formed early in life with their peer groups, often in competitive contexts such as athletics (or hunting, if you are a forager male) and as men grow older, the contexts within which friendships can coalesce become more rare for males that rely less on personal and emotionally based exchanges.[29] Women continue to form friendships throughout their lives and adapt readily to novel environments, whereas when men lose established friends, their total number of friends dwindles throughout their lifetimes.[30] It also means there are differential biological effects to the loss of friendships between men and women. A Johns Hopkins study showed that married men without friends suffer from heart disease 17 percent more than married men with a friend.[31] Whereas socially connected women prove more adaptive at adjusting to changing circumstances associated with aging.[32]

Over a ten-year study, cocooning was quantified and found to be pervasive even among pair-bonds that did not have children.[33] Married couples tended to feel as though they were relieved of the greater social ties they had relied upon beforehand. Specifically, married couples had fewer network ties to relatives, neighbors, and friends. This relationship demonstrates the way our society has put pressure on our social networks that were, for most of human evolution, a group—not dyad—context: "Weakening those other ties . . . may end up closing off another source of nurturance that is equally essential to the long-term health of a marriage."[34] The mechanisms in which this occurs is by way of *witnesses.* Witnesses benefit couples in several ways; as noted by Jacqueline Olds and Richard Schwartz:

> When any aspect of a life is seen by others, it feels more real to the participants. . . . When a cocooning spouse is outside of the cocoon, the marriage can seem very far away because it is not interwoven with other relationships. Witnesses also provide a married couple with an audience to perform for *as* a married couple. People try to perform their best for an audience. Some of that improvement lasts after the audience is gone.[35]

Evolution weaved the pair-bond into a larger tapestry that includes our kin and peer networks. It never intended it to play out independent of those systems and thus is a source of modern mismatch. In this instance, without a larger social group context within which the pair-bond can perform, there remain fewer social forces keeping them together. People can be at their worst when they are cocooned. Either in a nuclear family or marriage without children, when we are behaving outside the context of a larger group, our basest behaviors can manifest. One study found that social isolation was the driving factor that predicted child abuse in nuclear families.[36] *Cocooning can ironically create the very conditions of instability that the partners are trying to avoid by isolating.* Moreover, the greater access that friends have as witnesses to the pair-bond, the greater accuracy and more beneficial advice they can give to help it regain its stability when it becomes unstable. Friends can be a bastion of social support when individuals feel doubts or insecurities about their partner.[37]

In sum, it is in the best interest of the health of the pair-bond for both partners to be socially connected in a network of strong ties. A mother with face-to-face female support and a father with shoulder-to-shoulder male peer support are not only going to be more effective as individuals but also a better support to their spouses or pair-bonds and nuclear families. Because in times of duress (for example, unemployment, an accident, or death in the family) it is this very network that will improve the odds the individuals can return full strength to their social roles in a healthy amount of time. Cocooning with your partner is unhealthy, and it is social connectivity that wards against those negative psychological effects.

Reason Three: Social Connection, from Birth to Death, Augments Development

For mothers, the benefits of social connectivity start from the moment of conception and are only accentuated postpartum. Postpartum depression (PPD) has been described as "the thief that steals motherhood." After birth, an expectant mother has been anticipating the joy of a new infant for at least nine months, and many are shocked to experience instead of joy, a downward spiral of depression. According to the U.S. National Library of Medicine, PPD is characterized as moderate to severe depression often occurring within the first three months after birth. In Western cultures, up to one in five mothers will experience PPD, making it one of the most common stressors related to childbirth. Approximately four million births occur each year in the United States, and that translates to up to 760,000 women affected by PPD annually. The symptoms of PPD are serious; it is a major depressive disorder that includes feelings of extreme sadness, low energy, diminished pleasure, thoughts of death, and in extreme cases can even lead to suicide. In a 2020 umbrella review of the risk factors for postpartum depression, researchers found that *lack of social support* (listed alongside variables such as current or past abuse, partner dissatisfaction, and high life stress) was one of the greatest predictors of PPD.[38] In other words, socially connected mothers who have just given birth are more psychologically resilient.

What about after birth? Evidence suggests there is a powerful link between how strongly integrated you are into your local community and your children's development. A 2020 study published in a special issue of the prestigious *Philosophical Transactions of the Royal Society B* uncovered a fascinating link between alloparenting, ritual, fertility, and child cognitive development.[39] Using ten years of longitudinal data collected with over ten thousand participants, researchers tested the predictions that church attendance is positively associated with social support and fertility, and that social support is positively associated with fertility and child development. Results show that, when compared to non-church attendees, female parishioners had stronger social networks and aid from co-religionists and that aid from co-religionists is positively associated with increased family size. This effect was reduced if the women had groups of friends outside a religious context. Moreover, when compared to social networks outside the religious context, co-religionist aid remained constant. These findings suggest that religious and secular networks differ in their longevity and have divergent influences on a woman's social support, and subsequently their fertility. Cognitive development scores were greater in children with mothers who had greater aid from co-religionists and mothers' social support. As previously discussed, ideology gives meaning to the tribal tokens that enmesh true believers within a shared cultural matrix. When embedded in this matrix, one fitness-enhancing benefit is access to human resources of mutual aid and social support.[40]

A systematic cross-cultural review assessing several dimensions of the influence of a child's social context for development suggests alloparenting may extend to children's moral development.[41] For children exposed to alloparenting, the advantages are numerous—including greater opportunities for engaging in beneficial social interactions and the development of flexible social skills, psychological perspective taking, and social openness.[42] Research demonstrates that the more nonparental adult relations a child has the greater resilience they have in face of adverse conditions. Thus, alloparents are a protective factor for children of all ages.[43]

In one study, multiple adult caregivers positively correlated with effective parenting, lowering the risk of social, emotional, and cognitive problems that extended into adulthood.[44] Finally, adult studies of retrospective reports demonstrate that alloparenting and social support (including increased opportunities for peer-related play) is correlated with adult mental health and morality—which is likely due to an enhanced adult capacity for physiological, emotional, and behavioral regulation.[45] In summary, children born with greater access to a larger network of adults and other children outside of their nuclear family become more resilient, and this leads to an adult life with an ability to effectively regulate emotions, resulting in a more grounded moral sensibility.

But what about getting old? One promising reason to work toward a life of greater intentional proximity is a healthier, longer, more meaningful elderly experience.* The institution of old-age care in our society is seemingly morally bankrupt. Nursing homes tend to infringe on one's independence by treating the elderly like children, with some institutions not allowing residents to administer their own medications or to drink wine when they wish (even if in a relative good state of health and fully capable).[46] Research that monitored the physical health of grandparent caregivers has demonstrated stable or improving health while caregiving.[47] Some grandparents have reported becoming healthier after assuming the caregiving role, citing a more active lifestyle.[48] Independent of caregiving roles, one study also showed that people who develop high-trust relationships throughout their lives significantly increase their well-being, suggesting that the development of trust across the lifespan is a crucial resource for a healthy, happy life.[49] Finally, in a society that discards the elderly as a burden, what better way to provide dignity, purpose, and meaning to the later stage of life than enmesh them in the dynamism of a developing social network? In summary, social connection improves child and family development and buffers against some of the nasty effects of getting older.

Reason 4: Social Connection Reduces Consumption and Improves Living Quality

The primary way we survive in the modern world is by accruing money to exchange for resources. Unlike hunter-gatherers that foraged for all their calories, the vast majority of the food we consume is provisioned and purchased. Yet, there is a massive disparity in baseline access to resources. This mostly comes down to luck. Scott Galloway, a professor at New York University's Stern School of Business and serial entrepreneur, noted how much luck was involved in his success:

* Later, we will explore examples of intentional communities that are specifically designed for seniors, such as Elder Spirit and Silver Sage Village, offering attractive alternatives to the nursing home model. These communities emphasize living dynamic, socially connected lives with autonomy and purpose.

> Being born in 1964 as a white heterosexual male in southern California was like winning the lottery. You came into a great university system that was free. You came into a professional environment where more wealth was created in a seven-mile radius than had been created in the entirety of Europe since WWII. You had [and benefited from] discrimination—you made up 22 percent of the American populace yet 97 percent of all venture funding went to that profile . . . to not recognize how much of your success is not your fault . . . I've come to that [realization] later in life.[50]

As a society, we should want to cancel the worst forms of disparity for the unlucky by making the entry-level luck-floor greater for everyone. For most of human evolution, this was done via *central place provisioning*.[51] Our ancestors overcame unluckiness and environmental challenges by bringing back the spoils of their efforts and distributing some of them collectively to form a sort of camp social safety net. This strategy is a type of caloric insurance; if hunters failed to bring back game,* they weathered the failure by coming back to a camp where foragers were cooking tubers on the fire. Everyone still ate. Everyone survived.

But, most of us don't live in caves and camps anymore. For that matter, the wealthy can simply *buy the products of campcrafting.* As noted by David Brooks, "Think of all the child-rearing labor affluent parents now buy that used to be done by extended kin: babysitting, professional childcare, tutoring, coaching, therapy, expensive after-school programs. (For that matter, think of how the affluent can hire therapists and life coaches for themselves, as replacement for kin or close friends.)" As wealth inequality in American families has sprawled over the last two generations, two different financial portfolios have led to two different family regimes.[52] While rich households predominantly own business equity (by way of stock and commodity assets), middle-class portfolios are dominated by real estate. In postwar America the income and wealth distribution decoupled, leaving rich, highly educated families able to retain the stability they have enjoyed since the 1950s, while ev-

* And more often than not, they do fail. Thanks be to the gatherers!

eryone else has adapted (with varying success) to a more chaotic, rapidly changing twenty-first-century economic landscape.

It is a different story for the less fortunate though. Purchasing outsourced family support has rarely been an option for people of color and first-generation immigrants. Yet, now that the middle class is feeling the effects of income inequality, many find themselves in a similar situation. Immigrants and people of color have, in this regard, had the advantage because they have always been more likely to live in extended-family households on a grander scale than the typical American middle-class nuclear family. Compared to the 16 percent of whites that live in multigenerational housing, more than 20 percent of Black people, Asians, and Latinos are dwelling as cohousers.[53] Whites living in Middle America are now facing greater economic and social stress, yet culturally have idealized the norm of the traditional nuclear family, dislocated from larger extended-family households. Thus, while living alone or in a two-parent, single-home strategy can still benefit many people with the means to outsource the support needed to nourish a family, for many experiencing the economic effects of growing inequality, moving toward a new, more communal strategy may have many advantages.

Being poor makes people unhealthy. That may sound obvious, but the implications are far-reaching and run deep. The way scientists in public health speak of this important phenomenon is the "socioeconomic status (SES)/health gradient." The world over, a cross-cultural trend has been quantified—the poor have worse health than the rich, more disease (in both incidence and impact), and shorter lives.[54] It is no wonder there has been extensive research on the SES/health gradient, given its importance to public health and policy making.

The summary of this work is as follows: First, the gradient can't be fully explained by greater exposure to health risk factors, such as more smoking, lead paint, or toxins in the water; only a third of the effects of health are driven by these causes. Second, the variation in the gradient is, well, a gradient, because with every move lower in SES categories—that is, the poorer you get—health worsens. Third, the gradient is not a result of poor health causing people's SES to go down, as being born poor predicts being a poor adult and being born rich predicts being a rich adult. Fourth, and surprisingly, the gradient is not related to access to health care for the poor, as the pattern exists in countries with health care ranging from universal to near-total privatization. What, then, is the driver of the gradient's relationship between health and SES? It is less about being broke and more about feeling broke (subjective SES) that predicts poor health outcomes, according to Nancy Adler, chair of management at McGill University.[55]

In one study, subjective status (a proxy for social inequality) in older adults was analyzed against subjective health. Individuals living in an adult day care center were compared to a continuing care retirement community over a one-year period. The results reported higher levels of subjective social status were associated with lower levels of loneliness only in the continuing care communities.[56] Another study that used a randomized trial examining the effect of social isolation on subjective well-being in elderly Japanese people discovered that one of the most powerful predictors of how a person *feels* about their well-being is linked with interventions aimed at improving social support. Synthesizing this information, we see that poor health outcomes are not fully explained by actually being poor, but also by the psychosocial mediator of the stress of being poor; as social connection decreases, up goes psychological stress.[57] It stands to reason that having a healthy peer group buffers you from the worst effects of being poor.

That's not all, though—having a healthy peer group also buffers you from *actual* poverty. For example, Timothy Miller, professor of religious studies at the University of Kansas, has documented the history of intentional community-building in America in a trilogy of books. In a section of *Communes in America, 1975–2000* highlighting the emergent secular cohousing trend, he interviews Lydia Ferrante-Roseberry, of the Swan's Market Cohousing in downtown Oakland, California. She lists several of the benefits:

> Cohousing models an alternative to the consumerism and isolation of the "American dream." Since many of our amenities are shared, as a group we consume less than if we lived independently. Among other things, we share three washers and two dryers, a gas grill, a large, well-equipped children's play room, exercise equipment,

> power tools and crafts materials, and . . . a hot tub. In addition, our shared guestroom means that every unit can be built to fit the needs of the family, not the family plus the occasional guest. Most of the things we share are items that people use intermittently. As my husband simply puts it: Everyone doesn't need their own hammer.[58]

Thus, intentional proximity with some type of collective sharing of resources can add significant savings to cost of living. Surveys from 2005 recorded the impact of cohousing on individual budgets; compared to the average American, cohousers consume 40 percent less water, have 50 percent lower utility costs, and drive 30 percent less.[59] Not only is it often more affordable, but it also reduces environmental impact. The architect Graham Meltzer, who specializes in cohousing, has noted that approximately half of the cohousing communities that he has worked with have environmentalist mission statements. Cohousing communities shrink their environmental footprints in several ways and are typically characterized by smaller housing units, shared facilities and tools, recycling programs, fewer cars, alternative energy production and energy reduction strategies (e.g., water-saving plumbing and solar panel roofs), and benign and renewable waste disposal.[60]

Another benefit to intentional proximity is the increased capacity to converge on collective financial strategies that can improve quality of life for the individual and the group. For example, two of Canada's youngest retirees, Kristy Shen and Bryce Leung, started their path to financial independence (FI)* living in Toronto (the most expensive city in Canada) with a net worth of nothing coming out of college.[61] As a team, they managed to retire in their early thirties with a seven-figure portfolio. When asked about money's value, Leung said: "Life is so hard, and what I keep telling people is, 'it is only hard if you don't understand money. If you understand money, life is incredibly easy; if you don't, it is the most difficult thing to get out of bed every day.' But if you under-

---

* *Financial independence* is the status of having enough income to pay one's living expenses for the rest of one's life without having to be employed. There is a growing community that rally behind the motto "FIRE" (Financial Independence Retire Early). The term is viral among Millennials who are not banking on Social Security as a retirement strategy.

stand money—life is easy." The simplest account of the FI strategy is: (i) find ways to keep expenses low, (ii) look for ways to boost your income, and (iii) make saving and investing a priority.

Yet, for those individuals that have been attracted to the FI movement, one of the greatest challenges has been the social and psychological costs of stoically abstaining. Society endorses and incentivizes social consumption; that is, in our culture people feel they need to spend money when out with others to have a good time. As one blogger, tormented by the FI strategy to the point of quitting, noted: "As well as my diet, frugality also starts affecting my relationships. Getting midweek drinks with friends is off. . . . It does make me realize how many of my friends I only really see in the confines of a pub."[62] Jim Collins, one of the founding fathers of the movement, described how critical community is when you are bucking society's norms when he spoke at a conference for the first time and met others dedicated to the FI cause: "Finally for the first time in their life, they got to hang out with people who got it. As one person said to me, 'I found my *tribe* [emphasis mine].' . . . We are very much the odd men out walking this path in society and . . . building community . . . has turned out to be the most powerful element of the FI movement."[63] By living in proximity to a group of people with whom you share cultural norms, you are freed from the capitalist expectation that social capital can only be accrued by going out and spending money.* In summary, consuming in isolation is expensive, whereas consuming as a group reduces resource costs and risk from adversity, and increases the standard of living.

Reason Five: Social Connection Boosts the Odds of Personal Success

In the political and economic sciences, two forms of capital dominate the literature: *human capital* and *social capital*.[64] The concept of human capital pertains to individuals' knowledge and abilities that permit pos-

* In fact, it is likely the opposite, as a good board game or session of Dungeons & Dragons can be just as enjoyable, and accrue more social capital, than most expensive nights out at the bars. We'll delve into this more in the chapters on *campcrafting*.

itive change and growth; the central proposition of social capital is that networks of relationships consolidate and mobilize resources that can be used for the good of the individual or the group.[65] Often human capital is measured in terms of the level and specialization of individual education, physical condition, and overall economic well-being. Social capital is measured as the embedded resources directly tied to relationships with others. In sum, human capital is the expertise of the individual and social capital is the power of the social network an individual is embedded within to execute plans and achieve goals; it's who *and* what you know that leads to success.[66]

When social capital is considered at the group level, it is often viewed as the value the group has to direct its energies successfully when engaged in collective action.[67] These two types of capital, of course, are not independent of each other. For example, analyses on the macro level have demonstrated that social capital has the capacity to create human capital; moreover, regions with only high levels of social capital do not end up fully prospering when human capital is weak.[68] However, on both accounts, the analogy of *capital* can be somewhat misleading because unlike traditional forms of capital, neither are depleted by use. In fact, the opposite is true; both are depleted by non-use. In other words, human and social capital are "use it or lose it" resources.[69]

An example of a region with both high levels of human and social capital is California's Silicon Valley. California's GDP in 2019 was $3.2 trillion, representing 14.6 percent of the total U.S. economy; if measured as a country, the state would be economically larger than India and the United Kingdom and would rank the fifth-largest economy in the world. This is in large part because of Silicon Valley. Silicon Valley drives one-third of all the venture capital investment in the United States, resulting in it becoming a leading hub and startup ecosystem for high-tech innovation.* Economists have attributed the incredible levels of success and innovation of the region to the presence of an intensive flow of tacit know-how (human capital) of individuals among local firms and a culture directed at open communication, which ultimately

* Although, with the advent of remote work, Silicon Valley's monopoly on capital is quickly shifting, with Texas appearing to be the main beneficiary.

resulted in a steady process of incremental knowledge development and experience, elevating social capital for the entire region.[70] When both forms of capital synergize with each other, it not only serves the individual but society as a whole. Ultimately, the Silicon Valley example shows when you combine human capital in transparent, communicative, open social networks—driving up social capital—you have the perfect recipe to strengthen people's capacity to engage and act on both the individual and group level. All boats rise with the tide.

Let's bring this to scale for your own personal social network; human and social capital are factors that, if properly aligned, can improve the likelihood of your personal success and the ultimate attainment of your most ambitious life goals. Much of developing your own human capital rests in you focusing time and energy on specialization; for example, a certified project manager needs to prove 4,500 hours of project management experience before they can take their certification test. That is an incredible amount of human sweat equity dedicated to a hyperspecialized task. The gains are obvious on the individual level, as a Project Management Professional (PMP) certificate will increase a person's salary, status, and work opportunity. But that's not all. A group that has a single member with a PMP has now bootstrapped its capacity to work toward the group's goals.

In complex environments, with challenges arising from the unexpected, the greatest power to ward off the unpredictable future is the massive adaptive potential of a group of different kinds of specialists.[71] Consequently, social capital—by way of trusting, tightly knit social groups—can lead to developing resilience and learning how to reorient oneself to confront the unknown challenges of the future. A true team is a group within which the trusting relationships between each constituent member leverages the specializations and life experience of the individual toward shared, superordinate end goals of the group. Looking at this from an egoist vantage point, *you* as the individual benefit from having access to the human and social capital inherent to the group. This direct access enables you to achieve *your* life goals. In summary, isolation reduces your potential for personal success, whereas social connectivity enhances it by increasing access to human and social capital.

Reason Six: Social Connectivity Gives Your Existence Meaning and Purpose

Some say the purpose of life is to pursue happiness. In a now-famous publication from the journal *Psychological Science* titled "Very Happy People," psychology professors Ed Diener and Martin Seligman performed an intensive study of 222 college students who were screened for high happiness, surveyed for fifty-one days over several months.[72] Diener and Seligman then compared the upper 10 percent of consistently very happy people with average and very unhappy people and discovered that "very happy people were highly social and had stronger romantic and other social relationships than less happy groups. Compared with the less happy groups, the happiest respondents did not exercise significantly more, participate in religious activities significantly more, or experience more objectively defined good events. No variable was sufficient for happiness, but good social relations were necessary for happiness."

Yet, is happiness really the target here? Is a life that maximizes pleasure and avoids pain a life well lived? In *The Sweet Spot*, psychologist Paul Bloom describes a more nuanced vision of how to craft one's own meaning and sense of purpose. People have an intuitive sense of meaning, but Bloom scientifically describes its properties:

> A meaningful activity is oriented toward a goal, one that, if accomplished, would have an impact on the world—and this usually means that it has an impact on other people. This activity . . . has some structure—it's the sort of thing that one can tell a story about. It connects to . . . spirituality and often connects to flow (leading to experience of self-loss) and often brings you into close contact with other people and is often seen as morally virtuous.

Unlike "experienced happiness," which can be measured as the pleasure derived from any given moment, "meaning" is measured by asking the question "How is your life as a whole?" The key to maximizing this dimension of human wellness is taking on tasks that require struggle to overcome obstacles. One example is becoming a parent. Studies show that by nearly every metric, parents' experienced happiness—moment to moment—bottoms out and marital satisfaction doesn't return to normal levels until children leave home. That's a long, hard dip in happiness. Yet, there also seems to be evidence that having children is one of the most powerful sources of meaning a human can experience. Attachment, specifically the connection experienced with another form of developing life, is often cited as the overriding factor that trumps the in-the-moment experienced pleasure of happiness. There is a magic in the struggle, a social struggle, that adds the spice of purpose to one's life.

Depression is a mood disorder that is associated with sadness and a general lack of interest in life—in essence, depression can be described as when one's life loses meaning and purpose. In support of this idea is data that shows that despite our growing societal wealth, the lifetime risk of a mood disorder increases as GDP increases per capita. As noted by Brandon Hidaka, primary care physician and researcher at the Mayo Clinic, in his publication in the *Journal of Affective Disorders*, "Depression as a Disease of Modernity":

> The growing burden of chronic diseases, which arise from an evolutionary mismatch between past human environments and modern-day living, may be central to rising rates of depression. Declining social capital and greater inequality and loneliness are candidate mediators of a depressiogenic social milieu. Modern populations are increasingly overfed, malnourished, sedentary, sunlight-deficient, sleep-deprived, and socially-isolated.[73]

The biomedical approach to treating depression has heavily relied upon the chemical imbalance paradigm. Yet, this rarely tackles the root problem. This is like treating depression as though it were an acute brain aneurism that requires immediate surgery, followed by swift recovery. However, depression is often more like diabetes; effective treatment can require a systemic lifestyle change that takes time, investment of energy, and commitment. Psychologists have been attempting to adopt the latter paradigm in a treatment for depression called *social proscribing*.

The major aim of the therapy is to remediate two forms of disconnect—from people and nature. An example of social proscribing is "therapeutic horticulture," where a group of depressive patients is tasked to change a dilapidated plot of land in an urban environment into a garden. The task is oriented around committing to cultivating something natural that requires patience and cooperation.[74] One Norwegian study found that participants moved on average almost five points on the scale of depression—nearly two times more effective of a treatment when compared to antidepressants.[75] The key to the success of these programs is that by being part of a team engaged in a long-term superordinate goal, patients began to feel as though they had purpose, and that with the help of others, they could not only be a part of, but also effect positive change in, the natural world.

Technology may be contributing to an increased risk of depression in a variety of ways.[76] Technological innovation has been linked to the lessening number and strength of ties in social networks, due to increased internet use driving less communication among families and greater feelings of loneliness. Specifically, weak or absent social support has been shown to exacerbate individual vulnerability to depression, whereas strong social support wards against the onset of depression.[77] This may be why interpersonal therapy—which focuses on enhancing connections between friends and family—has been demonstrated to not only be an effective treatment against depression but also reduce the risk of relapse.[78]

Consumerism correlates strongly with anxiety and depression.[79] For Americans, the effects of advertising on spending and happiness are particularly acute, as people are exposed to a daily average of five thousand advertisements. The goal of these corporate tactics is to implant and elevate *extrinsic* values over *intrinsic* ones. Intrinsic motivation is the kind of reasoning acting from within. Its opposite, extrinsic motivation, is the kind that originates in external factors. Imagine that you enjoy playing the guitar. You play every day because you love to do it. On days that you don't play, you just feel off. This sort of natural pleasure derived from playing the guitar would be called intrinsic motivation. Now imagine that you play not because you like to do it, but because you work at a bar that you hate. It's your job to play to entertain the patrons. Or you may be learning to play because you have a romantic partner you wish to impress, and the reason you play is more about courtship than the actual joy of playing—that is an extrinsic motivation. The science of happiness shows us that those of us who are motivated more by intrinsic values appear to be much happier than those who are motivated by extrinsic values.

In fact, one study showed that with small, group-level interventions that led individuals to assess their intrinsic and extrinsic values, participants experienced a significant reduction in their levels of materialism and higher self-esteem.[80] A leader of the study, Nathan Dungan, believes the group discussion was the most critical component of the elevation of intrinsic values: "There was a lot of power in that connection and that community for people—removing the isolation and the fear . . . [so] you could actually get to the meaning, to the heart: their sense of purpose."[81] Thus, we see another positive feedback loop, where people who are in a group that openly discusses their intrinsic values are more resistant to advertising and less likely to resort to consumerism to reduce feelings of loneliness. Being socially connected is cost saving on many levels.

In the modern world, we are two times more likely to hate than love our work; 87 percent of people feel disengaged or negative about their jobs.[82] In large part, this may be the result of the antisocial ways in which we organize workplace institutions. One study carried out in the British civil servants' sector showed three factors are associated with negative physical and mental effects of the kinds of work we perform: (i) loss of autonomy (feeling like a deindividualized part in a larger machine), (ii) imbalance between work effort and reward (feeling like working above and beyond while never getting noticed), and (iii) a sense of low status (feeling low on the hierarchy).[83] A study that investigated small businesses has shown that alternative work systems that model cooperatives with greater democratic properties—like workers setting their own agendas and having greater input into problem solving than the traditional top-down control models—are not only more effective in the long term (growing on average four times greater) but also provide workers with meaning and sense of purpose by having a significant role in the collective effort.[84]

The *ultimate causation* of the more democratic work model imparting greater purpose to workers is that it is more closely aligned to the social suite of successful ancestral groups—where everyone was needed and everyone had a significant role to play in the camp's survival, and therefore had a purpose—to help the overall success of the group. In the words of a worker employed in the cooperative work model: "Everybody wants to feel useful, and have a purpose."[85]

Rumspringa is the Amish tradition in which when a child turns sixteen, he or she is encouraged by the group to go out into the modern world of the "English" (the term the Amish use for North American society) to "sin," experiment, and explore. One young Amish woman commented on the experience: "God talks to me in one ear, Satan in the other. Part of me wants to be Amish like my parents, but the other part wants the jeans, the haircut, to do what I want to do."[86] Remarkably, the individual empowerment of Western society does not ultimately hold muster for most Amish youth, as 85 to 90 percent return to permanently join the Amish church.[87]

This is fascinating on several counts. The Rumspringa is a naturalistic experiment, where participants can choose between two competing models of social organization. Perhaps the Amish approach to social organization is more appealing to those who have sampled both because it is closer to the ancestral social state, which was more aligned with our body's biology and psychology. They have adopted a life that emphasizes intrinsic (over extrinsic) values to the extreme. They also appear to be warded against many of the psychological traps of the modern world, as studies have found their mental health problems to be significantly lower than those of other Americans.[88] This is likely because they lead lives embedded in community and buttressed by strongly defined identities. It's this orientation between themselves and their group that engenders meaning and purpose to their lives. This is something they apparently cannot get in the English world. And so a vast majority of them, after experiencing their exodus into our hyper-individualist modernity, return to their communities.

Yet, there are serious costs that each Amish person weighs upon making this choice. As we recall, one of the most powerful tribal signals that enhances coalitionary alliances is ideology, with a special emphasis on religion. The Amish, although in many ways an inspirational group, are also glued together by religious fundamentalism that is appalling when viewed from a modern, post-Enlightenment ethical lens. For example, LGBTQ+ people are treated as abhorrent anomalies, child abuse by way of corporal punishment is extoled as a moral virtue, and women are decidedly subordinate to men.[89] Yet, despite these harrowing trade-offs, the Amish have a choice. Do we have a choice? A main argument of this book is that we do—but before making it, one must seriously consider the costs, trade-offs, and disadvantages of working toward intentional proximity. In other words, what do you have to give up to become more connected?

The Costs of Community

Careful readers may be skeptical of the claim that intentional proximity is a panacea to the ills of modern life. And for good reason—it's not. Before we hold hands and sing "Kumbaya," let's be clear: intentional proximity—and its most expressive form, intentional community—is not utopic. Also for many people—especially those with a measure of societal success, intellectually satisfying work, and significant financial means—the costs of putting energy into intentional proximity or active intentional community building may be too much to warrant engagement. So there are several drawbacks, some of which can be transplanted into modern group-living contexts.

The biggest challenge of intentional proximity is Ernest Gellner's *tyranny of the cousins.*[90] A romantic view of the egalitarian and pluralistic societies of our ancestors, created to meet the challenges of past landscapes, holds that hunter-gatherer life is idyllic. Camps are not subject to any polity, disputes are solved communally with everyone having a voice, people do not let each other sit in sadness, and in general day-today life is peaceful. In many ways, individuals in these societies live freer lives than those in larger agrarian groups that have been the dominant society since the agricultural revolution. But there are several trade-offs of living in a camp, embedded in one band of many that share language, ritual, and custom. The external-oriented cost is the propensity for in-groups to pseudospeciate out-groups by dehumanizing the "other." The internal-oriented cost is the continuous threat of exile or capital punishment if one does not adhere to the social norms and rules of the group. Whereas most post-agricultural societies relied on domineering leaders to survive—in what Gellner called *the tyranny of kings*—without an alpha, hunter-gatherer societies adhered to the often-draconian tyranny of cousins. Tyranny of cousins "is a metaphor for the group of adults whose decisions held sway. Their power was absolute. If you did not conform to their dictates, you were in danger."[91] In essence, the trade-off is this: when people group together, individuals limit their personal freedom by their willingness to conform to group expectations.

As discussed earlier in the chapter, over time, intentional proximity in the form of cohabitation or colocation in housing can lead to the use of fewer resources and cost saving with shared facilities. Yet, the initial buy-in can be expensive in several ways. Graham Meltzer recounted the difficulties that an intentional community being built in the 1990s in suburban Seattle, Washington, encountered when the construction bids came in: "The cost per household was way too high for most participants to bear. Within a time all but five families had departed. . . . We attracted people . . . not earning enough money to buy in."[92] Other sacrifices can carry costs, such as relocating to live with or closer to a friend. Not only is relocation expensive, but one risks losing a source of income during the process. If a friend is helping, it can not only strain the friendship but the social network itself as people adapt to changing circumstances. The trade-off is this: the startup costs (both fiscal and social) for pursuing intentional proximity can be high.

A benefit highlighted in this chapter is the increased resilience that intentional proximity confers upon its practitioners. An obvious cost to this resilience is the loss of privacy. As observed by David Brooks: "[Extended families] can . . . be exhausting and stifling. They allow little privacy; you are forced to be in daily intimate contact with people you didn't choose. There's more stability but less mobility. Family bonds are thicker, but individual choice is diminished. You have less space to make your own way in life." If your privacy is part of your identity, group living may be nearly untenable. Two types of identities in particular will find intentional proximity challenging. If you identify as someone who is autonomous and mobile, without any ties—a lone wolf type—you may want to intentionally live with weak bonds and social distance. If you identify with your perfectly crafted and controlled environment and hold extrinsic values of material wealth and status in high esteem, note that the loss of these things is a likely compromise necessary for cultivating stronger social bonds through intentional proximity.

As noted by Olds and Schwartz: "If small matters of taste [are] central to . . . identities . . . decorating a little abode with knickknacks that represent one's identity gets transformed from self-indulgence to self-definition. [In this condition] people seek partners who will fit seamlessly into the lives they have already created." If you know that compromise and consensus building are conditions you cannot live with, then intentional proximity will likely not be worth the effort.

Working toward intentional proximity is not for everybody. If you generally eschew social consensus, are weary of ritualizing symbols to enhance in-group cohesion, and find that autonomy, mobility, and control of your environment are central to your identity, then the social and financial costs of working toward intentional proximity are likely too high to make the endeavor worthwhile for you.

WE NOW UNDERSTAND THE COSTS of both social isolation *and* living a life of intentional proximity to other people. Perhaps there is a balance to be struck between the two. Even modest steps toward living a life of intentional proximity can have lasting and profound effects. Again, social mismatch is more like diabetes than an acute brain aneurysm. A small change to your diet, like adding or removing a few processed foods, can lead to a tipping point that significantly alters a complexly evolved gut biome and either strengthen or decimate your body's ability to effectively fight off disease. Similarly, a behavioral change where you adopt wearing blue light–blocking glasses after sunset or refrain from eating right before bedtime (resulting in earlier onset of melatonin production and therefore deeper, higher-quality sleep) can lead to thousands of added hours of slow-wave sleep (SWS) and rapid-eye movement sleep (REM) over a lifetime. This can make the difference in whether you suffer from Alzheimer's disease later in life.

Campcrafting circles and protects; it keeps you safe and removed from physical danger; it buffers you in a natural world that is stressful, precarious, and unpredictable. Yet camp life demands a robust, stress-tested trust among its practitioners and requires sacrifice of individual needs for the needs and norms established by the group. It is ultimately up to you whether you find the investment worthwhile.*

Summing it all up, intentional proximity can help you by (i) improving your psychological well-being, (ii) protecting you from the negative effects of dyadic withdrawal (cocooning), (iii) improving conditions for your family and children to grow with cooperative alloparenting, (iv) improving your long-term finances and quality of life while exhausting less resources, and (v) increasing your access to human and social capital. This means the "what's in it for you" is a life where your dreams can be more easily attained. And perhaps most important of all, the science is showing that even small efforts to move in this direction can give your life added purpose and meaning and even make you happier and more fulfilled.

The Roseto Solution

By now perhaps you've solved the Roseto Mystery.

The reason the town was named Roseto was because in 1882 a group of ten men and a single boy left Roseto, Italy, their impoverished medieval-style village in the province of Foggia. After finding work in slate quarry mines near the town of Bangor, word was sent back to Italy and the first trickle of immigrants became a flood that nearly left the streets of the Italian town empty. After only a few years, the hundreds of Rosetans pooled together resources to purchase land. This fateful act empowered the fledgling community with *the power of proximity* and bound them together in the shared fate of *intentional community* in their efforts to survive in the Americas. Modeling their homes after the closely clustered two-story dwellings built on the narrow, brick cobbled streets of

* Choose your own adventure time. If you are reading this book, you are likely up to the challenge, and I highly recommend continuing to read the section on *camp-crafting*; if your answer is "no," then feel free to skip to the book's societal conclusions in the final two chapters, where we'll tackle how political tribalism is affecting the world of tomorrow.

Roseto, Italy, the clans lived intergenerationally—often with three generations in a single home. As depicted by Gladwell, the secret of the Roseto Mystery that Wolf and Bruhn discovered was the *community* itself:

> They looked at how the Rosetans visited one another, stopping to chat in Italian on the street, say, or cooking for one another in their backyards. They learned about the extended family clans that underly the town's social structure. They saw how many homes had three generations living under one roof, and how much respect grandparents commanded. They went to mass at Our Lady of Mount Carmel and saw the unifying and calming effect of the church. They counted twenty-two separate civic organizations in a town of just under two thousand people. They picked up on the particular egalitarian ethos of the community which discouraged the wealthy from flaunting their success and helped the unsuccessful obscure their failures.[93]

The Rosetans were, by every anthropological definition, a contemporary tribe. This tribe was culturally bootstrapped with powerful signals and tokens. Their green-bearded traits were proudly on display, with shared language (it was said that even in the 1950s people predominantly spoke Italian), a shared immigrant origin story, and even shared tribal heroes—the main street, Garibaldi Avenue, was named after the Italian hero of unification. During the foundational years of the town, a dynamic shaman guided and cultivated their collective spiritual health. Father Pasquale de Nisco was the priest who organized spiritual societies and festivals for the town to rally behind; he encouraged farm-to-table food (by passing out seeds to parishioners at their church), with a common tactic being residents clearing land behind their houses where they grew crops, raised pigs, and cultivated wine. Given the German, Welsh, and English immigrant towns that dotted the region, they even had outgroups with which to juxtapose and keep their identities singular and strong. In other words, there was a strong sense of the Rosetan "us" and a clear sense of everyone and everything outside Roseto being a "them."

When Wolf and Bruhn first presented their findings, as well as their explanation for the Rosetan outlier, the medical community was skeptical. The science of the day—and to a large extent, the way we still view health and wellness—was dictated by a conventional wisdom that this or that gene, or a particular physiological process, or our access to medical systems, or our diet and exercise predicts how much quality life we live. Nobody was used to thinking about health from the lens of *community.* But the new applied science of evolutionary anthropology tells us that to ignore community and dwell in isolation is to live in a state of precarious mismatch.

As we continue forward, I want to invite you, like the late nineteenth-century Rosetans leaving their broken homes of the past, to walk your first steps away from our Old World of modern technological and social isolation and move incrementally toward a more fulfilling, truly socially connected New World. The world of *intentional proximity.* A world of your own making. But how can one build such a world? Let's go find out together.

7

TRIBAL FRIENDS

An (Evolutionary) Guide to Forging a Meta-family

Just as our capacity for love has antecedents and parallels in other species, so does our primal, unshakable capacity for friendship.

–Nicholas Christakis, 2019

We need to take a moment's pause and learn how to be with each other again in pairs and groups before we venture further out.

–Douglas Rushkoff, 2019

You are the average of the five people you spend the most time with.

–Jim Rohn, 2012

Choosing Family

It is the late morning hours of the winter solstice of the year 2012, and I am about to become a *blood brother*.

My arm aches. It is swollen and inflamed, my body responding to the several preceding hours of continuous puncturing of its skin. The outline of what, over the next several years, would become the tattoo that would sprawl across my left arm, shoulder, back, and chest, was freshly inked upon my flesh. The act is the beginning of the rite of passage I am embarking upon with my best friend.

I have been researching blood brotherhood sacred ceremonial texts for months, using my anthropological training in an applied way to piece together the thoughts, philosophies, and actions of past peoples to best emulate with precision the most universal elements of the sacred friendship ritual. The culture that my friend introduced me to, and whose tales of friendship impressed beyond all others,* was that of the ancient nomadic horse-warriors of the ancient steppe—the *Scythians*. The tattoos we now shared were from the same source: a mummified male body from a 2,500-year-old kurgan burial mound archaeologists had named Pazyryk 2.[1] The kurgan, located in the steppe close to borders with China, Kazakhstan, and Mongolia, faithfully preserved the skin of the mummy so that infrared light from archaeologists could reveal the tattoos. The archaeologists could not only reconstruct the tattoos after thousands of years, but also determine that the man, likely a chief among his tribe (tattoos were symbolic of high station), had died at around the age of sixty.

These mythically infused "koulans" that now adorn my left and his right arm are an amalgam of stylistically twisted horses, griffins, deer, and other animals common to the Siberian plains. After four hours of tattooing and the quaffing of several brews to celebrate, we are now back at our loft in downtown Indianapolis. My stomach is filled with a nervous joy—a type of boundless energy that can barely contain itself. The time is here, the moment is now. With this ritual I would gain what would prove to be one of my life's greatest possessions.

Engineering the Social Suite—The Primacy of Proximity

Our understanding of the evolution of relationships has recently been thrown into sharp relief with the idea described by Nicholas Christakis as the *Social Suite*. Christakis articulates in his book *Blueprint* the power of the Social Suite to not only influence individual health and well-being but drive the ultimate success or failure of societies.[2] A group's success depends on the degree to which it adheres to eight social laws. It stands to reason that those individuals who situate themselves within the parameters of the

* The ancient Greek novelist Lucian wrote a second-century text that set up a conversation between an Athenian called Mnesippus and a Scythian by the name of Toxaris. The two men have a fireside debate where they compete by telling stories of friendship from their own cultures limited to those they have themselves witnessed. After five tales of friendship each, the Athenian yields to the Scythian. The two are so impressed by one another after the heated exchange that they are happy to call each other friends henceforth.

social laws will be in social alignment, and thus better able to ward off all the bad mental and physical mismatches that can arise from being out of alignment. What, then, makes up a *good* social niche? The Social Suite, summarized in chapter two and broadened here, is composed of the following laws:

1. *Individual identity* (without which tracking long-term patterns of reciprocal kindness is not possible)
2. *Love* (commonly oriented toward kin or reproductive partnerships)
3. *Friends* (long-term, non-reproductive partnerships)
4. *Social networks* (mathematically universal ways in which groups assemble themselves)
5. *Cooperation* (bias toward, and not against, cooperative behaviors)
6. *In-group bias* (a preference for one's groups of residence)
7. *Mild hierarchy* (hierarchy based more on prestige than dominance)
8. *Social learning and teaching* (where prestige leaders engage in skill-dependent social teaching)

As Christakis notes: "We live in a sea of genes, and other people's genes may be even more critical to our destiny than our own." There are precise predictions about the optimal parameters for each of the eight social laws that can influence the effectiveness of the relationships within and among groups. Groups that adhere closely to and do not defy major tenets of the Social Suite thrive, whereas groups that defy key principles go extinct.[3] In the following pages we will illustrate how to use others' identities to search for and cultivate *partners* (long-term, emotionally committed relationships, both platonic and romantic) as the foundation for friendship and cooperation. This cornerstone of relationship theory is critical because identity provides us with the capacity to track who is who across time and space and to repay kindnesses (or punish offenses) offered to us by others. Then we will consider the building blocks of your social perimeter by exploring the elements that are characteristic and predictive of high-quality *friendship*. Identity and friendship, and how to balance them with romantic relationships, are the primary foci of this chapter (we will do a deep dive into laws 4–8 in later chapters, where we will discuss how to craft your own camp, how to safely cultivate in-group bias and leverage evolutionary psychology to maximize cooperation within your group, and share some ways to contain its more nefarious effects when groups tribally collide).

Here, we will begin a voyage into the new, untapped realm of *applied evolutionary anthropology.* We will traverse novel terrain; as noted by the evolutionary biologist David Sloan Wilson and his colleagues, "applied evolutionary science exists, but just barely." As a university professor who teaches human evolutionary biology, some of my most powerful tools to stimulate my students' imaginations are concrete demonstrations of how, if properly leveraged, evolutionary knowledge can be *applied* in their lives to positive effect. My students are intrigued by evolutionary forces that still influence us today. They are skeptical of society's answers of how to live a good life but come up short when attempting to formulate alternative strategies of their own. In fact, as an anthropologist with a penchant for heterodox thinking, I, too, am skeptical of the standard answers society gives when asked the question "How do we live good, healthy, fulfilling lives?"

For years I have been exploring, both personally and professionally, how to mold the best of the old with the new to maximize my fulfillment. I've been test subject number one in the experiment of my life. I have been recording these experiments, the details of which go beyond the scope of this book, on a blog that serves as a kind of "Darwinian Workshop"—a how-to guide of campcrafting for the twenty-first century. It's anecdotal, but I believe it is because I live a life consistent and "in sync" with the Social Suite that I wake up every morning bursting at the seams with a sense of purpose. I invite those who are curious and wanting to consider my case study and more examples, in greater resolution, to go to the blog compendium of resources (see QR code in the appendix of the book).

THE SOCIAL SUITE ALLOWED OUR ancestors to carry with them a type of protective social shell that made them hyperadaptive in any environment. I want you to think of your camp as a shell, a guiding metaphor for our aims.

Snail shells, unlike typical animal structures, are not made up of cells. The exoskeletons are composed of three distinct layers, a matrix of mineral and protein. Think of laying down steel (protein) and pouring concrete (mineral) over it. Shells grow from the bottom up, or by adding material at the margins. This pattern of growth results in three distinct shell layers: an uncalcified outer proteinaceous layer, a calcified prismatic layer, and an inner layer of pearl. These proteins tend to bind calcium ions while guiding and directing calcification, enhancing crystal formation according to precise hierarchical arrangements that constitute the entire protective shell. A shell implies a bound matrix—a unity of elements that is greater than the sum of its parts. In our shell metaphor, the matrix of protein and mineral is the social network within which we live.

What should be obvious by now is that without a social shell, you are a less resilient and more vulnerable animal. The most socially isolated among us are snails without shells. For most of us, with the weak social bonds afforded to us by the jobs we work or the unknown neighbors we nod to out of politeness and nothing else, our shells are fragile. Even for those of us with strong shells, there are steps to take to fortify and reinforce structural weaknesses. If the most important human beings in your life are geographically spread in a diaspora beyond reach, this is akin to being a snail with its shell shattered and dispersed. A broken shell does little to serve the snail.

The Social Suite cannot adequately be achieved by people who do not share geographical space. This is what I call the *Law of Intentional Proximity*—the idea that being spatially collocated with your *honor group* is a precondition of living in sync within the Social Suite. Your *honor group* is to be distinguished from your *sympathy group.* Recall when we discussed the channel capacity constraints on our cognition that makes sympathizing with others extremely energetically expensive; your *sympathy group* is the list of people about whom you would be psychologically devastated upon hearing of their death. This could be a rather long list for some, but on average has about twelve people.[4] Your *honor group* is the short list of your sympathy group that is the in-group of people you are honor bound to share your emotional resources with. Creating this short list can be challenging, and we will end the chapter on the best ways to go about cultivating your own honor group. If you do not possess a large enough or strong enough sympathy-honor group, according to social baseline theory (see chapter four), you will be in an elevated state of energy consumption that leaves you drained and prone to both mental and physical illness. This explains why humans tend to get really sick in isolation, because it is so far misaligned with the Social Suite. But if social support is so important, how do we know who to let in to the concentric circles, the protective calcified matrix shell of our in-groups? In other words, how do you know someone is going to be a good life partner?

Sometimes it's like capturing lightning in a bottle. You move to a new school and the first person you sit next to in the cafeteria is destined to become your best friend for your entire life. Other times, a person is introduced to you by a friend and at first seems antagonistic, but over the course of years you end up developing a begrudging respect that evolves into the most trustworthy relationship you have. Other times you think you've got the greatest friendship in the world, and later they end up betraying you. Are there any ways to improve our ability to identify the people we want playing on our team?

In essence, we're trying to predict the future and determine if a person is in-group. Thus, the critical question is, does science tell us whether there any differences between people in their ability to predict the future? Fortunately, with the help of the psychologist Philip Tetlock and decades of research that culminated in the book *Superforecasting*, we know precisely the kinds of temperaments and manners of reasoning that make people either abysmally bad at predicting the future, or what Tetlock calls *superforecasters*.[5] Tetlock and his colleague Barbara Mellers tasked themselves to find the best predictors through forecasting tournaments held by the research organization of the federation of American intelligence agencies called the Intelligence Advanced Research Projects Activity. Based on their own research, Tetlock and Mellers correctly picked out the people who not only performed better than random, or professional, pundits, but also significantly better than professional intelligence officers that had the advantage of access to classified information. The superforecasters out-predicted "prediction markets" and some edged in close to the theoretical maximum of prediction. It is noteworthy that "public experts"* are among the worst predictors, as they tend to be ideologically driven, which leads to oversimplifying complex problems; they use absolutist words like "impossible" or "certain." Tetlock's superforecasters were the opposite in temperament and approach:

- They reasoned in probabilities, not certainties.
- They were more likely to admit they were wrong and change their minds.
- They were intellectually curious.
- They were anti-impulsive and distrusted their first gut feeling.
- They constantly reassessed their reasoning.
- They were conscious of their own biases.

Ultimately, they were Bayesian. Recall the Reverend Bayes who formalized the concept of generating probability distributions—that is, instead of binary answers like yes or no, the answer itself is a distribution of likelihoods. Another way of putting it would be that Bayesian thinkers consistently update their degree of credence of a proposition in light of new evidence. The way they do this, as noted by Steven Pinker, is "they begin with the base rate for the event in question: how often it is expected to occur across the board and over the long run. Then they nudge that estimate up or down depending on the degree to which new evidence portends the event's occurrence or non-occurrence. They seek this new evidence avidly, and avoid both overreacting to it ('This changes everything!') and underreacting to it ('This means nothing!')."[6]

Our task, then, is to utilize these strategies to become superforecasters of human relationships. Adhering to the Law of Intentional Proximity is costly. If only a handful of people can be your closest and most trusted partners in life, then it's worth framing your mind to be more Bayesian when considering your relationships. In one respect, we al-

* The very traits that put them in the public eye (unfaltering ideology) make them the worst predictors of the future. Take note, the next time you watch a political pundit predict an election.

ready do this instinctually. There is a branch of thinking in primatology that theorizes that emotions serve as a kind of cognitive accounting for social exchanges over longer-term interactions in the wild. This results in us feeling good about others in proportion to how nice they are to us. In essence, if you light up when you see someone, it's likely because of the sum of the previous emotional balance sheets of remembered past interactions. All we are trying to do here is be more intentional—more mindful—about how we record the balance sheet. Each interaction with another human being is a datum—a single pass worthy of awareness. Interactions can be judged as either negative, neutral, or positive. We don't need to use a spreadsheet or anything like that, as you likely can guesstimate based on feeling, but it can be helpful.*

When we encounter human beings, we begin to generate internal data sets that serve to create *predictive models of behavior* for that individual. When your fitness depends on carving out a bit of certainty in an overwhelmingly chaotic, dynamic, and uncertain world, these predictive models are all you have and perhaps nature's most sophisticated tool gifted to our big-brained species. The magic of this phenomenon is that as your data maps a higher-resolution picture of another human being, so, too, does the model's power to accurately predict their behavior improve. In essence, you become a mind reader on that individual's behavior. Importantly, when this happens, the brain begins to change on a fundamental level, and the *self* and *other* become intertwined. The closer the relationship, the more the lines blur between the concepts until the brain loses the ability (or perhaps utility and will) to distinguish between the two. Once you harness this power, you can begin

* Or you can use a spreadsheet . . . When I was first dating my wife, I recorded the first thirty interactions and formally ran the stats in a Bayesian perterior distribution. It was highly predictive, as we have been together years—and now we're happily married. I look back at the results with some measure of fascination at its eerie accuracy. In fact, my mate framed the figure plot I made of the analytical prediction. I'm reminded of a Carl Sagan quote: "It is sometimes said that scientists are unromantic, that their passion to figure out robs the world of beauty and mystery. But is it not stirring to understand how the world actually works—that white light is made of colors, that color is the way we perceive the wavelengths of light, that transparent air reflects light, that in so doing it discriminates among the waves, and that the sky is blue for the same reason that the sunset is red? It does no harm to the romance of the sunset to know a little bit about it."

intentionally recruiting partners in your life journey that truly add depth of meaning, and eventually shared purpose.

For the bold among us, we'll explore the idea of intentional community building in the following chapters. But I recommend approaching these chapters as if you were adopting a new diet. Alternative diets aren't for everyone. Moreover, some people can tolerate certain foods better than others. For example, your personality may be strongly extroverted or introverted; people will have different social networking needs along this continuum. This is why we will survey the core set of tools to assess what diet is best for you. We'll explore how the *Big Five* personality traits, *attachment styles*, and an application of *optimal stopping theory* can help you find the right social diet for you. The Big Five, commonly referred to as OCEAN (openness, conscientiousness, extroversion, agreeableness, and neuroticism), are measures of personality traits that can give some insight on yourself, the people actively in your life, or the kinds of personalities you believe could enhance the quality and compatibility of your relationships. With respect to changing your intentional proximity diets, you'll likely find bits you like, while others you try don't taste good and are jettisoned. This is good. Metaphorically changing your social diet to consume even modest amounts of intentional proximity will likely have a lasting, positive impact on your life.

Darwin's Handbook of Friendship

Before we get to the task of "finding the others," we must first define their function. The concept of good friends may seem self-evident, but it is worth exploring how evolution crafted these roles to help us on our quest for survival. So then, what does science tell us about the nature of friendship? Friendships are special. They are a unique form of relationships, extremely rare in nature. This special type of connection embodies the most precious of human commodities—a trustful mutual relationship of lifetime support that defies tit-for-tat reciprocity. From an evolutionary perspective, what makes this relationship singular as a phenomenon is that it is neither influenced by genetic similarity* (in

* Although friendship networks do have above-average genetic correlation.

most cases) nor the need to engage in the project of reproduction.[7] For all social animals, kin are irreconcilably bound by genes, and when animals invest cooperatively in their progeny, they can be held hostage by offspring. That is not to say in humans these types of relationships are not capable of creating beautiful and profoundly durable interconnection—they classically do; but friendship is distinct.

Scientists have historically underplayed the role of friendship in favor of reproductive partners.* On this issue, I am more aligned with Ralph Waldo Emerson, who in 1849 wrote: "A friend may well be reckoned the masterpiece of nature." Recall Maslow's hierarchy of needs. The basic needs of the bottom of the pyramid are food, water, warmth, and rest, and then above them are the psychological needs of friendship and love. But in humans who evolved in Paleolithic camps and bands, it was the perimeter of friendship that served as the outer layer of the social shell that protected and provided the security that could enable a pair-bond to safely unite for the more long-term project of reproduction. Christakis acknowledges this when he writes: "The friends we choose affect our prospect for survival."[8] It may run counter to the popular sentiment that romantic love is the primary force that predicts our happiness, but discoveries made in the relatively new scientific realm of positive psychology—the study of the nature of human happiness—show that friendship, not romantic partnership, is foundational to determining wellness.

For example, Daniel Gilbert, a professor of psychology at Harvard and author of *Stumbling on Happiness*, argues that a more powerful predictor of an individual's happiness and well-being other than marriage is friendship.[9] Furthermore, Daniel Kahneman, a Nobel Prize–winning psychologist and the founder of the positive psychology field, has said: "What determines how happy you are is spending a lot of time with . . . friends more than children."[10] It makes sense that evolution crafted our minds to be most fulfilled when surrounded by a protective perimeter of friends; otherwise, why invest in a mate or have a child that is simply

* I believe that is because of the cultural bias, endemic to societies in the Global North, that emphasize individualism over communalism, coupled with the typical tendency to experimentally measure psychological phenomenon in individuals, rather than individuals within groups.

going to be eaten or killed in an unsafe environment? In other words, no variable is sufficient for survival, but friends are necessary for survival.

Friendship is perhaps the rarest of all unique types of relationships because it is a gift only privy to those (few) highly social species with the cognitive capacity to hold in their long-term memory stores previous interactions with others. As described by Christakis:

> Friendship is rare in the animal kingdom. The human propensity to do this has been shaped by natural selection and is written in our DNA. Friendship in animal species serves the stunningly useful purposes of mutual aid and social learning. And it's the foundation of the capacity for an enduring culture that transcends individuals and transmits information across time and space. Many further aspects of human psychology are also related to friendship, like the joy and warm feelings we experience in the company of friends and the sense of duty we feel toward them.[11]

The way friendships manifest in other species is incredibly informative to why human friendships exist. So much so that "once the concept of friendship takes root in a species, there might be just one basic way to become socially organized in this regard."[12] But how do we measure friendship, especially in animals other than humans? The most basic comparative measure of friendship is the *association index*. Say a primatologist is observing a chimpanzee network. After the course of a week, the primatologist had dedicated thirteen hours to observing chimp A and his friend chimp B. The chimp friends spend six hours together, where chimp A is alone for three hours, and chimp B is alone for four hours. The primatologist would then calculate their association index and approximate that both friends spend about 46 percent of their time together. We can logically surmise, then, the level of friendship this pair has relative to all other relationships in the network.

HUMAN FRIENDSHIPS ACROSS CULTURES TEND to share the same developmental milestones.[13] From ages five to nine, children focus on extrinsic, superficial awards when engaging in shared activities with peers.

Critically, by the age of ten, children cognitively gain the capacity to process abstract concepts of loyalty and trust (recognizing both responsibilities and transgressions) in friendship. The final stage on their way to an adult capacity for friendship is adolescence, where they find intrinsic value in the friendship beyond material gain, and invest in them as an end, and not a means to an end. With the neocortex reaching full development in the midtwenties this tells us a crucial function of adolescence is the forging of friendships. There is never a period throughout life where the potential to find your allies and partners is more fertile. The fact that friendship develops along a similar childhood and adolescent trajectory speaks to the degree to which the adaptation is innate. There is something truly *natural* about the way friends, and thus, social networks that surround these keystone relationships work.

Yet, more informative is the striking similarity of adult social network structure. Human networks the world over—whether in the Global North or South, or in postindustrial agriculturalist or hunter-gatherer societies—resemble other social animals' networks. Friendship is an instinct.[14] How, then, do we define friendship? To get an operationalized definition, we turn to a comparative study among sixty representative cultures* by Daniel Hruschka, whose analysis showed that friendship is a human universal.[15] Table 2 shows what friendship is, and is not, among these societies.

TABLE 2: FRIENDSHIP CHARACTERISTICS ACROSS CULTURES

CHARACTERISTICS	PERCENT DESCRIBED	PERCENT DISCONFIRMED
BEHAVIORS		
Mutual aid	93	0
Gift-giving	60	0

* In a larger sample, five out of four hundred cultures may have possibly lacked friendships. The common denominator: extremely collectivist communities with totalitarian bends that see an individual's preference for some small group of people over the community to be a threat to social order. Any governmental regime that suppresses the expression of friendship is an enemy to a flourishing human condition and in direct violation of the Social Suite.

CHARACTERISTICS	PERCENT DESCRIBED	PERCENT DISCONFIRMED
BEHAVIORS (cont'd)		
Ritual initiation	40	4
Self-disclosure	33	10
Informality	28	0
Frequent socializing	18	55
Touching	18	0
FEELINGS		
Positive affect	78	0
Jealousy	0	0
ACCOUNTING		
Tit-for-tat	12	71
Need	53	0
FORMATION AND MAINTENANCE		
Equality	30	78
Voluntariness	18	64
Privateness	5	66

If a trait has a high percentage of being described, in combination with a low percentage of being disconfirmed, it can be considered a fundamental characteristic of friendship. Four traits stick out as important. Behaviors of *mutual aid* (93 percent confirmed, 0 percent disconfirmed) and *gift-giving* (60 percent confirmed, 0 percent disconfirmed) are seen in most cultures. *Ritual initiation* is also noteworthy, and was confirmed in 40 percent, and disconfirmed in only 4 percent, of cultures. Seventy-eight percent of cultures report feelings toward friends should have positive affect, with none describing jealousy as a component of friendship. Cultures report a consistent story with respect to how friends account the cost-benefit of the relationship over time—as 71 percent disconfirm tit-for-tat strategies and 53 percent confirm *need* as the primary context for altruistic behavior. In summary, friendship functions to give aid when needed, and the relationship is glued together by positive feelings that relax a kind of *quid pro quo* typical of acquaintance relationships. A friend is the best thing humans have to serve as a social insurance policy and safety net against chaos, misfortune, and an unpredictable world.

Thus, friendship can be operationally defined as a long-term relationship that is characterized by mutual support, especially when a person is in dire need of assistance, resulting in feelings that the person is family. This properly demonstrates the way in which friendship is less about transactional utility in the moment, and more about the emotional connection and feelings of mutual goodwill toward one another. These manifest emotionally as sentiments of closeness, trust, and affection. Ritual and gift-giving* are powerful methods of publicly announcing friendship bonds, as well as maintaining the bonds throughout space and time. Simply put, a true friend is the first person you turn to when in need because you trust them.

In the quest to find those that walk among us who can become part of our meta-family, one study in particular is elucidating. Subjects were asked to spend time in a painful posture either for a self-reward or the reward of someone else. For themselves, on average, they spent 140 seconds in pain. For immediate kin (a parent or sibling), they spent 132 seconds. For cousins, it was 107 seconds, and for a children's charity it was 103 seconds. For their best friend, the average clocked in at 123 seconds—somewhere between immediate and extended kin.[16] On average, good friends endure pain for each other more than for a cousin. Some other concepts that help us circumscribe friendship are how you (or they) feel when you are in pain or in joy. If you are jealous of a "friend," that's a good indicator that you aren't actually friends. Friends, by definition, are part of the social self, and so if you feel bad about someone's success, it is likely you don't consider them close to you on the self-identification scale. Scientists use a social intimacy scale that can measure self-identification by way of perceived inclusion of the other in the relationship. Friendships can vary in intensity, and a

* One experiment in the hunter-gatherer Hadza showed that when gift-giving anonymously (honey-filled straws, a favorite food item of this community), individuals were more likely to bequeath non-kin friends than kin with the honey.

stronger friendship is one where you perceive things that benefit the other friend as though they benefitted yourself. This is a simple visual way to assess the quality of your friendships; when your best friend succeeds, it should give you the same serotonin and dopamine rush as though you yourself were the victor.

In summary, whatever it is you're looking for in a friend, you'll know they are a really good friend when you're willing to sacrifice for them; and the more you self-identify with them, the more likely you'll be to sacrifice for them. Science provides some other pointers to help us flag good friendships among the list of people we know. In general, the people in question are not your friend if: Exchanges are tit-for-tat. In other words, if they only provide services as friends simply to keep a balance sheet score even, they aren't friends; they are acquaintances. You don't feel a sense of joy when they succeed or notice they do not express joy at your success. They are not there for you when you experience serious challenges. Fairweather friends are no friends at all. They aren't willing to sacrifice anything on your behalf. The logic of this can be reversed—you can flip these questions around to test whether or not *you are a good friend* to the partners in your life. Now that you know what characteristics define friendship, you can actively work on raising your friendship value by continuously improving and striving to develop the characteristics you would want in an ideal friend within yourself.

Find the Others—Identifying Partners and Improving Relationships

You are born into your family. You did not choose them.

As with a legion of other factors that served as preconditions for your existence, you didn't choose the time, place, culture, and conditions of your birth; nor did you select your parents, whose gametes collided, thereby bestowing upon you your genetic legacy. You likely had little to no say as to whether you were an only child or one of a brood of a dozen children. As we recall from chapter three, our ancestors were born into even greater constraints—nested camps, with loose affiliations with other camps that shared a tribal identity. If you are reading this book, by chance you likely were born into a culture that has vastly more mobility and geographic spread than any previous ancestral condition.

Imagine for a moment being born into a Paleolithic hunter-gatherer camp: Your parents would have been a pair-bond among a dozen or so other adults. Your gaggle of peers would have numbered about the same, and your potential mates you could have counted on one hand.* From one vantage point, most human relationships are held hostage to something. You don't choose your biological kin, and some of your relatives may even possess traits that you would never seek out in strangers or friends, yet you share genes, and so by forces unbidden, you are driven to cooperate. Your pair-bond you have some choice over, yet you are held hostage in some measure by the offspring you create together, as their chances of survival are enhanced by reproductive cooperation.

Humans, though, unlike many other species, have a particular penchant for *imagining kinship*. Anthropologists call this "fictive kin" or "meta-family." We have special ways of thinking about non-relatives we like that co-opt "kin cognition," and it isn't always rational. Typically, we treat people like relatives when they *feel* to us like we are related. The instinct to do this is hardwired. The Westermarck effect is one example of this: non-related children who grow up together are sexually repulsed from each other because they grew up as children in the same environment. This was an evident effect of the Israeli kibbutzim, where rates of non-related children becoming romantically involved were extremely low.[17] This is the incest-avoidance strategy, leveraging fictive kin cognition, in action.

So how then do we understand and use fictive kin cognition to the betterment of our twenty-first-century lives? As North Americans steeped in democratic tradition, we tend to think of real change and reform as coming from the top down: "If only the government enacted this or that policy, or supported this or that initiative, or if only our political team won this or that election, then everything would get better," is the common line of thought. As an individual, you have little

* Imagine swiping right on your dating app, and it telling you your deck is empty after three swipes!

influence on which candidate wins the presidency or what type of government policy is enacted. However, you do have significant power over whom you choose to associate with. There are few factors that are better predictors of lifetime fulfillment than the makeup of your social network. Unlike your ancestors, who had extremely limited options to their "family of choice," you have the option to build your family from among tens of thousands of humans you chance upon throughout your highly mobile, complex life. This ability to construct a *meta-family* is incredibly powerful, and a source of hope.

JUST LIKE WITH OUR SELF-ASSESSMENTS for personality, to better understand how to resonate with others, we need to take initial stock of our own social health. In broad terms, how can we take a social health exam? In the latter twentieth century, several surveys were developed to quickly assess people's social support. One such survey is a simple, validated six-item questionnaire, which I have adapted to give you a quick social support checkup.[18] Consider the statements in Table 3, and then in the empty boxes rank from one (you completely disagree and are unsatisfied) to six (you completely agree and are satisfied).*

TABLE 3: SOCIAL SUPPORT SURVEY

SOCIAL SUPPORT STATEMENT	SCORE: 1 = COMPLETELY UNSATISFIED AND IN DISAGREEMENT 6 = COMPLETELY SATISFIED AND IN AGREEMENT
You have someone to count on to distract you from your worries when you feel under stress.	
You have someone to really count on to help you feel more relaxed when you are under pressure or tense.	

* For a more in-depth survey, with accompanying graphs and more detail on how to interpret your and your camp's scores, feel free to use the QR code in the book's appendix to check those and other additional resources to assess your social fitness.

You have someone who totally accepts you, including both your worst and your best points.	
You have someone to really count on to care about you, regardless of what is happening to you.	
You have someone whom you really count on to help you feel better when you are feeling generally down in the dumps.	
You have someone to console you when you are very upset.	
Total score	

Now add up the scores. The average respondent scores somewhere around thirty. Most people fall within five points of thirty, meaning the closer you are to thirty, the more average your perception of social support. If you score a perfect thirty-six, you are a rock star of social support and in prime social health; between thirty and thirty-five you are above average, and between twenty-five and thirty you are below average. For those who are above average, if you were at the doctor's office, they'd say you were in good shape. If below average, they may give you some references on how to improve your diet and increase exercise and reduce stress. If the score is below twenty-five, the doctor may prescribe some medication and refer you to a specialist. The lower you are on the scale, the more steps you can take to enhance your social networks, which could potentially lead to significant increases in security and psychological well-being. That being said, just because you're in good shape doesn't mean you can't be even healthier.

Now that we have a general sense of our social health, let's get a feel for the makeup of our in-group social network. Because of the individualistic nature of North American culture, we tend to confuse acquaintanceship with friendship. I witness this often when someone introduces their best friend to me; I ask how long they have been friends, and they respond with, "Oh, a couple months." As we will see, that's not enough time to be able to determine another human's compatibility, nor true grit, loyalty, and resilience in the face of challenge. Acquaintances can be the pool from which we select life partners, but the bar should be much, much higher. Whether a mate or a best friend, the roster for "life partner" is small and each relationship precious. The science of social networks has uncovered strong patterns of relationships across all the world's cultures. How do researchers identify someone's relevant social connections? The tool is called a name generator, and they ask people the following questions:

- Who do you trust to talk to about something personal or private?
- With whom do you spend free time?
- Besides your mate, parents, or siblings, who do you consider to be your closest friends?

Let's outline the outermost concentric circle—the sympathy group. This is the furthest psychological extension of your social self and is, in effect, your in-group. Get out a piece of paper and write a list of everyone whose death would truly and utterly devastate you. It's not a fun exercise, but it's crucial to the task at hand. Meditate, and in a state of open awareness, envision those you love most perishing. Then list those persons (best to use initials for anonymity) whose imagined deaths caused anguish.

It could be dozens of people or only a handful—the ranges can be wide for this list. If your list is around a dozen people, this is about average. This list likely has both kith (friends) and kin (relatives) and maybe even affines (relatives by marriage). These are the people that you view as embedded within your social network; likely you have spent most of your social energy in your life dedicated to cultivating these relationships. Because of channel capacity (discussed in Part I) the emotional energy you've dedicated to these relationships is likely very expensive. It also makes these relationships precious and irreplaceable.

Now it's time to narrow things down even more. Although you truly care about these people, not all are able or willing to be anything more to you than they are at this very moment, for various reasons—perhaps they just had kids and are focusing on their nuclear family, or they moved away and live on the other side of the country. You want more, though—you want an *honor group*. This is your forged family, an inner concentric circle within your in-group. Your principal value as a group should be tethered to *honor*—because whether it's praising a tribal god or standing up for a social norm that you all have deemed important, honor is the core value that naturally conveys group loyalty and signals to your coalitionary alliance's status. You want a team to take on life together.

Here we'll lean on social networking science to help us accomplish this important task. Name generators have been used in cross-cultural research and reveal that people typically identify 4.4 close social contacts with whom they can confide personal and privileged information. On average, respondents list 2.2 friends, 0.76 spouses, 0.28 siblings, 0.44 coworkers, and 0.30 neighbors. In sum, the average person has four to five close *partners* (strong ties that can include one or two siblings and one or two friends), that also can include a mate. Importantly, these patterns have remained consistent over decades of analysis.[19]

If life were a sport, then what would be your team's roster of players? Typically, in our society we can only juggle a single romantic relationship—this is the "mate" position. This often results in our only being able to dedicate enough energy to a single best friend—this is the "blood sister or brother" position. Typically, the number of strong social ties that are close enough to qualify as a confidant is 4.4, with most Americans having between 2.6 and 6.2—this is the "friend" position. In sum, this is your life partner roster and constitutes what I call your "Fireteam."* You can call it whatever you want; in fact, if you have your own special name for it, it's likely to work its magic much more effectively. All in all, if you find compatibility between and among the people on this short list, you are in a powerful and secure life position, as these are your A team and the core concentric circle closest to you.

Now, let's run an exercise to see if we can't narrow your long sympathy group list down to an honor group. The following diagnostic

---

* *Fireteam* is a term used in the U.S. military. It is a small subunit of infantry designed to optimize tactical capacity on the ground. Militaries adopted the four- to five-person team subunit structures because they prove to be the most trustworthy groups. Any more than this, and trust begins to falter because the social ties weaken. I understand military analogies may not resonate with many of my readers, but consider this: What better testing environment is there to reveal the greatest levels of success and cooperation among peers than one where literal lives are on the line?

will help you crystalize the short list of your strongest relationships. Regarding the sympathy list, consider the following: *List the number of individuals you would reach out to for help in the case of a crisis.* If you're struggling, do your best to identify up to three. An alternative exercise, inspired by the "self-other identification scale," may be more helpful for the more visually inclined. List your sympathy group on a piece of paper. Next to each initial representing that person, draw two circles: circle one is the "self," while circle two is the "other." Imagine and visualize that person before drawing the circles. If you were to transpose your body onto theirs, how much of it would overlap and blend in?

To the degree that this happens is the degree of emotional proximity you feel toward that person. The amount of overlap between the circles should represent your feeling of closeness with the other person, with the person you feel closest to in the entire world having the most overlap in the circles. Draw these out for each person, visually inspect the circles, and rank them by closeness. The set of closest self-other circles are your honor group.* Given psychologists call this list a *support group*, it should not come as a surprise that these can be people who provide you with emotional, physical, and mental company in times of crisis. (If you can't think of enough people to get to three, then your social health is in serious risk, and it's time to do some critical care work to make new friends and begin the process of developing and cultivating your sympathy group.)

It should become apparent why this exercise is beneficial; not only does it allow you to think intentionally and take an inventory about the kinds and quality of relationships you have in your life, but it could also become acutely clear who is in your honor group. For more introverted people who find socializing energy expensive, the top three to four individuals can qualify, whereas for more extroverted people who have a large capacity for socializing (which for them is overall mentally less taxing) between five to six people can qualify. These numbers are backed up by the scientific studies that link social network structure, health, and prosocial behaviors.[20]

These exercises have a purpose. They help us figure out our highest-

* Additional visuals and resources provided using the appendix QR code.

quality social ties. Having more friends is good because it drives improvements in health, but there is a trade-off between the degree and closeness of social contacts; as the number of close social contacts increases, the average closeness of each individual contact decreases. In other words, more friends equals cheaper-quality relationships. At the end of the day, there's only so much love to spread around. That's why there is a sweet spot to the number of people in your honor group.

How do romantic relationships fit into the context of a group? Is this something that can or should be factored into the equation? First, let's clear some terminological brush: I will rarely use the terms *marriage* or *spouse*, as these refer to long-term political, legal, and institutional recognition of two people's union. We're less concerned with this, and more concerned with *pair-bonds* and *mates.* A *pair-bond* is a long-term, emotionally committed romantic relationship. Those pair-bonds (either heteronormative cis or LGBTQ+) that engage in procreation by any means (by birth, adoption, or artificial insemination) we'll term *mates.* Thus, for our purposes, mates are simply pair-bonds that engage in child-rearing.

In the attachment literature, there is a principle known as the *dependency paradox.*[21] This is the idea that people are only as needy and insecure as their unmet needs. When effective communication meets the individual's needs, they can then turn their attention outward. Therefore, the paradox is that the more effectively dependent people are with one another, the more independent and daring they become when facing the outside world. Although often contextualized within the context of pair-bonds, this applies to relationships of any type, and as we will discuss in the next chapter about campcrafting, it can apply to groups too—the more effectively we depend on each other, the greater our capacity to reach a state of true independence. With this in mind, consider how much more attractive and valuable you can be as a pair-bond when you are bolstered by a strongly defined, supportive honor group. If the dependency paradox means that having an active, strong, social network enhances your mental and physical well-being, instilling you with confidence to engage the world on your own terms, it also follows that those with stronger, prebuilt honor groups will be able to attract higher-quality mates.

MUCH OF THE SIGNALING WE'VE discussed in the context of the Tribe Drive also applies to assessing mates. Geoffrey Miller has linked our preferences of mates with moral virtue signaling. He writes: "In evolutionary terms, a moral person is simply one who pursues their ultimate genetic self-interest through psychological adaptations that embody a genuine, proximate concern for others."[22] He likens courtship in most cultures to a moral obstacle course that is a ritualized test of diverse moral virtues, where kindness in gift-giving, conscientiousness in keeping promises, empathy in listening, and sexual self-control are key signals that can be emitted to those individuals seeking long-term attachment with a pair-bond and who do not adopt a quick (diversified) mating strategy.

The lesson to keep in mind is it's best to date after you have the support of an honor group, to ensure a measure of compatibility between them, your prospective pair-bond, and the rest of your in-group. This philosophy is echoed by several leading researchers and communicators on the challenges of modern relationships. Psychotherapist Esther Perel, in her book *Mating in Captivity,* writes: "We live miles away from our families, no longer know our childhood friends, and are regularly uprooted and transplanted. All this discontinuity has a cumulative effect. We bring to our romantic relationships an almost unbearable existential vulnerability—as if love itself weren't dangerous enough."[23]

Historian Stephanie Coontz observed: "Until a hundred years ago, most societies agreed that it was dangerously antisocial, even pathologically self-absorbed, to elevate marital affection and nuclear-family ties above commitments to neighbors, extended kin, civic duty, and religion."[24] And psychologists Jacqueline Olds and Richard Schwartz write: "Resilient marriages usually achieve a balance between restorative intimacy and outward-looking engagement; the couple is alternately a self-contained unit and a building block in a larger social network."[25] In Western culture, since there is such a hyper-focus on finding the "perfect spouse," we get paralyzed when a mate doesn't fill every single desire that we have. But if you date and court with your honor group and healthy friendships intact, then your mate doesn't have to be *all things* but only a *few things*—because you have a healthy group within which to situate a pair-bond. This is a much more realistic and humane bar for a potential mate to meet.

This network of witnesses to your pair-bonded relationship may also help your sex life. By being an active part of a larger social network, you also have a distinct role and identity, and this in-group identity helps create a healthy distance between you and your mate. Perel notes that it is this distance that is essential to the erotic pull to your pair-bond: "Too much merging eradicates the separateness of two distinct individuals. Then there is nothing more to transcend, no bridge to walk on, no one to visit on the other side, no other internal world to enter. When people become fused—when two become one—connection can no longer happen. There is no one to connect with. Thus separateness is a precondition for connection: this is the essential paradox of intimacy and sex."[26] In other words, it's the complete loss of identity and overlap that is endemic to cocooned mates that is the long-term killer to the lusty erotic, sexual pull that bonded you as mates in the first place.

YOU'VE TAKEN THE TIME TO assess and gather a group of humans to be your meta-family. These are your life's greatest allies and champions. These are the people I challenge you to engage with and adhere to the Law of Intentional Proximity. *Be* with these people. *Build* with these people. Now we have our team. But how do we strengthen it? For relationships new and old, time-tested or never tested, now we go to one of the most powerful methods to maximize and solidify the bonds that humans have ever invented—*challenge* and *ritual.*

Challenge Unites and Ritual Bonds

Charles Fritz was a pioneering scientist whose main interest was the effect of disaster on human psychological well-being. He was part of the United States Strategic Bombing Survey that evaluated the effectiveness of the bombing campaigns on England during WWII. Fritz became a critic of the reasoning behind the bombings; shockingly, instead of demoralizing a population, they did the opposite—civilian resilience and sense of unity was raised in response to air raids. After these initial observations, Fritz began a decade-long study on the effect of calamity and challenge on communities. His "community of sufferers" thesis, which he fully developed in 1961 in his paper titled "Why Do Large-Scale Disasters Produce Such Mentally Healthy Conditions?" clearly demonstrated the power of outward challenge to reinforce social bonds. For example, in 1970 in central Chile in the city of Yungay, 90 percent of the population died instantly when seventy thousand people were killed by a rockslide. The disaster had a status-leveling effect, where for the remaining survivors, the only thing left was a sense of kinship that drove unprecedented levels of collaboration.[27]

British psychologist John Drury, who studies group-level responses to disaster, recounts response patterns typical of emergency situations. First, panic and overreaction are rare. Second, even in a group of strangers, survivors help support one another. Third, and critically, the key mechanism by which this help is generated is an instantaneously created shared identity; the group is banded by the new label *We are survivors*.[28] In essence, humans are wired in a way that makes a group experiencing adversity stronger and mentally more resilient—where the adversity itself is recollected with fondness and nostalgia.

How do we use this feature of the human code to bootstrap our honor groups? The answer is surprisingly simple—do challenging, hard, adversity-spurring activities together to forge stronger bonds with our life partners. Throughout most of human prehistory, in small-scale societies around the world, creating kin outside of people you are biologically related to happened wholesale. Often people were pushed together through suffering, challenge, or tragedy that went deeper than most experiences we face in our "safety first" society. In Micronesia, the Chuukese people have a saying: "My sibling from the same canoe." That is, if two people survived a dangerous trial at sea, then they were kin. Human migration is a challenging survival task in any era, but for the Ilongot people of the Philippines, those who migrated somewhere together became family. On the Alaskan North Slope, the Iñupiat of Alaska channel death to create new fictive kin; by naming children after dead people, those children are considered new members of their namesake's family.[29]

Thus, if you are trying to find friends, a solid approach to assessing people of like mind is to look in places where such activities take place—whether it is a spin class, hot yoga, or martial arts. These activities need only to simulate adversity to start a chain reaction of bond building. For a non-physical example, adversity activities can be completely imagined. Cooperative video games, board games, and even tabletop role-playing games (RPGs) are ways to imagine and play out how a group could overcome adversity together. Nothing beats the real thing, though; camping, hiking, or any type of group activity that takes people out of their comfort zone generates valuable data of someone's true nature.

Life will also provide ad hoc tests. When disaster strikes, as it often unpredictably will, take note of those who are there to help you pick up the pieces; in essence, ask yourself: Do the people in my life fulfil the evolved function of a friend? These activities are meant to be combined with more formal assessments of personality, such as the Big Five, to help you generate predictive models of your potential honor group member's grit. If these tests begin bringing you closer together, then it's time to take it to the next level and formalize and solidify these bonds.

THE OXFORD DICTIONARY DEFINITION OF ritual is "a religious or solemn ceremony consisting of a series of actions performed according to a prescribed order." Since we now know that religious actions are, at their core (see chapter five), signals of tribal allegiance, we can modestly modify this definition to make it a bit more operational. For our purposes, we'll define ritual as *performing a pattern of actions that conveys meaning and significance using identity signals of coalitionary alliance.* This covers a bit more conceptual turf, because it includes behaviors like a ritual kiss goodnight to your pair-bond signaling your continued partnership, or Notre Dame's football team's tradition of touching their "Play like a champion today" sign while heading out of the locker room before home games.

Ritual practice is one of humanity's oldest tools for deepening the bonds of a relationship or a network of relationships. As Jim Clarke notes in his book *Creating Rituals*, ceremony honors what has occurred, but ritual is a symbolic tour of the inner demands of your new life stage.[30] The forging of the bonds between people by way of ritual weaves together powerful forces of attachment—shared history, experience, and overlapping life goals glued together by sacred values such as honor and loyalty. Friendships, transcending kin selection and not held hostage by reproduction, are an incredible moral achievement by our species; rituals can cement these bonds in something beyond words, in action itself.[31] As described by Sasha Sagan in her book *For Small Creatures Such as We:*

> It's really science that's been inspiring rituals all along. Beneath all the specifics of all our beliefs, sacred texts, origin stories and dogmas, we humans have been celebrating the same two things since the dawn of time: astronomy and biology. The changing of the seasons, the long summer days, the harvest, the endless winter nights, and the blossoming spring are all by-products of how the Earth orbits the sun. The phases of the moon, which have dictated the timing of rituals since the dawn of civilization, are the result of how the moon orbits us. Birth, puberty, reproduction, and death are the biological processes of being human. Throughout the history of our species, these have been the miracles, for lack of a better word, that have given us meaning. They are real, tangible events upon which countless celebrations have been built, mirroring one another even among societies who had no contact.[32]

Rituals can be private or public. They can mark time, season, and transition. They can help us face fear or explore new possibility. They can atone and make pure. But most important, rituals connect. They connect us with ourselves, with nature, and with others. The universality of rituals speaks to their power to bond. Cross-cultural analysis demonstrates that rituals fall into several categories, all of which you can co-opt for your friendships, your pair-bonds, and your camps and communities:

- **Seasons**
  - Spring and fall equinox
  - Summer and winter solstices
  - Feast and famine

- **Developmental life history**
  - Birth
  - Sentience—development of moral sense
  - Puberty and entrance into adulthood
  - Death

- **Time**
  - Daily
  - Weekly
  - Monthly
  - Annually

- **Transition and consecration**
  - Rites of passage
  - Moving/traveling
  - Purification, forgiveness, and atonement
  - Blood brother/sisterhood
  - Wedding

We will investigate the various types of rituals as we move toward different aspects of campcrafting in the following chapters. Rituals for *seasons* and *developmental life history* are best performed publicly; from Thanksgiving feasts, to birthdays, to funerals, these are community- and family-focused celebrations marking important milestones or reminders of core community values. Of particular interest to the solidifying of friendships are *time* and *transition* and *consecration* rituals. A great starting place for friendship development is a weekly and annual ritual. This is a consecrated period of time solely dedicated to meeting with your friends. For example, Sunday samosas and brunch or Thursday night poker, but with a dash of tradition and sacredness to protect it from being atrophied by life's other obligations. Annual rituals are truly sacred as they not only mark significant periods of time, but also permit the participants to pool together significant resources to execute a retreat or reunion. These are especially important for those who are geographically dispersed, as they are compelling reasons to bring an extended social network together.

Once a friendship breaks a critical threshold of closeness, then it's time to consider consecration by way of ceremony. Perhaps the most powerful historically verified friendship ritual is the practice of blood bonding. Becoming a blood brother or sister implies the scaffolding of kinship. You are forging kin and choosing family. This threshold was traditionally met after years of courtship and was as important as any wedding, only it commemorated friend alliances and friendships—this is the standard we should aspire to when choosing our own rituals.

Of course, modernized humans might articulate several objections against the reinvigoration of blood bonding, a seemingly outmoded behavior. First, it must be conceded that the ethnographic evidence suggests it is typically a male practice; with this heavy male bias, it could ignore half the population. There are exceptions—such as the blood sisterhood practiced thousands of years ago among the Mongols or Amazonians mythologized by Herodotus—but regardless, I am less interested in how it was applied in the past and more intrigued about how it *could be* leveraged today to augment the future. In my view, the function of the practice supersedes concerns of gender; anybody along the feminine-masculine spectrum can benefit from ritualistically forming connections with others.

Second, there are legitimate health concerns that are raised by blood bonding. Modern medicine has exponentially increased our understanding of disease transmission to a level that was simply inconceivable to the majority of cultures that utilized this practice, and it's clear that direct fluid exchange can be very dangerous. I am not a health care professional nor am I advocating one specific blood bonding cultural practice over another; much of the value gained from the ritual is symbolic. In fact, some blood bonding traditions do not involve fluid exchange at all, such as the Norsemen who mixed blood into the earth. Many previous practices called upon the participants to mix blood into alcoholic beverages (with natural antiseptic properties that would reduce risk of disease transfer) and to together drink the elixir.

But blood transfer can be substituted by countless symbolic gestures, such as the exchange of meaningful gifts, sleeping side by side, making oaths over shared real or imagined sacrifice, exchanging names, or undergoing acts of co-creation. Tattoos—already a popular Millennial pastime—can also be used to solemnize bonds. Finally, to add some perspective, most cultures had strong norms against making blood bonds with too many people. Any more than three or so pacts was considered a dilution to its significance. There is no singular approach, and the details of the ritual are only limited by the participants' imaginations. It's yours to envision, craft, create, and execute.

This is the beauty of ritual; its magic is that you and your friends can author and customize the perfect rituals for your own lives. I challenge you to get together with your honor group and fill in the following table of rituals you will hold together as sacred:

RITUAL TIME	RITUAL TYPE	ACTIVITY FUNCTION
Week	Check-in	This is the weekly check-in with your honor group, where you protect your relationships from atrophy and disuse and strengthen bonds; the focus here is communication.
Month	Group activity	This is the monthly activity that is focused on building resilience as an honor group. Integrating some type of challenge that pits your group against the elements or other groups is ideal to forge unity.
Rite of passage	Group identity	This is some type of ceremony that elevates individuals into your honor group. Change and transition, and a type of rebirth around a shared identity, is the focus.

Randi Griffin, a data scientist, experienced an exceptional example of how to use both challenge and ritual to forge lasting human bonds. Randi and I met while working in the same lab during my postdoc at

Duke University. She's the kind of person that instantly distinguishes herself upon first meeting. A coding wizard, she sees the world with mathematical precision that is awesome to inspect.* But she holds many credentials beyond being the smartest person in the room. Before she pursued a career as an evolutionary anthropologist, she was a ferocious attacking forward who attended national camps run by USA Hockey, which helped her get into Harvard, where she played on the women's hockey team. She is an Olympian who took part in history in the 2018 Winter Olympics.

As documented by author Seth Berkman in the book *A Team of Their Own*, the South Korean women's hockey team featured members from across the globe and drew from incredibly diverse backgrounds (concert pianist, actress, data scientist, high school student, and convenience store worker). Twenty-first-century school textbooks still promote the idea of "one blood" among Koreans as a source of national pride, which has led to a history of mixed and adopted children being treated as inferior for not being "pure" Korean. Though nobody on the Korean team openly expressed such sentiments, the idea of "imports" coming in to, as worded by Berkman, "change the face of the tribe" was not taken to without reservation.

After a year of wavering (she thought the initial slew of emails from the Korea Ice Hockey Association were spam), Randi—a North Carolina native born to a white father and Korean mother—eventually became one of the team's "imports" and ultimately one of its leaders. It was difficult at first; the linguistic and cultural signals of the Westernized imports crossed wires with the native Koreans. Randi recalled how, despite the tribal differences they each brought to the team, they began increasing team cohesion. Although a cultural gap existed, they also shared the lived experience of being female athletes and women who don't follow gender norms in their countries of origin. This, combined with the

* When defending her thesis proposal, she was challenged by one of the department's senior faculty on one aspect of the model she had developed. Watching her dismantle the attack with laser precision and doing so without breaking a sweat was a marvel to observe. Randi's hockey coach at Harvard said she was one of the most trusted players on the roster, who could withstand the pressures of a big moment or a game-winning penalty shot as if it were a mundane task.

shared challenge and the goal of overcoming adversity, radically changed their diaspora of hockey players into a sisterhood and tight-knit secret society. In an exchange with me, she recounted:

> We truly forged bonds despite remarkable cultural differences and initial hostility. There are girls on that team who don't speak English and I don't speak Korean, but we forged alliances partly out of necessity and partly out of our gradual recognition of a deep connection that was even more profound than language. We all shared the experience of feeling alienated and unvalued by our birth communities because we are females with unconventional talents and passions; we found ourselves ignored by men and reviled by traditional women. The tribal alliances formed on women's hockey teams had been transformative for all of us in our lives, to the point that we would sacrifice material wealth and broader societal acceptance in exchange for the opportunity to spend the majority of our time surrounded by twenty-five women who truly understand and value each other, toiling toward dreams that the world doesn't validate or care about, but *we* care and that's all that matters. In the process we built trust and understanding that I don't share with 99.9 percent of the English speakers I interact with in America. It also helped to have a common adversary . . . since it is the most recent and extreme example of tribal lines being redrawn that I've ever experienced, and I am still kind of in awe that it happened.

Ritual, pre- and post-game, was an important component of cohesion building. As a team, they always hit the ice in a ritualized way; after skating in unison for a few laps, they would gather together around the net for the final huddle before the puck was dropped. This "unfurl ritual" was a call-and-response style cheer, where the skating circle would unfurl as everyone high-fived the starting goalie (named SoJung) in a line. Paying homage to the special role a goalie plays on the team, they would chant things like "Let's win this for SoJung!" Unlike other, more arbitrary patterns of behavior that usually were performed with coaching staff, this one felt more powerful to them because the net huddle was just for the players, and the location—surrounding the goal and goalie—had an air of sacredness that was just theirs. After the game, the team had a somewhat unique ritual where they formed a circle at center ice, and on cue of the captain, tapped their sticks and bowed to each other. Next the circle unfurled and they formed a line with the captain at one end, and on her cue, they tapped their sticks and bowed first to the opposing team's bench and finally to their coaching staff. It was a noteworthy ritual because it singled out their unique group identity and injected some Eastern-style respect into typical Western hockey traditions. Moreover, for the Westerners on the team, this helped reinforce that they were representing Korea. It was singularly theirs—and so it was especially powerful in leveraging the Tribe Drive to enhance their coalitionary alliance. By the end, they were commonly referring to each other as "sister" and one "family." This blood sisterhood, this sodality of women separated by culture, geography, and language, became fictive, forged, chosen family. Let them serve as inspiration to all those seeking something with deeper meaning and purpose beyond themselves.

If they can do it, so can you.

A Blood Brotherhood Ritual—Kith Become Kin

We begin the ritual with the taking of sacred vows. I have difficultly uttering the words, for I have never uttered words with such honesty nor with greater fidelity, up until that moment in my life. I wipe tears of truth from my face, and I steel myself to break my skin and draw my blood. So, too, does my friend, and we dip our hands into a large drinking horn that is filled with dark, bloodred wine. We take our swords and dip their tips into the wine, a symbol that we will protect and defend each other's honor with our lives if need arises. For the Scythians, as for many of the world's past cultures, the drinking of the wine is the act of consecration. He brings the vessel to his mouth and drinks, and when he passes it to me I am reminded of an old Scythian verse that spoke of their custom to drink a beverage until it is completely empty. I tip the end of the horn to the heavens as I drink the last drop of the magical elixir. I am reminded of the verse by Rudyard Kipling from *The Ballad of East and West*:

> *They have looked each other between the eyes,*
> *and there they found no fault,*

They have taken the Oath of the Brother-in-Blood
on leavened bread and salt:
They have taken the Oath of the Brother-in-Blood
on fire and fresh-cut sod,
On the hilt and the haft of the Khyber knife,
and the Wondrous Names of God.

This ancient, sacred, and culturally universal rite of alliance has proven itself to be among the most significant moments of my life. Today, it stands as one of my most valuable experiences. It is a gift I would never give up—nor return—at any price. It is interesting to consider that the simple act of a ritual can elevate something that almost everyone has—a friend—into something far greater: a walking, sentient sacred value, imbued with shared identity that binds those with different blood into the same family.

8

CAMPCRAFTING (PART 1)

An (Evolutionary) Guide to Forging a Twenty-First-Century Camp

To have a good life, you don't need kids or a husband or a picket fence. You just need to find people who matter and keep walking toward them.

—Mandy Len Catron, 2018

We can't rely solely on the traditional ceremonies of the past. We must have the courage to create new ceremonies in the present so there will be traditions for our children, grandchildren, and our great-grandchildren to follow.

—Bernice Falling Leaves, Métis/Lakota Sioux Elder

Ancestral Homes—The Power of Intentional Community

Benjamin Maxmillion Charles is telling me a story.

His words are measured and precise, but also lyrical and rhythmic. Setting intent and speaking from one's heart is a sacred value among Ben's tribe—the Yup'ik. Compelled by the power of his narrative, my heart is open and I listen closely as he speaks to me of the *coming-of-age ceremony* that welcomes young men and women into Yup'ik adulthood. Ben is the descendent of one of the last remaining lineages of bear shamans of his people, who for thousands of years have survived and thrived in Siberia, Saint Lawrence, and Diomede Islands in the Bering Sea and Strait, and Alaska. He is also an environmental researcher and activist and at the time managed the museum at Alaska's Bethel Cultural Center. In this moment, he is sharing a crucial dimension of culture with me at the center. It was my 2019 summer field research season, and I was investigating how circumpolar environments influence sleep across the dramatic seasonal changes of polar nights (that never see the sun) and midnight suns (that never see the sun set). He was generous enough to give me a tour, and we found ourselves gazing at a series of pictures depicting a young girl's coming-of-age ceremony.

The word *Yup'ik* means "real human being." As Ben tells me, one of the aims of the ritual is to ensure that society is entrusted to real human beings—those who have been properly trained and possess the character and capacity to behave responsibly to serve their family, their community, and humanity. This ceremony is among the most significant in one's life, and is ranked in importance with birth, marriage, and even death. The girl's ceremony, he tells me, involves incredible preparation on the part of the participants. It takes several years for all the aunties and grandmothers to be notified and make decisions that circumscribe the time and place of the ritual. Offerings are made to bless and consecrate the site. Each would-be woman during that time is made a ceremonial medicine vest, created out of the fabric and color of her choice, alongside a medicine bag, beaded eagle feather, and leather moccasins.[1] Days before the ritual, the girls are taken into the woods to silently meditate and to craft the goods they will gift the participants upon entering the women's circle. A sacred fire is lit and the dance begins. The girl dances, in front of her elders, pouring her heart and soul into the act, which ultimately leads to her ascension in the eyes of her people to becoming a real human being—a Yup'ik woman with full rights and responsibilities of her community.

It is the telling of the dance that hits me. Ben's words are serious and true, and as he speaks this story, so central to Yup'ik identity, our eyes lock. I am welling with emotion and overwhelmed with the images of beauty and power prompted by the telling of this rite of passage. I feel grateful for this moment and his sharing of the wisdom of his people that was forged through thousands of years of engaging together in the project of survival. I think of the "sweet sixteen" birthday party, and its subtle yet underwhelming acknowledgment of a young girl's transition into becoming a woman—an atrophied whisper of a rite of passage. The Yup'ik possess something so much more poignant. I offer an internal prayer in that moment that if I am ever blessed with a daughter, she, too, may experience a ritual ceremony such as this, surrounded by a rich community of people that channel strength from their shared identity.

It occurs to me, though, that creating the daughter may be the easy part in this hope of mine. Where will the community of people that lend the ceremony its power come from? Without that community, how would the Great Spirit and our ancestors be able to afford unto my daughter blessing, support, protection, and guidance in her life?

THROUGHOUT THE BOOK, WE'VE SURVEYED the anthropology of how groups, for better or worse, past and present, rural to urban, have oriented themselves. For those of us living in a state of mismatch, we feel that the way our parents did it, with the emphasis on a nuclear family at the expense of friends and community, is less than ideal. Nor do we have the means and motivation to strike out into the wilderness and attempt to live like a romanticized camp of nomadic hunter-gatherers. In the following pages, we'll explore a distinct, third option—of harnessing the best of the Social Suite that is embedded in our DNA to harmonize with the suite of technological and institutional innovations that prop up our current societies. *Gemeinschaft* is a German word that describes the sense of group identity and solidarity born of personal interactions. People past and present who have sought out ways to live communally seek gemeinschaft—they seek the authenticity of face-to-face, human interaction by reducing the scope and scale of their social networks.

As opined by the architect Jay Austin, "I think we've only begun to reexamine what it means to be successful in life. . . . I think people are beginning to realize that they've been tricked. They are realizing they have more agency and options than they thought they did."[2] Our goal here is to help provide a scientifically grounded guide on how to build an *ancestral home*—a space with the capacity to transcend generations and to bring the people you care about most together in both space and time. This chapter describes the parameters of what constitute stable, cohesive, cooperative, prosocial groups—but the question remains: Do you have what it takes to build one? If you have read this far, chances are, you do. In previous epochs of the human experience, campcrafting would have been the meat and bones of survival, but modern society, with its built-in securities, has robbed us of this gift and this challenge. This chapter, and the next on *tribal practice,* are the low-hanging fruit that, if applied, can begin the process of improving people's lives. This chapter is for the true trailblazers and pioneers of modernity. This will take work. It's a life's project. But it's good work . . . it's *human* work.

How to Use Group Psychology to Build Successful Camps

The astronauts that land on Mars will have something in common besides being the first humans to dwell on the Red Planet—they will also share a specific personality trait. It has long been recognized by many institutions that a carefully selected team is vital to the success of attaining their goals, while a poorly chosen team invites chaos, disaster, and poor performance. It's not just having high-performing people assigned to work together; there is a magic "sum that is greater than its parts" effect when the right constellation of humans synergize together.[3] In recognition of this fact, the National Aeronautics and Space Administration (NASA) sponsored research to determine which personality traits are the best selection criteria for high-stakes expeditions when lives are on the line.

Conscientiousness is defined as wishing to do what is right, especially to do one's work or duty well and thoroughly; in the context of a team, it "can be thought of as a pooled team resource." Julia McMenamin, the study's first author at Western University in Canada, said, "The more conscientiousness a team is, the better they will likely be at accomplishing tasks."[4] In a way, campcrafting is a bit like a Mars expedition. It requires some level of courage to leave the comfort of our isolated nests and voyage into the social unknown. Fortunately, it's a lot less risky than surviving on another planet, and so another trait in particular will be helpful for groups to grease the social wheels—*agreeableness.* These two together, as we will see, can pack a one-two punch that leads to teams winning and achieving their goals. But there are also some traits that, if expressed by certain individuals, can be beneficial despite the wider group not collectively scoring high for them. Extroversion and openness are good examples. Although neither predicts group success, they do permit novel routes by which a team can access new resources. Thus, for some traits, a little dash of diversity is complementary to the meat and potatoes of playing nice and being responsible.

The personality makeup of a team influences its success. Therefore, camps that have an awareness of this makeup have a decisive advantage in managing their groups to increase the odds of attaining their goals. Just like in the previous chapters, we'll list each personality trait and describe and interpret the results of the science that shows how they influence the success of groups.*

- **Openness**, as a group-level trait, is the capacity for a team to be creative, broadminded, and willing to experiment or to try new things. Open groups would be expected to adapt easily to new situations, build upon each other's ideas, and look for alternative ways to solve problems they encounter.[5] They would also be expected to foster a creative atmosphere in which team members have opportunities to learn and to experience satisfaction.[6] Although previous research has shown that higher elevation of a team's openness results in better decision-making performance and higher overall team performance, broader meta-analyses showed that elevation (a high group average) of openness to experience was not positively related to team performance.[7]
    - **Lesson:** Openness is a bit of a mixed bag, as it does appear that groups with open people are also more proactive. But the tradeoff can be those groups being less organized at completing tasks because of their propensity to get lost in the creative and abstract. In sum, having low- or high-average group openness won't make or break a group. Groups with high openness will likely be characterized as more creative than proficient at task completion. Diversity in this trait could be a good thing, as a few individuals could be the wellspring for new ideas, while others step in to execute those ideas deemed worthy to pursue by the group.

* Marriam Khan, my research assistant, did a marvelous job of coalescing and synthesizing the vast literature on personality.

- **Conscientiousness** is a critical team trait. In a study that had a multifaceted measure of work performance ability, conscientiousness was the strongest predictor of individual-task proficiency, and it was overall the greatest driver of how good one was at their role.[8] Amazingly, it drives focus and commitment to the task, cooperation, and predicts how adaptable one is in face of role changes within the team or task.[9] The lack of these characteristics may lead to social loafing or free riding.[10]
  - **Lesson:** This trait may be the most important one to consider at the group level for two reasons. First, as a team average, high overall conscientiousness drives key success. If there are crazy levels of variability in this trait, it could lead to some nasty conflict; the team members with a high achievement orientation may become frustrated with others who are more relaxed in the execution of tasks that lead to goal success. It's the classic "freeloader problem" and for good evolutionary reason. Second, nobody wants to risk their lives going out to hunt and forage only to have people who do not contribute benefit from their personal sweat equity. Therefore, feelings of frustration or hostility toward other members can potentially distract them from the team's ability to perform at an optimal level.[11]

- **Extroversion** may predict group-level success, and there has been a long history of researchers adhering to this theory. One could easily assume it promotes a smooth functioning of the social mechanisms within a team. Researchers hypothesized that the specific characteristics of being talkative, outgoing, enthusiastic, energetic, optimistic, and assertive would result in a positive attitude toward teamwork and high performance expectations.[12] Furthermore, extroverts in a team are expected to stimulate discussion. It turns out most tests of these predictions do not bear fruit. Meta-analyses show that neither elevation of high group averages nor greater variability influences team performance. There may be a trade-off in that extroverted people within teams appear to experience personal success, but this individual success doesn't translate to the teams themselves.

- **Lesson:** Although a team full of extroverts performs no better than a team full of introverts, teams that have extroverts should leverage their specific talents as "special agents" to accomplish goals on behalf of the group. Think of extroverts as great point people to interface with the outside world as diplomats for your group's cause.

- **Agreeableness** is, without exception, positively related to team effectiveness. Team members high in agreeableness are friendly, tolerant, helpful, altruistic, modest, trusted, and straightforward and compete less with each other.[13] Studies show higher levels of agreeableness lead to higher team performance.[14] Variability in agreeableness, or even the presence of one single disagreeable team member, is expected to disrupt cooperation, and these results bore out in a meta-analysis that showed teams with members with both elevated and similar scores (i.e., high scores with little variation) on agreeableness outcompete all other teams.
  - **Lesson:** Teams that have both high team averages *and* fewer disagreeable members are the super cooperator teams.[15] With their open communication, they can have smooth conflict resolution and work easily to target team goals with high levels of task cohesion, as well as group cohesion on the most effective way to work together as a team.[16]

- **Neuroticism** plays an important part in team output. Team members lower in neuroticism and therefore higher in emotional stability have more self-confidence about the team's chosen goals and decisions.[17] Research supports that lower neuroticism typically fosters cooperation, a relaxed team atmosphere, stability within the team, coordination of work behaviors, and task cohesion. Some researchers have tested the idea that the presence of one or a few unstable or neurotic team members will have an adverse effect on team effectiveness by disrupting the cohesion within a team, but this has not been supported by large meta-analyses.[18]
  - **Lessons:** The more emotionally stable and less neurotic a team, the more easily they will be able to attain their goals. Their suc-

> cess will be buttressed by the relaxed atmosphere and stability around behaviors that support task completion. But it is important to note that having a few neurotic people in the group need not adversely disrupt team cohesion. Similar to attachment theory, where there is a link between anxious attachment styles and neuroticism, here, too, groups can be supportive of their more anxious members. Communication on this front is key, as it is important for neurotic members to voice concerns and insecurities, and for team members to be supportive. This patience and support is typically rewarded tenfold because when anxious people feel accepted, they become super loyal compatriots to groups that "get them."

For those readers going on the campcrafting journey, it could be a worthwhile—and awareness-raising—exercise to take the Big Five and workshop it with your honor group.

If you have a group Big Five profile, how would you use it? It's critical that there should be no judgments toward individuals that score outside the "average" ranges. Although there are patterns that could be beneficial to work toward, having individuals in your group outside the average can strengthen the group by way of variability. Every group will benefit from everyone being aware that being disagreeable or lacking conscientiousness *when acting in the context of the group* can hurt your honor group. But it is also important that individuals can feel free to express these traits as they see fit when outside of a group context. (The discussion on *tight versus loose* cultural norms later in the chapter will help with this.)

For example, if someone is low in conscientiousness and has roommates that are high in this trait, they can be strategic with the level of conscientiousness they put into keeping public spaces tidy and organized and not lapsing on group or public chores. Meanwhile, their rooms can be as messy and disorganized as they like, because it does not affect their roommates in the group context. When working in group contexts (for example, landscaping a backyard, building a collective gym, or simply doing the dishes with your roommates), it is beneficial to keep in mind that these are the times where conscientiousness and agreeableness are important, and not moments to let insecurities and anxiety manifest in ways that hurt group cohesion. At its core this exercise is about both individual and group awareness—the more of each, the better for the self and everyone in the network. Groups that do this will be in a good position to achieve their goals.

It's also important to realize that the percentile scores you recorded take into account the individual's age and sex—there are differences inherent to both sex and age that mean it's not always fair to directly compare two people who differ in these categories. For example, it should be acknowledged that a young, disagreeable male is the most extreme expression of that trait they will likely be in their entire lifetime—because all things being equal, young men are more disagreeable. For those individuals who seek to change one of their traits in one direction or the other, there are some key points to keep in mind. A study performed by psychologist Rodica Damian and colleagues titled *Sixteen Going on Sixty-Six* suggested that personality has a stable component across a life span, yet it is also malleable, and people mature as they age. They reported that the average person matures to become more open, conscientious, extroverted, agreeable, and less neurotic over time. The average change is about eight points, which can be a substantial difference in personality simply due to experience and maturation.[19] People are most malleable to such changes between thirty-three to forty-two years of age, the period of time they grow and mature into adults.[20]

Ultimately, these exercises help lay the foundation of understanding how each person fits into the whole to help the group achieve its goals. Research has shown that teams that understand each other enhance their cohesion dynamics in a way that forms a positive feedback loop.[21] Each person brings to the table something unique and can use their special talents to help groups provide accepting, prosocial environments. The science shows that certain traits objectively improve group performance, but there is also support for the *special chemistry hypothesis*—the idea that successful teams may be nothing more than an effective meshing of a set of personalities, and that it's *compatibility* of the group that matters, not the elimination of certain traits.[22] At the end of the day, sticking with and encouraging each other to be the best version of ourselves is one of the key functions of being a member of a team.

Identity and Purpose—The Rocket Fuel of Group Cohesion

Have you ever been part of a group where you felt a pure sense of trust and closeness with its members? A group where you felt you truly belonged? A group with a sense of commitment and purpose so strong you would never even consider leaving? Atkins, Wilson, and Hayes remark: "Being part of a group with strong bonds of commitment, trust, and shared purpose is deeply, deeply satisfying for human beings."[23] Such occasions, while rare, often stand out as peak human experiences. Fortunately, with applied evolutionary science at our disposal, we can use a tried-and-true formula to *create* these experiences. We've explored some of these highs in chapter eight.

I experienced such highs fighting alongside my medieval martial arts group, The Brotherhood of Steel, at Pennsic; I imagine Randi Griffin also experienced it during her time as an Olympian playing for the Unified Korea hockey team. The commonality among these two groups was that their identities were strongly defined, membership clearly delineated, and purpose crystal clear. And every day the groups worked together more perfectly, cooperating more effectively as they adapted, fought, and negotiated their way through extreme challenges. Although difficult to put a name to this type of feeling, scientists use the term *group cohesion*. The classic definition from 1950s group performance science defines group cohesion as "the total field of forces which act on members to remain in a group." This is the degree to which the group sticks together and is unified in the pursuit of its goals.[24]

Cohesion can be enhanced with *shared identity* and *purpose*. Groups that have this property build a sense of "us" beyond each individual member's self-interest. Groups that don't share a clear sense of purpose and identity are doomed to conflict and splintering; the mechanism driving the wedge is a lack of cohesion. There are two parts to group cohesion: *social cohesion* and *task cohesion*. The former answers the question "Do people within the group like, trust, and feel affinity toward one another?" The latter answers the question "Do the people in this group really believe in the goals of the group?" Either believing in group goals but not getting along with its members or vice versa will create substandard group performance. When groups perform well together, they also enhance cohesiveness, and in a virtuous positive feedback loop, enhanced cohesion improves group performance.[25] Members of a cohesive group are more likely to work hard and be motivated to perform that work and be loyal.[26]

Cohesion takes hard work to build. Like trust, it takes much effort to cultivate, and it is easily broken. Yet, when groups have it they become highly motivated to work hard in the name of the group. In fact, studies have shown that cohesion is such an intrinsic value, it is a more important motivator than money.[27] One cross-cultural study showed that in war, even more important than physical and material strength is *spiritual strength* measured as the will to fight, which is the individual's willingness to mobilize and seek success in violent conflict regardless of the personal risks. Group loyalty, leading to cohesion, was the driver of spiritual strength.[28]

Our purpose, as set out in the beginning of this chapter, is clear—to create a camp with your honor group. Fortunately, it's a powerful one that is written into the very fabric of our DNA. Shared purpose creates motivation, cooperation, and personal benefits, and elevates collective interests as inherently valuable.[29] Most important, this is a "living purpose" that is embedded in everything—every conversation, agreement, process developed, ritual created and experienced—and is used as the ultimate bar of evaluation of success. Because you are campcrafting with your honor group, you have the highest-level "living purpose," which should also be aligned with your group's *sacred values.* As discussed briefly in Part I of the book, a sacred value is one that an individual observes as absolute and inviolable—conceiving of breaking one is a social taboo. Intriguingly, values correlate also with personality.[30*] Sacred values differ from material values in that they incorporate moral beliefs, such as the welfare of the family, commitment to country, or identification with a particular religion. Think of them as your group's primary social norms, the spir-

---

* The social psychologists Velichko Fetvadjiev and Jia He discovered that agreeableness correlates with values of wanting to engage in positive actions for society. Conscientiousness was correlated with values of conforming with societal norms. Extroversion is linked with values of enjoyment, and openness was related to values of self-direction and actualization.

itual compass that helps map and guide all other norms.* For example, one popular sacred value for many modern intentional communities has been environmentalism. As noted by Miller: "Those who flock to the new ecovillages are as passionate about their environmental convictions as were medieval monks and nuns about their spiritual beliefs."[31] Developing your own group's sacred values will help symbolically crystallize your collective purpose.

With a strong sense of shared identity and purpose within your honor group, it will be easier to create a sense of *psychological safety.* Google performed a fascinating internal study called *Project Aristotle,* where the goal was to determine the optimal mix of personality traits and group dynamics that predicted team success.[32] The results were astounding, as they found almost no significant patterns except for one group dynamic: *psychological safety.* The greatest teams provided a supportive environment in which people felt secure enough to provide alternative perspectives without fear of negative consequences. Socially adept leaders use a velvet-glove approach to cultivating social environments that reduce the cost of productive dissent, which ultimately helps people feel they are in a "safe space." It is all too common to think that psychological safety means that you don't have the capacity to criticize; in fact, real safe spaces are the opposite of the silencing effect that many modern social environments encourage.

True psychological safety in group dynamics means that debate is welcomed and embraced, and people can disagree respectably and come back together without hurt feelings because they feel they are all working toward the same goal. The acid test for your group is whether you hear minority opinions voiced often and without judgment during group conversations. If everyone is saying the same thing and there are no dissenting views, your group is likely suffering from a lack of psycho-

* These norms can even be exercised by playing games. A *New York Times* article titled "Why the Cool Kids Are Playing Dungeons & Dragons" describes the advantages afforded to people that play games together over tabletop rather than in virtual spaces. One thing to keep in mind while playing games with friends—the real game is the metagame that demonstrates it's not about winning the game, but *how* you play it. RPGs are perfect to practice and play out sacred values and social norms of the group, because whether you win or lose, if you play in accordance to your group's values, you gain esteem within your group.

logical safety. This will ultimately hurt its capacity to make high-quality, informed, and innovative decisions.

When your honor group comes to see themselves proudly as *who they are*, then it can be said they share an identity. When they come to happily believe that the group represents something they care about, they share purpose. Behaviorally speaking, this shift occurs when individuals begin speaking less about "What's in it for me?" and begin to consider "What's in it for us?" In other words, when the "us" of your group includes "me" as a key part but is no longer limited by the "me." To create, reinforce, and bake group identity into your honor group, keep in mind the following strategies:

- **Motivate people to get involved and be proactive.** At first, simply motivating people to be in the group can be challenging. Individual members must feel like they get something out of it. This is the "What's in it for me?" problem. Fortunately, we spent the opening of Part II of the book outlining the extrinsic and intrinsic benefits to intentional proximity and community.
- **Use the inherent power of small groups.** Small groups are more cohesive than large groups. Larger groups are more prone to social freeloading and it's harder to monitor agreed-upon behaviors within them. (Chapters eight and nine outline both how to select your small group and what parameters best facilitate the success of a small group.)
- **Define and ritualize membership.** Clear standards for membership are critical to forming and reinforcing group identity. Specifically, it is important to be clear about who is inside and outside of the group. This may be easy for smaller groups, but if you get bigger and more ambitious in your campcrafting, you will eventually need to be explicit about membership. If you decide to create a sodality—a topic we will discuss in more depth in the next chapter—this step is crucial. Groups naturally form inner circles, which correspond to greater rights and responsibilities. Outer circles may join and bolster your network, but correspondingly have fewer rights and responsibilities. One strategy is to require contributors to be sponsored by at least one member, with con-

tinued membership needing approval by all members. Clarity about belonging and clear rules for how membership is attained will enhance cohesion and make the group perform harmoniously. Finally, making sure there is on "onboarding" process for new members by way of ritual can profoundly enhance their feelings of shared identity with the group. Initiation ceremonies are powerful tools to this end.* One key element to proper initiation is to emphasize the unique history of the group and tie in membership as part of that tradition so that initiates feel a connection to it from the beginning.

- **Create a space for personal, meaningful communication rich with psychological safety.** Cultivate a time and space for meetings. Practices and norms within a group that give members a chance to share about their personal lives outside a group context can be helpful in building cohesion. Ritualized meetings with team building, goal setting, and built-in time to reflect on completed projects are all opportunities to accomplish this task. One strategy is to set a meeting time for the honor group that is consistently observed. *Camp meetings* can be the nexus for all projects and related activities to be planned and executed. We'll discuss the importance of the camp meeting later in this chapter. Opening these meetings by asking each member how their personal life is going is a great way to enhance group cohesion and productively share individual struggles or victories.
- **Enhance similarity in the context of diversity.** Groups that are more similar are more cohesive. The science behind tribal signaling was detailed in Part I of the book, and anything that serves as a tribal symbol or token can enhance cohesion when applied to small groups that share identity and purpose. Shared uniforms (for example, T-shirts, hats, or other clothing), rituals, nicknames, ways

* Readers may be worried about hazing. The function of hazing is to create a community of sufferers by causing suffering unto prospective members. Yet, it can also be used as a tool in hierarchical groups to establish pecking order. I want to emphasize here that the ritual needs to be, overall, positive and not harmful to the well-being of the members. The ultimate goal is shared identity and purpose, not just entertainment or dominance. Challenges can be excellent techniques to create a sense that "we all had to overcome something" to be a member of this group. Hazing is when these challenges cause undue harm to the members.

of talking and insider jargon can bolster cohesion as well. Creating a camp symbol that can be put onto clothes, or even a flag, taps into the deeply rooted Tribe Drive and can also be an incredibly fun way to create an identity with your honor group.

An important caveat to enhancing similarity is that if it is overdone, it can also disservice your camp with groupthink. Ultimately, you'll want to cultivate an atmosphere in discussion that enhances your group's productivity by instilling what psychologist Todd Kashdan calls *principled insubordination* in his book *The Art of Insubordination*. In the most effective groups, psychological safety is so engrained that individuals can readily voice counter opinions to the group orthodoxy. If you find yourself in a group you love and identify with, but feel it could be better aligned with its values, take a page out of Kashdan's book by following these principles:

1. Use and emphasize your status as an in-group member. Emphasize "us" and "we" terms, focusing on in-group status. Then elevate the group's attention to your shared identity before rebelling against its orthodoxy.
2. Don't spark fear; cultivate curiosity. Present new ideas in non-threatening ways. Consider and frame the current members of the orthodox views as future allies.
3. Always project an aura of objectivity. Cite studies. State when you are basing something from hard data, and flag and label hot takes and speculation on your part.
4. Communicate your self-sacrifice as an act of courage. Signal the ways you've personally sacrificed to attempt to change the group for the better. Don't be a disagreeable naysayer. Seem uncomfortable speaking out against the group. Be direct: "I feel seriously uncomfortable disagreeing; here is why I am doing so . . ." Candidly evoke how hard it is to act rebelliously.
5. Be consistent yet flexible. Know the hill you will die on, but besides those mission-critical arguments, bend on the less important issues.

If you remain consistent without being rigid and never forget to not *push* your values upon others, but live the values you espouse in action, you'll be amazed at how many previous members of the orthodox opinion will begin to seed and eventually stand for the principled act of insubordination.

Ritual and Religion Repurposed

On March 29, 2021, Gallup, one of the most trusted polling institutions in the United States, released a survey that illustrated a pivotal moment in an eight-decade-long trend—the emptying of places of worship.[33] For the first time, church membership fell below a majority. In 1999, 70 percent of Americans regularly attended a church, synagogue, or mosque, whereas in 2020, only 47 percent of Americans said they belonged to and practiced a religion. The decline is principally driven by Americans who no longer express religious preference. The evidence of this is everywhere; from abandoned places of worship being revitalized into breweries, arcades, or even residential homes, organized religion has seen a drastic reduction to its ranks.

That doesn't mean, however, that North Americans are becoming less spiritual. In fact, there is evidence to the contrary. As noted by the Pew Research Center, while religiosity is decreasing, "at the same time, the share of people across a wide variety of religious identities who say they often feel a deep sense of spiritual peace and well-being as well as a deep sense of wonder about the universe has risen." Fascinatingly, this has happened for both the religiously affiliated and unaffiliated. For instance, between 2007 and 2014, both Christians and non-religious people showed a 7 percent increase, yet for religious "nones," there has been a radical rise of seventeen percentage points in the same time period.[34]

The spirit drive, a key subroutine of the Tribe Drive, is too powerful an evolutionary adaptation to discard, mute, or fully suppress. Religion was the invention that channeled the spirit drive, but it's only a social tool. And as we well know, tools can be used for both good and evil. To move forward in the twenty-first century within an ethical framework, we need to be able to protect against the worst outcomes of misapplied religious expression. In chapter ten, we will forward an ethical precept that will ensure groups can work prosocially with other groups—a philosophical move I call *the concentric circle.* For now, we simply want to safely use some of the basic levers of spirituality to increase group cohesion and prosociality.

But if we don't believe in a God or gods, then how can we use religion to help ourselves or our groups? The philosopher Andy Norman puts it this way: "What if we could have the benefits of religion without pretending to know things we don't really know?" We absolutely can. "The valuable aspects of religious faith can be had without willful self-deception. . . . The same healthy attitudes that religions have long been in the business of cultivating can be induced in other ways." Religious wisdom can be used functionally, even if it is not literally true. We can craft sources of values that adaptively replace religious functions without abandoning an evidence-based approach to the world.[35] Tapping into the prosocial functions of religion, Norman expounds: "A religious claim doesn't need to be true to feel deeply right. The feeling of rightness can stem instead from its power to *express allegiance* [emphasis mine]." That's the core of religion's role as a subjugate signal of the Tribe Drive. We can tap into the power of ritual to express allegiance, while consistently maintaining a worldview derived from the scientific method.

The components of the religion tool can be broken down into two parts: sacred values and rituals. The reason why religion has been such a widespread social institution is because it effectively leverages spirituality to coalesce groups around sacred values enacted by ritual. This is likely why religious people tend to benefit (as discussed in chapter seven) from the coalitionary alliance buttressed by their tribal signals. Religionists tend to have higher overall levels of social support; this is especially true for moms, as the number of congregational social ties increase when they have a child, and mothers receive more social support from co-religionists than non-mothers.[36] Thus, religions are good at creating strong in-groups, but it comes with a significant societal cost; they also create out-group suspicion and distrust. This may be in large part why highly religious societies tend to be characterized by greater levels of social dysfunction (e.g., murder, poverty, divorce, teen pregnancy, and general crime).[37] We can conclude then that religions are good for the individual, but can be bad for societies. The psychological rewards of religious faith often ignore the indirect cost to others, which is selfish and may turn out to be ultimately unethical.

What are our options, then? Option one, stick with your traditional religion and leverage its time-tested sacred values and rituals to bind your group. Option two, adopt new (or in some cases, ancient) religious practices to the same end. The latter is an increasingly popular approach, as seen in Europe by the contemporary restoration of ancient indigenous religions (aka, paganism). Paganism has grown in popularity during the last hundred years, a growth that coincides with a decline in Christianity in Europe.[38] It could be a trendy, spiritual bandwagon your honor group could hop on.

A third and distinct option is to build your own sacred values and rituals with your honor group, eschewing formal religious practice altogether. This may be particularly attractive to those people in the ever-growing category that feels uncomfortable with religion's top-down, centralized control role in society, but are growing an increasing awareness of their own spirituality. My own honor group has adopted this option—I detail our cocreated spiritual journey elsewhere,* which is aided by entheogens (chemical substances that are ingested to produce a non-ordinary state of consciousness for religious or spiritual purposes).

The application of psychedelics to improve individual and group well-being is undergoing active research by world-renowned institutions like the Johns Hopkins Center for Psychedelic & Consciousness Research. Roland Griffiths, the director of the institute, and his team are demonstrating how the guided use of prodrugs like psilocybin can profoundly help people overcome a range of challenging conditions, including addiction, distress caused by life-threatening disease, treatment-resistant depression, and even the existential fear of death.[39] Yet, it is clear that psychedelics are not a panacea for creating utopian, peaceful, prosocial societies.

The Aztecs, due to their Central American geography, were generously surrounded with entheogenic plants and had access to and used nearly every known psychedelic. It permeated the society. Much akin to the drug use of the United States during the hippie movements of the 1960s, the setting was often "ravelike" recreation with individuals thrown into an experience with little preparation, intent, or attention to their environment. Without restriction, the use of hallucinogenic drugs

* Which can be tracked down on the website linked in the QR code in the appendix.

informed the activities of every niche of the Aztec civilization, yet their cultural lack of respect for human life is infamous. They had assembly-line beheadings for ritual human sacrifice, ball-game sports that incorporated murder, and near-continuous tribal warfare. In a way, it strikes one as analogous to modern Western societies, where psychoactive drug use is everywhere, but does not positively impact the goals of governing the society.[40] While it's true that the 1960s was a period of great experimentation with communal living, most of these experiments failed once the realities of living off the land without resources tore at the fabric of weak, transient social bonds.

However, history and anthropology show us many instances where psychedelics augment human individual and collective well-being. The practice was used in the ancient world, and (as described by Brian Muraresku in *The Immortality Key: The Secret History of the Religion with No Name*)[41] began in recorded history with the Eleusinian Mysteries practiced in ancient Greece and is still practiced among small-scale indigenous communities and modern religious movements today. One powerful example is the *Hoasca Project* of the 1990s. This was a major study that followed a scientific protocol including a control group and long-term follow-up examinations, which revealed that users of ayahuasca (a psychoactive beverage containing dimethyltryptamine [DMT] that is prepared especially from the bark of a woody vine) possessed more stable personality measures, keener neurophysiological functioning, and greater strength of interpersonal relationships than those who didn't use ayahuasca at all. This led the researchers to conclude that through the sometimes life-altering personal experiences, entheogens have the capacity to significantly improve the well-being of both individuals and groups.[42]

Psilocybin is a naturally occuring compound (called a "prodrug") that is produced by more than two hundred species of mushroom. It has generally similar psychoactive effects to LSD, mescaline, and DMT, which are clinically described as euphoric visual and mental hallucinations and produce fundamental alterations in perception; these include a distortion of space and time and spiritual experience. There are also risks, which are modulated by the type and dosage, the "set" (the preparation one undergoes before the experience), and the "setting" (the environment in which the entheogen is taken). When these elements are aligned, the individual increases the likelihood of an easy, positive experience. The more of these elements that are out of alignment, the greater the risks of a challenging experience with effects ranging from nausea to panic (although the possibility of overdose is negligible and almost never lethal). In addition, even challenging experiences are often reported by participants as proving beneficial and therapeutic in the long run. Those who take psilocybin in controlled studies have reported years afterward that the experience ranks among the top three most significant of their lives—up there with marriage and the birth of children.[43]

The ability of these compounds to improve well-being in our mismatched Western societies is as astounding as it is untapped. Psychedelia sources seem unambiguous in descriptions of group encounters with entheogens enhancing close communal bonds.[44] The Huichol (with peyote) and the Huni Kui/Ashaninka (with ayahuasca) tribes have, as observed by anthropologists, leveraged entheogens for spiritual practice in group settings that enhance social networks alongside their tribal identities. The key component of successfully accomplishing this is the ritual code of having experienced shamanic guides of psychedelic initiation that wisely control dosage, set, and setting. At Eleusis—which featured optimal preparation and attendant ceremonies, just as they are among the Mesoamerican Indians—where their use was in the hand of and controlled by hierophants (ancient Greek religious shamans), this type of drug had profound effects upon its participants. As noted by an expert on psychedelia, Patrick Lundborg, on the subject of the ancient Greek Mysteries ceremony:

> Having prepared their bodies and minds, the climactic hours were inaugurated with a psychedelic potion, and what followed inside the hall was an artful multi-media mix of sound effects, song, flickering light, smoke, fire and joint celebration, all aimed to maximize the effects of the *kykeon* drink. The Great Mysteries were the earliest Acid Tests, the original all-night dances. As the ceremonies continued beyond sunrise, each initiate would experience a moment of profound, often life-altering, personal revelation. This private experience, rather than some scripted climax in the rituals, was the ultimate goal of the Eleusinian Mystery, the end for which the preparations had been made.[45]

In Plutarch's *Parallel Lives,* Cicero writes that the Mysteries not only strengthened Greek identity, but the group of individuals that shared the experience were forever bonded. The Greeks used entheogens at scale, but your group can leverage them for several ritual purposes under a communal banner. Depending on the primary factors of dosage, set, and setting, one group member may come out of a communal night of psychedelics with a life-altering shift in perception, while their companions can mainly enjoy the perceptual sensations and sanctified mood of the shared ceremony. Entheogens can bring joy and melancholy and facilitate soul-searching and shamanic healing of participants who have experienced trauma, but at higher doses they can open a visionary realm through which individuals can explore the group's sacred values. During the experience, there can be moments of deep bonding along with moments of euphoric relaxation, as the social tension dissolves by the communal prodrug experience. Given their capacity for personal revelation, therapeutic healing, enhanced creative expression, and the bonding of group participants, with the right ritualistic configurations the potential for entheogens to be used to increase group cohesion remains a fascinating area of study.*

Ritual as the actualization of your sacred values that engender your purpose is a powerful driver of group cohesion. Rituals are a kind of language; they capture the ineffable with action. They are a way of expressing meaning that is too deep for words. One study demonstrated this by experimentally comparing work experiences, such as the successful execution of tasks, among a control group and a group that ritualistically used physical, psychological, and communal associations with their work. For the participants practicing rituals, results showed that their work became more meaningful, satisfying, and enjoyable.[46]

We've discussed rituals in the context of solidifying friendships and strengthening a pair-bond, but at their core, rituals were meant to be a community practice. They can encompass many dimensions of the

* The Native American Church, also known as Peyotism and Peyote Religion, is a Native American religion that teaches a combination of traditional Native American beliefs and Christianity, with sacramental use of the entheogen peyote. They have 250,000 members and growing, and they use the all-night eight- to ten-hour peyote communal experience as a central aspect of their modern religion.

collective experience: groups can burn something—signifying sacrifice; they can put something on—signifying the taking on of a new identity and role; or they can pass something around—signifying the creation of new community. Below is a group ritual table, where you can cocreate the most meaningful rituals that will be held sacred by your group:

RITUAL TIME	RITUAL TYPE	ACTIVITY FUNCTION
Week	Check-in	This is the weekly check-in with your camp. Here is where the executive decisions that will guide your camp can be deliberated.
Month	Camp activity	This is the monthly activity that is focused on building strength, cohesion, and resilience as a camp.
Rite of passage	Camp identity	This is some type of ceremony that elevates individuals from outsiders to insiders of your camp. Change and transition, and a type of rebirth around a shared identity, is the focus.
Coming-of-age	Celebrating the growth of a child into an adult	This ceremony is ideal to bolster children's sense of self as they transition from childhood to adulthood. Typically, this ceremony is gender-specific and happens around puberty. The end goal is to adopt a new role in life and the camp.
Feast and famine	Honoring the life cycle as a camp	This annual ceremony can occur around the solstices or the equinox. Both fasting and feasting are moments that bring people together as collective acts of appreciation for life.
Camp anniversary or birthday	Camp survival	Just as birthdays can acknowledge individual survival, so, too, can they celebrate the group's survival. Celebrations can occur on an especially important date, such as the origin of your camp's identity.
Death	Celebrating legacy	Just as with birth, death is a once-in-a-lifetime event that can be both a private and public recognition of the legacy of someone in your camp.

Many of the rituals you cocreate with your camp can be drawn from the legion of examples recorded from societies both past and present. There is no greater homage to the survival of our species than acknowledging the transition of a camp member from child to adult. The coming-of-age ritual often coincides with the developing human releasing gonadotropin into the pituitary gland, thus reaching significant levels of testosterone or estrogen. Biologically, it is about celebrating the milestone of reaching sexual maturity, but practically it is an acknowledgment from the adults in the camp that one of their own now faces new possibilities in life. The fledgling adult must now wrestle with the complex behaviors of sex, pregnancy, and the powers and responsibilities of becoming a parent.

Mormons have their teenagers go on missions. The Baha'i at age fifteen is offered a chance to join the faith. For Jewish adolescents, the male bar mitzvah and female bat mitzvah celebrate coming-of-age. A watered-down version that United States society practices for girls is the sweet sixteen, which as noted in this chapter's opening pales in comparison to many indigenous cultures' approaches to celebrating this momentous moment in a human's life. As we have the capacity to take on multiple identities throughout our lives, we can liken this process to what industry technologists call a "technology stack." This is the data ecosystem that serves as a list of all the technology used to build and run one single application. In this instance, it takes many kinds and degrees of identities to run the *identity stack*. Ultimately, if we are not actively involved in helping our children create prosocial, positive identity stacks, then this can leave them vulnerable for others with more malicious, antisocial identities to fill the void. Be active in warding your beloved future adult humans from predatory identity stacks.

Many societies have chosen the fall equinox (when the days become shorter rather than longer) as the harbinger of change to a colder, darker time. This can be a moment to recognize the harvest yielded from a productive summer or for many periods of human history, the famine that followed an unproductive one. Recall that when humans adopted agriculture, embedded in a single crop was the chance of a whole society's prospects of survival. Thus, many ceremonies have been created to this end. The Green Corn Ceremony, practiced by many Native American tribes, was a festival featuring dance and ritual forgiveness. For the Masaru ceremony among the Paiwan, an indigenous people of Taiwan, rice is the primary yield of survival and so is elevated in their rituals. The same can be said of the Shinto harvest in Japan. The New Yam Festival, called the Iwa Ji, is critical among the Igbo of West Africa. In Argentina, La Fiesta Nacional de la Vendimia celebrates the transition of grapes into wine. In theme with all of these, America's best-loved secular harvest celebration holiday is Thanksgiving.

As equally important is the acknowledgment that human evolution has been riddled with cycles of boom and bust—but mainly bust. A near-universal practice in religions across the world is to observe a fast for some varying period and frequency. Yom Kippur, the Jewish Day of Atonement, is one such sober holiday where one refrains from eating. Ramadan, as an example of ritualized intermittent fasting, has Muslims fast every day from sunrise to sundown for thirty days. Not only is fasting for certain periods of time a way to gain an appreciation for what you have, it has the power to collectively bind groups that do it together. Abstinence builds a kind of "community of sufferers" that can serve as a bonding agent. Finally, fasting is also good for you. The human body can only undergo certain types of repair, a process scientists call *autophagy*, during a fasted state. These benefits include the removing of waste products from cells, dropping insulin levels, and activating gene expression that can produce a number of beneficial changes by releasing molecules that increase longevity and protect against disease.[47] One of my favorite annual rituals is a three- to five-day fast I perform with my honor group. Food, and hence life, is never more appreciated than the moment you break a fast—and we do this act together.

Survival and death are intimate bedfellows. An anniversary is the celebration of survival and is not only the purview of the individual but of the group as well. Tying in the origin story of your camp to an anniversary can be a worthy ritual to celebrate the persistence of your group through space and time. Death, while a time of great mourning and loss, is also a type of celebration of the legacy that a camp member created during their tenure as one of your own. There is a growing movement in the Global North of a new profession: the "death doulas" (also "death maidens" and "deathwalkers"). These are end-of-life planners and ritual providers, and have been disruptors to the whole "business of death."[48] Much of the increased interest in these movements stems from fears of a ritually sanitized and "repressed" version of secularism associated with the medical establishment and funeral homes and the need for some autonomy in expression for how an individual's death ritual is enacted. From the "death positive" movement to natural burial advocates, this diverse group of practitioners exist within different religious and cultural traditions and geographical locations and work to redesign and reclaim death care.

Creating your own camp's procedural rituals for loved ones lost can be an act of reclaiming. A specific example of honoring your past ancestors borrowed from the Roman, Greek, African, and Chinese traditions is the pouring out of a favored drink into the ground. Another example in the Jewish and Muslim traditions is the casting down of dirt; in the former, the community takes turns shoveling dirt, and in the latter, three handfuls of dirt, to help fill the grave of the dead. This could have an even greater significance and impact if performed in some way within the confines of your camp's land.* The spreading of ashes, either real or symbolic, in a camp space can also convey the appropriate symbolism. This can magically transform a piece of land where people share a network of homes into a true *ancestral home*, where the DNA of ancestors, both forgotten and still fresh in memory,† comingles with communal earth.

In summary, working ritual and transcendental practice into your camp's culture could have incredible potential to enhance meaning and purpose in one's life.

Campcrafting—Actualizing Your Social Suite

Nancy and her twelve-month-old son, Roland, enter a toy-filled room. A smiling research assistant shows them around the room, helping them acquaint themselves with their environment. Roland eagerly explores this novel place and crawls around to survey the toys, testing

* In places where it is legal.
† The Ovambo of Namibia have different words for the concept of ancestors that are beyond memory and the recently deceased. *Aathithi* refers to forgotten ancestors, where no living member of the group can personally remember the deceased.

their properties by squeezing, biting, pounding, and throwing them around. Periodically, Roland looks up at his mom while he goes about his adventures. Then Nancy is instructed by the research assistant to leave the room, and she does so quietly. Roland keeps playing until he realizes that his mother is gone, and upon the realization he bursts into tears. The assistant attempts to show him some more toys, but he is agitated and he entirely dismisses them. After a short time, Mom comes back into the room, and she consoles Roland until he appears at ease once more; fascinatingly, in only a few short moments his interest in exploration reawakens. He continues his quest to find every toy in the room.

This experiment, created by Mary Ainsworth, is called the *strange situation test*. One of the crucial scientific studies of attachment theory demonstrated that children's exploratory drive, which predicts their ability to play and learn, could be muted or enhanced simply by the departure or presence of their mother in the room.[49] This presence is what is known as a *secure base*, and it's a prerequisite for a child to explore, develop, and learn. Although adults rarely play with these kinds of toys, a great measure of our fulfilment in life is created by going out into the world and dealing with novel situations and overcoming challenges.

The key insight of attachment theory is that if we feel like we have a secure base, then we are decidedly much better at taking the risks necessary to pursue our dreams out there in the real world. As described by *Attached* authors Dr. Amir Levine and Rachel Heller: "Our partners powerfully affect our ability to thrive in the world. Not only do they influence how we *feel* about ourselves but also the degree to which we *believe* in ourselves and whether we will attempt to achieve our hopes and dreams." I believe the groups we choose, and the intentional community we build, are a part of the adult secure base. In other words, when we've got our honor group working together cohesively, we are best able to overcome hard times and challenges that ultimately propel us to focus our attention on that which gives our lives meaning.

But once we've found that group and we begin the undertaking of creating intentional community, how do we not only keep it healthy, but exercise and improve its capacity to be the best secure base it can be? The Mandarin word *yuanfen* means "predestined relationship;

natural affinity among people; lot or luck by which people are brought together." This section describes how to take the luck out of creating affinity toward your in-group, and what science says about how to *create yuanfen* in your camp.

As DETAILED BY THE THEORY of the Social Suite, of the myriad imaginable forms of human societies—that is, forms that *could* be taken by humans in groups—only a few have arisen in natural systems. This is why despite the insistence by the postmodern branch of the humanities that there is no human nature, there is strong evidence that the Social Suite is universal.* In 1966, paleontologist David Raup published a paper titled "Geometric Analysis of Shell Coiling," where he defined three axes that can possibly vary among shells, giving rise to a *morphospace* of potential shell shapes.[50] The dimensions are (i) *size*: the rate at which the shell's cross section gets larger moving through the shell; (ii) *coiling*: the rate of shell coil as it spins away from the axis; and (iii) *elongation*: the rate at which the coil moves up and or down along the axis. Only a small part of the morphospace has ever had an example of a shell in nature occupy that space, whereas the vast majority of potential shells have never been occupied. Raup's work was a profound discovery that extends to all life. Looking at the myriad forms of life-forms, one can be overwhelmed with the variation one sees, yet what is truly remarkable and astounding is how, for many of these forms, there are exponentially more forms that

* In 1991, anthropologist Donald Brown wrote a book titled *Human Universals*, which he defined as "those features of culture, society, language, behavior, and psyche for which there are no known exceptions." In this work, he coined the term *The Universal People*—a kind of hypothetical human society that had a fundamental social order. He argued that these people would play music, create and adhere to rules of taboo, gossip about each other, believe in magic, participate in rites of passage, give speeches in public occasions, and pay strict attention to how males assert aggressive impulses. Although cultural relativists and those who fetishize human variability tend to cringe at the concept of a universal standard for which humans organize themselves, if we do not define and circumscribe the real, useless, and impossible ways humans can exist together, then by what empirical benchmark could we understand how different or similar we are? When we fully understand the social morphospace and witness how infinitesimal the range of possible ways we can organize ourselves in existence is, then this actually reinforces our similarities and brings out our common humanity.

could have come into existence but did not. Seashells have many more useless forms than useful. So, too, it is with human groups; only certain forms are physically possible and adaptively valuable.

In 1999, Lee Cronk, a professor in the Department of Anthropology at Rutgers University, forwarded the idea of the *ethnographic hyperspace*, which mathematically described human societies in a type of social morphospace. Using the Human Relations Area Files (one of the most comprehensive codified documentation of features in over eight hundred cultures), he considered political, economic, religious, social, and reproductive practices in a comparative analysis of these societies and based on his calculations, estimated the potential combinations of such societies would be only a sliver of a fraction of the number of social structures that have ever existed.[51] In other words, relative to the non-functional ways humans can group together, there are only a few configurations that actually work and permit fitness on the individual and group levels of organization. Our goal then is to build a camp that doesn't break the core rules of the Social Suite—the rest is up to you to invent and adapt to best fit your preferences and those of your honor group.

The first step in what I call *campcrafting*—the act of building intentional community—is assessing the key parameters of social groups. By now you may be working on cultivating your honor group of people that together want to fight *social drift*—the social and environmental factors that drive people into isolation—by living with intentional proximity. How do we go about doing this? First, we need to set the bar for the measure of success for a group. The metric of success is *cooperativeness*, which is the likelihood of a behavioral interaction between people within a group resulting in cooperation over conflict. Recall our public goods games, where game theorists quantified the chance that two people would cooperate with each other (between zero to one hundred, with larger numbers indicating a greater chance for cooperation). Although the chance of cooperation between strangers differs somewhat across societies, the typical chance of cooperation is roughly 65 percent. When behaviors lean toward conflict, we call this *antisocial*, and when they result in cooperation, we call this *prosocial*. Thus, the measure of success for our intentional communities is cooperativeness, and these cooperative groups can be said to be prosocial groups. Close your eyes and use your prior experience with your honor group to imagine one thousand simulations of encounters with each other. If you cooperate more than two-thirds of the time, you're doing pretty good; you guys are prosocial! In sum, primary factors that drive cooperation in groups are as follows:

- Size
- Fluidity
- Transitivity
- Hierarchy

Specifically, good, thriving social networks have a specific size, degree of freedom in movement, a level-specific structure to their social ties, and a type of hierarchical style.[52] This data has been analyzed in intentional communities across the globe and even in the digital space—whether it be online communities in World of Warcraft, Israeli kibbutzim, the Quakers, or communes of the 1960s stemming from the Free Love movement—those groups that fell within the factor's parameters were successful (and still around today) and those groups that crafted social norms that deviated from these parameters went extinct.

Size, as discussed in the previous chapters, is an important variable that regulates the channel capacity. As we recall, the channel capacity is our cognitive bandwidth that regulates how many close social ties we can juggle without our brain's processing power tapping out. We know, after taking measure of our own social health, that high-quality ties can number between three and six people, and usually average to about four individuals. We also know that Dunbar's number—the number of people we *know* face-to-face—typically extends to 150 people. We have a range of from 4 to 150 then, and within this value are concentric circles with you at the center and subsequent circles radiating outward, with each farther layer representing less and less channel capacity of sympathy, until you get to the circles of "them." The camp, which is the primary level of concern with respect to building intentional community, is your honor group situated within its local environment. This can be incredibly varied, but studies from artificial (online environments), intentional (historical and contemporary communes), and accidental communities (island shipwrecks) show that the average stable group is about seventeen adults, and 90 percent of groups are under thirty-five people. With historical and current communities, such as the Hutterites, that specific Goldilocks zone works best, with fifty people being optimal for cooperation.[53]

For campcrafting, start small and build up with a channel capacity cap in mind. You don't want things to get too big. Too many people means too little face-to-face experience, which weakens ties and destabilizes your camp. In an experiment, researchers determined the extent to which the benefits of cooperation had to exceed costs to predict when people would cooperate.[54] The study showed that cooperation is harder in larger groups, and so the payoff had to be significantly bigger; that is, the cost/benefit ratio of cooperation must exceed the number of friends by a rate of two-to-one.

For example, social groups became cohesive units if the benefits were twice the cost when a person had two partners, four times the cost when an individual had four partners, with the same pattern increasing as partnerships got larger. To ground this with an example, consider if you have only one good friend and you both make approximately the same modest salary. Before you roomed together, you both lived on the other side of town in two rundown, small, cramped one-bedroom apartments. You decide to move in together—to cooperate—and so now you've got a decent three-bedroom in a downtown condo. Things are good, and your friendship grows and attracts a few other like minds. The work you put into these relationships is significant, but now the five of you decide to go in on a million-dollar house together in an amazing neighborhood, and the bank is happy to take your mortgage because there are four people on the line. Even more ambitiously, your honor group starts making real ties within your community, and some of you are having kids and starting families. With the increased need in size, you get together with another couple groups and decide to take on the project of building a cohousing community together. The key being, at each subsequent stage, for there to be success, the benefits must outweigh the cost of investing significant stores of energy into building relationships in a cohesive network.

Social fluidity and *transitivity* are other important factors that drive group cooperation. Social fluidity can be described as the rate with which ties may be rewired. That is, how much autonomy does an individual have within a group to rearrange their connections? Transitivity is a social network measure that quantifies the chance that a person's friends in the network are also friends with each other. Cooperation depends on the types of rules that govern the formation of friendship ties in a network.

The way social networks connect is important to their cohesiveness. In an experiment with over a thousand people arranged over ninety groups, researchers assessed social fluidity drive on cooperation. There was a Goldilocks zone discovered for fluidity, where rigidity in social ties disincentivized people to cooperate because they couldn't escape bad actors and so shut down their cooperativeness.[55] On the other end of the spectrum, too much fluidity reduced people's cooperation because there was never enough social glue to bind them together. A revolving door of social partners does not lead to reciprocating groups. The Goldilocks zone was 0.7 (where 0 was never rewiring ties, and 1.0 was always rewiring ties), and so a system where people can vote with their feet seems to be optimal to group stability. For transitivity, in both real life and online, this property appears to be relatively consistent at 0.25.[56] Meaning there is a 25 percent chance that a person's friend is also friends with other people in the network.

In sum, what this means in practical terms for your camp is that there should be a modest degree of high-quality ties among one another in the camp. Fortunately, not *everybody* has to be friends with one another; in fact, this is simply unrealistic. But there should be enough transitivity to glue together the honor group. Moreover, there should be some type of built-in social fluidity release valve or escape route for people to access in case things need to get mixed up and rewired for a time. Real-life examples of this can be seen in the Hadza, who benefit from their band-level structure of distributed camps. They often camp in one place for a season and then break down their huts to hang out with another camp during another season. When they are "feeling it" and have found some stability, they have the autonomy to stay at a camp for several seasons.[57] How does this apply to your camp? Perhaps you all go in on a rural cabin that can serve as a vacation spot for individuals in your camp, or even more ambitiously, you are a band-level community of camps where each community has a visitor center where people can come and stay for a time. The possibilities are only restricted by your innovation, your creativity, and the enthographic hyperspace.

Hierarchy is another important axis in the morphospace of prosocial groups. This is not the type of hierarchy we typically think of in the animal world—the kind where individuals dominate others with brute force and coercion. It is important to strictly define this within the context of campcrafting, which is essentially the distribution of wealth within a group and how that drives or inhibits cooperation. In a study that explored this issue, 1,462 subjects were randomly assigned to eighty different groups where real money was given to each individual. Different environments were engineered by the researchers, with some groups having total equality of wealth and others being played in either a mildly unequal or hyper unequal distribution of wealth. By mediating how much money subjects were given at the onset, they were either classified as rich or poor. The researchers measured how the levels of prosociality within groups changed depending on the distribution of wealth.

One might predict that unequal wealth distribution would be corrosive and decrease cooperation, but fascinatingly, they discovered this was not the case. However, one measure in particular was quite corrosive to cohesion; *visual displays of wealth* were highly subversive.[58] In the groups where money was on display, subjects were half as likely to cooperate. This point is corroborated by analyses looking at real-life intentional communities that intentionally try to conceal status differences to foster greater level of prosociality.

Finally, a foundational anthropological concept, discussed briefly in chapter three, is that of residency. Typically, there is a sex bias to residence. Social anthropologists call a male bias to residence patrilocal or virilocal. The inverse is a female bias, similarly called matrilocal or uxorilocal. Of all the world's cultures, approximately 70 percent are virilocal and hence male biased.[59] In many postindustrial countries, including the United

States, the predominant pattern of residence is *neolocal*, where a nuclear family moves away and forms their own residential core. Likely, the end result of what type of camp, and hence whether this is a male or female bias in residency, will be shaped by the strength of the social bonds of the original honor group. If the strongest bonds in the honor group are female, and the pair-bonds and mates are latticed into the network, then it will likely end up being matrilocal; whereas the opposite is true if the initial strong ties are male. Independent of sex, the beginning core relationships that have the strongest ties in the honor group will serve as the foundation and likely the residential gravity of the camp.

THE HARVARD PSYCHOLOGIST B. F. Skinner, giving a nod to Henry Thoreau's move to Walden Pond in 1848, published *Walden Two,* a novel describing a fictional rural community of a thousand people that lived according to his own theory of *behaviorism*. The theory, in vogue during this era, deemphasizes people's own feelings and motivations for behavior and lessens the role of their genes in contributing to their behavior. He thought any social arrangement was possible and that if set up in the right way, it would "run itself."[60] Although initially a flop, by the 1970s *Walden Two* was selling 250,000 copies annually and spurred a wave of intentional community building that resulted in dozens of fascinating experiments highlighted in *Communes in America: 1975–2000* by Timothy Miller.[61] Almost all of them failed, insomuch as they countered and fell outside the parameters of the Social Suite, but several still persist today. Here I will juxtapose two ultimately successful experiments—*Twin Oaks* and *Los Horcones*—to illustrate the camp-crafting parameters that we have so far explored. The former barely survived and experienced a challenging and troubling early era, while the latter grew slowly and steadily to be a working example of a thriving intentional community.

In 1967, with copies of *Walden Two* in hand, eight Twin Oaks founding members perused the pages within to find some guiding structure for the intentional community they wanted to build. The group adopted a communal childcare system, a labor point system, and used Skinnerian positive feedback to attempt to instill cohesion. But nothing seemed to work as they planned. One example being the childcare system, which failed because it went against the Social Suite parameter of *love* toward kin—parents simply didn't want to be away from their children (a running point of contention with the Israeli kibbutz communal childcare system that also collapsed).[62] From Hadza camps to urban jungles, the pair-bonding drive for parents to be in close proximity with offspring is too embedded in our genes and thus the Social Suite. After the setback, the community experimented with several childcare programs, eventually adopting the system where parents spend more time with their own children and make the decision on their own whether to send them to public school or to homeschool. Moreover, people wanted autonomy over their lives and hated the planner-manager system of governance described in *Walden Two*.

The decisions were made from the top down, yet people wanted to *take part in the decision-making process*. As one member of Twin Oaks recounted: "I don't know of any decisions I would have wanted to be different. The decisions were fine, I just wanted to be part of it."[63] This also struck counter to a relatively egalitarian hierarchical structure, which needs some level of democratic balance. With such a rocky start in the Twin Oaks early years, about a quarter of the population moved away, and even though they recruited new arrivals, the continual reshuffling moved social fluidity too high and transitivity too low, causing even greater levels of instability within the community.

Another member recounted the emotional trauma of the lack of social cohesion: "The disruption of friendships this almost always entails is emotionally wrenching for those left behind. . . . Departures tend also to be demoralizing and to call community belief systems into question."[64] Size seemed to vary a lot through time, and with boom and bust cycles (that could be attributed to more lax admissions standards during bust cycles) that today results in the community having approximately one hundred people.

Los Horcones (meaning "the pillars" in Spanish) was founded in 1973 in Mexico. The founders were an honor group of four people who were already married with strong social ties and optimum transitivity (that remain stable until this day). The members didn't have a dogmatic view of Skinner's hypothetical community, but built their community around the science on which the novel is based.[65] The form of communal childcare they enacted was cooperative and revenue building—as the community founded a school that was open to outsiders in the adjacent city. Property was shared and the governance system was what they ended up terming "personocracy," which aimed to inspire, by way of positive reinforcement, participation of all members in decision making. The community's size is optimal as it today has thirty members (interestingly, the same size as your average hunter-gatherer camp!) with a charismatic and benevolent prestige-oriented leadership. Finally, membership is not easy to attain, as the community grows through procreation and the layering on of friends and family, ensuring the social ties remain strong within the community. This begs the question then—What are your options when it comes to building a Twin Oaks or Los Horcones of your very own?

9

CAMPCRAFTING (PART 2)

An (Architectural) Guide to Forging a Twenty-First-Century Camp

Create a community, offering friendships with people who have similar values to yours, people to exchange ideas with, people to commiserate with, people to succeed and fail with, and people to fast and feast with. It's been crucial to our survival for eons. And no matter what comes next, we need a tribe to face it with us.

–Sasha Sagan, 2019

As long as people dream of a better world, some of them will try to create it.

–Timothy Miller, 2019

Perimeters, Houses, Huts, and Hearths

When I lived with the Hadza, I always knew when I was within the confines and safety of the camp and when I was venturing outside of its protective perimeter into the world of the savanna. I knew because of the properties of sound—once I was out of earshot, I was on my own.

If you were to walk out of your house right now and scream at the top of your lungs, how far would the sound of your voice travel? Math can give us the answer with some level of precision. Let's assume the following: it's a clear day with little wind and therefore minimal dampening effect. Let's also assume an average, low-octave male voice. Under these conditions, a scream would emit approximately 90 dB. Note that for every doubling of distance from the place where the scream occurs, the sound drops by 6 dB. Also, let's assume the ear that hears it is healthy and can hear sounds that approach 0 dB—about the sound emitted from a falling leaf. Therefore, our man, standing outside his home, screaming at 90 dB would be audible for 32,764 feet, which is almost 10 kilometers or 6.2 miles.[1] But, our dynamic, ever-changing environments are variable in both density and pressure, so sound usually gets absorbed and is only audible for several hundred feet. With these factors corrected for, the range of this scream could cascade through waves for anywhere between 100 to 300 meters, but on good days up to a single kilometer. Let's land then, with this back-of-a-napkin calculation, on about half a mile.

Now, go to Google Maps and type in your home address and zoom out to about the scale of five to ten miles. Print out the map and, using a compass, draw a circle with a mile diameter, with your home as its center. For some cultures in time, like the ancient Chinese walled villages called *hakka*, this was a literal perimeter. For us, this need not be so literal, but it does serve as a solid conceptual perimeter of your camp. Who lives in it? Who would hear your alarm if you were in peril? Would it fall on the deaf ears of neighbors who don't even know your name, or would it be one of your most trusted friends? What makes an honor group a camp is whether its members dwell within or outside this parameter. If within, your honor group is also a camp. Thus, a camp is where you can live intentionally with your honor group.

Your camp is the first step toward *intentional community*. With intentional community, we are emphasizing the importance of *colocation*. Cohousing is defined as housing comprising of individual apartments or homes with shared spaces and facilities designed to create a community, oriented toward collaboration among residents.[2] Putting it all together, our working definition of a camp is a colocation (approximately a mile in diameter from the centermost house) that is composed of a house or cohouses that dwell in an intentional social network. In other words, if you can walk there, you're close enough to circumvent social drift. We are taking Jacqueline Olds and Richard Schwartz at their word when they say: "Someday technology may indeed make physical proximity in living arrangements irrelevant. But it would be a terrible mistake to start to live as though that day has actually arrived."[3]

Timothy Miller—who surveyed the North American history of

communal living, starting with the first intentional communities that migrated from Europe—states that for a community to be intentional, it needs these four attributes: (i) shared purpose, (ii) colocation, (iii) a pooled resource, and (iv) a minimum of five adults.[4] A few of these attributes—the people and the purpose—we've already discussed, so now it's time to focus on the logistics and architecture of campcrafting. The ways in which a camp can come to fruition are only limited by your imagination, as long as they attend to the parameters of the Social Suite. As Miller wrote on the nature of the communal impulse: "I am . . . reluctant to predict just what the future communal world will look like. . . . Although the basics remain the same—people living together on the basis of a common purpose, practicing some level of economic sharing—the details are elastic." The following is an exploration of how we stretch the "elastic" details around in a twenty-first-century-friendly way that synergizes with our Paleolithic impulse for intentional community.

What is a *house*?

A simple definition is: a building intended to provide shelter for human habitation, typically holding people and their possessions.[5] Both the interior and exterior of a house can be altered through decoration in a way that an otherwise empty space is transformed into a residential space. Some societies may value the privacy of the house as an intimate space for nuclear families, with distinct rooms each serving a purpose, and a strong locking door to separate the house's interior from the outside world.

Now consider what it means to feel *at home*. In other words, when does a *house* become a *home*? According to the anthropologist Fran Barone, "Without the family a home is 'only a house.'"[6] Houses, then, are transformed into homes by embedding them into the social matrix of our lives:

> *Home* can encompass the built structure of a dwelling; an extended community; a safe haven to defend; and/or a place for intimate family moments. It can be settled, or unsettled; but it is most often in a constant state of being made and remade. Emotionally, home pulls

on the heartstrings and its potent memory can follow us wherever we travel. The symbols of food or landscape may trigger homesickness, even though those things can have little to do with an actual house or any bounded territory that can be circled on a map.[7]

Because of these social connections, home is the place where we sense and experience true belonging and is the source of both joy and pain throughout our lives. It can provide protection, comfort, challenge, and even serve as a symbol of aspiration—as many strive to build and maintain a good home for themselves and their families. Intergenerational homes with roots can serve as a symbol of social relationships that last generations. Over time this process unfolds and incorporates many voices, locations, and even negotiations or contestations of how the space within a home can be used. At its core, home is the physical manifestation of the biological and chosen family of your honor group in the form of a camp.

We have defined the perimeter of our camp. But what are the building blocks of a camp? Let's explore residential behavioral architecture to find our bearings. In the book *Life at Home in the Twenty-First Century,* a team of researchers tracked thirty-two middle-class Los Angeles families around their homes to see how people actually live within their house. Primetime living was tracked based on "the location of each parent and child on the first floor of the house every ten minutes over two weekday afternoons and evenings."[8] The houses were typical in size, if not a bit smaller than the average new home in the United States—which in 2013 was 2,662 square feet—and radically more space* than prehistoric dwellings.[9] Notably, the distribution of activity was heavily dominated by the kitchen. A key property of a camp is the *hearth*—which is where food is prepared and cooked. For many societies, especially nomadic ones, hearths can be either inside or outside. For postindustrial societ-

* Comparatively, in 1950, the same number was 983 sq. ft., with an average of an extra occupant in those homes. Archaeologists have done the calculations on the difference of the average total floor area of residential dwellings in prehistory. On average, in small-scale sedentary societies, a single human requires 65 sq. ft, whereas nomadic mobile communities averaged about 10 sq. ft. That certainly makes my living in downtown Toronto in 700 sq. ft. a relatively palatial experience.

ies though, like the one where houses like this are common, hearths are typically indoors. We can say people are *cohabitating* if they are sharing a hearth within a house.

Traditionally, houses crafted in postindustrial societies are comprised of several rooms with differing functions. But for many human societies, the room and the house are identical. For our purposes, we will define a *hut* as a room where either an individual, pair-bond, or mates and children reside. Cohabitation—also known in the intentional community lingo as *coliving*—is a form of condensed colocation where multiple individuals, pair-bonds, mates, and offspring share a house and hearth. Hearths are important because they serve as a hub of activity, and for many cohabitation scenarios the cohesion of the group and how much social pressure it experiences correlates with the number of hearths, relative to the number of social units within a camp.

Imagine, for example, seven roommates sharing a single stove and the conflicts of time, maintenance, and access. Now imagine how that pressure would shift if there were two stoves. With this in mind, if you are considering a cohabitation scenario with your camp, you can easily calculate the *hearth-to-hut* (HTH) ratio to capture and predict this core pressure. Mathematically, it is the proportion of the number of hearths (cooking, stove, and food refrigeration areas) to the number of active rooms within the house. For example, if you had a single two-story house, with two hearths and five active rooms, the ratio would be two to five and the HTH would equal 0.4. As the HTH gets closer to one, the less pressure there will be around this fundamental human activity.

For camp members just getting to know each other in a living context and shared space, it is recommended to keep the HTH as close to one as possible at first, until you learn to understand each other's personalities (especially conscientiousness and agreeableness) and iron out governance strategies that will keep the camp cohesive. For people who haven't lived with roommates often, it may be wise to test out their compatibility of cohabitating in the same space with a low-stakes living situation. You may have a very strong honor group, but going into a situation where you collectively take out a loan on a house without having any previous experience living together may be a recipe for disaster. Before sinking tons of resources to this end, there are many ways to get a feel for living with someone; for example, traveling and sharing rooms together for a prolonged period of time, or simply renting a space together as a trial run before purchasing a mortgage together. This is especially important for those who are more introverted or who were raised as only children. Sharing space is a muscle that many of us in our mismatched societies haven't exercised, so just like you would approach tweaking your running form or working on your mechanics for practicing a new lift in the weight room, take it slow and steady before trying the full range of motion with peak power in execution.

THE 1960S WITNESSED AN INCREDIBLE surge in communal experimentation. In many ways, it was an inevitable collision between four centuries of the North American communal experience with a massive change in American society at large. The 1960s was a period of great social upheaval where long-accepted social institutions were challenged to their core. Religious practice, food consumption, and even sexual norms were taken to task. The back-to-the-land communal movements, which were the roots of today's ascendancy of organic foods, began in this period. Intentional communities whose purpose was to liberate their members from the institution of traditional heterosexual marriage were formed and saw the coming of groups—such as LGBTQ+, celibate, polyamorous, and serially monogamous communities—that originated to challenge orthodox sexual politics. Thousands of religious communes with incredible breadth were established, with settlements that adhered to New Age theologies, Asian philosophical traditions, paganism, and radical Christianity all simultaneously running their experiments.

This is a fascinating period for several reasons, yet for our purpose it was also a natural selection of intentional community models that by the mid-1970s was winding down. Of the tens of thousands of these experiments, most failed, with only a few—those that best adhered to the parameters of the Social Suite—remaining intact. The universal thread that weaves among all these efforts is the "joining with others to live together in pursuit of a high purpose . . . and a dedicated minority of human beings at any time or place [that] still seeks to make a difference—together, because many are stronger than one."[10] From this critical data set, we can see a pattern emerge that guides successful modern communal efforts in a twenty-first-century world.

While massive upheaval was being experienced in North American society, quietly, four thousand miles away, one such pattern was being conceived and applied by a Danish architect named Jan Gudmand-Høyer. He and his wife, disaffected with urban dwelling, began a series of discussions with their best friends. The core themes should be familiar, as they dissected the feeling of isolated mismatch they experienced in day-to-day living in the city. They talked about the lack of community that most contemporary housing, with single-family houses and apartment buildings, offered their tenants. The group correctly identified the trade-off between the isolation of modern life and the need for community to be balanced with private life. After mulling over the dilemma, they realized the essence of what they wanted—to craft a type of rural village, but one that had access to urban facilities, cultural resources, and jobs.[11] They sprung into action, purchased a plot of land not far from Copenhagen, only to be challenged by neighbors' objections to their seemingly radical idea, which ultimately ended in failure and sale of the property.

Some were disillusioned, but a few persisted, and two groups working together purchased sites of land in 1968 without conflicting neighbors. Challenges remained. The construction bids came in higher than expected and they had issues with financing, but eventually, after a series of adaptations, their hard work paid off. By 1972, four years after the renewal of their second attempt, both groups ended up bringing their modern concept of intentional community into reality. The *Danish Cohousing* movement was born. By the early 1980s, new communities kept springing up, and laws within the country were enacted to make financing for cohousing simpler, easier, and more affordable, such as the Cooperative Housing Association Law.[12] Elsewhere in Europe, the Dutch in the Netherlands were quick to follow. Their model tended to have smaller units and incorporate more sharing, with a smaller HTH ratio, with fewer hearths serving more than one hut. They also tended to design for flexibility; for example, it would not be uncommon for a bedroom-hut to be placed between two houses in such a way that it could be part of one or the other, such that as a family needs change, so, too, do the assignments of huts. Since these early adopters began running their experiment, the cohousing model has prospered throughout Europe, yet Denmark today is and remains the world leader in cooperative housing. Perhaps, in some part, this is why Denmark consistently ranks as the happiest country on Earth. Denmark is known for its common goods such as social equality and universal health care and postsecondary education, yet it also has the greatest proportion of its population living in cohousing communities of any other modern industrialized nation.

A modern example of one such community was highlighted in the documentary *Happy* by Academy Award–nominated director Roko Belic. Twenty families live together in a multihousing complex that sprawls on a plot of land. In one striking scene, the children (six girls and two boys) of the collective homestead are interviewed to get their perspective on their living arrangement: "It's like a big family," a statement that is followed by laughter from the group. "It's nice to have grown-ups who are always looking out for us. If I've hurt myself down in the hall someone always comes running, and it doesn't necessarily need to be my mom." Upon turning fourteen, the adolescents cook for the entire community at least once a week. Chores are divided equally. One woman with several children moved in after her divorce and had, up to that point, lived there for twelve years. She remarked: "I like the elderly people that are living here because they are like grandparents for the children. . . . I feel like it is a gift for me and my children to be living here . . . it is like a miracle." In a North American context, adolescence, especially for young girls, has typically been characterized by anxiety and stress. When I look at the group of young Danish girls sitting together with their friends, ensconced in a multigenerational community of prosocial people, I cannot help but notice a kind of confidence effusing from them. I could only wish that my daughters, if I am ever blessed enough to have them, would be instilled with such confidence.

It didn't take long for the United States to notice what was happening on the other side of the pond. It also helped that two of the architects of the original Denmark group, the wife-and-husband team Kathryn McCamant and Charles Durrett, came over in the 1980s to hold workshops describing their experience that culminated in a book: *Cohousing: A Contemporary Approach to Housing Ourselves.*[13] By the 1990s, the first American cohousing project was established in California and called Muir Commons. Today, the Danish model inspires the most successful and trendy intentional communities that exist in North America, with over two hundred cohousing clusters and many more in development. The details of how the communities come together are varied and, as we will discuss, so are the challenges. All residents know each other and cooperate in various ways day to day but retain private premises, and common activities are frequent but not compulsory.

American cohousing developments, on average, are characterized by two dozen units over several acres of land; typically, car parking is assigned to the periphery of the development, and the property has a pedestrian-only common space, often with playground facilities for children. Many communities leave a portion of their land forested and underdeveloped. Critically, a universal and defining commonality is a centrally located common house structure that serves a variety of purposes for all members. In essence, the huts remain private abodes for individuals or families, with their own hearths, but their home is a constellation that revolves around the communal house structure. Communal houses usually have a kitchen hearth and common dining, which is used weekly, and for some communities up to five times a week. Often, shared childcare is centralized around this structure, with recreational amenities for both children and adults. Many cohousers also share laundry facilities and some even have a guest suite, which removes the pressure from each individual house to have a vacant room simply so visitors can stay a few days a year.

For individuals, there is a tension between the desire for autonomy and belonging; the difference for many is the relative proportions they seek—with the more introverted needing greater autonomy and the more extroverted seeking social environments with a greater sense of belonging. For some, cohousing in which people have a private living space but a range of shared common spaces has become an appealing option. Studies have shown one of the appeals of cohousing is that if properly executed, it can provide both a sense of connection without being controlled, and conflict-free exits.

The cohousing model creates camps with a sharing structure reminiscent of the hunter-gatherer central place provisioning discussed in Part I of the book; it strikes an ideal balance between privacy of dwelling spaces and public spaces that facilitate community. Surveys show there is extremely low turnover rate, with less than 4 percent of people leaving communities over time and very high rates of satisfaction. Lydia Ferrante-Roseberry summarized her experience as such: "My housing now reflects my values more than at any other time in my life." This seems particularly to be the case for multigenerational and elderly residents, as 97 percent of the seniors that have been surveyed would recommend it to other seniors.[14] Moreover, it's not just for rural settings, as successful communities have retrofitted large commercial buildings in urban areas with a revitalizing effect to the neighborhoods they invest in. Thus, for those with the most campcrafting ambition, sinking your teeth into a cohousing project could be an option worth exploring. To this end, many resources exist to help nascent groups of people start the process.* Comparatively, this campcrafting strategy is one of the most successful models and serves as a kind of gold standard of intentional community in the modern world. But it's not without its challenges.

Communities that pull this off typically start from a place of privilege. It takes a starting place of extremely committed people, and good leadership with some measure of financial stability and a temperament for years of planning and overcoming roadblocks. The average cohousing development from conception to founding takes four years.[15] It requires incessant communication with your fellow campcrafters to answer the questions: *Where will the community be located? What should our design be—what will it look like in the end? How will we fund the project? Who will lead meetings? Who will manage renovation or construction of the space? Are there any kind of environmental goals, such as renewable energy, low carbon emissions, or efforts for organically grown foods? In practice, what is our balance of privacy and sharing? Do we share meals;*

---

* Two principal organizations that have spearheaded the North American cohousing movement are (1) www.cohousing.org and (2) www.ic.org, which provide directories of existing communities, online courses, classifieds, and ways to connect with other communities.

if so, how many? Do we have some way to facilitate shared childcare? Should we subsidize those in our community with lower incomes? These are incredibly complex questions to answer. Other challenges include:

- Assuaging the fears of neighbors of an unfamiliar communal development
- Adhering to codes of construction in your zoning area
- Funding initial construction and land purchasing costs, which can be high, and securing cohousing project–friendly sources of financing, which can also be hard to find
- Rallying behind a strong ideal or identity-defining purpose that need not be religion-based moral innovations

For many, these challenges can seem too complex and overwhelming. That's OK. The principles are relatively simple and can be adapted to fit the differing needs of many honor groups on multiple scales on the city to country (urban-rural) gradient. Moreover, if you don't yet have a fully developed honor group but want to jump ahead to living within an intentional community, organizations such as CohoUs and the Foundation for Intentional Community can facilitate finding such communities that are under way or already in existence. This could also serve as another option for your honor group; instead of building a camp from scratch, you could leverage the hard work of like-minded others to join a broader cohousing community.

For those of you who want to take smaller, measured steps toward campcrafting, let's inspect some basic, simple models to help navigate the complex landscape of modern living. With the aforementioned definitions of house and home in mind, let's explore some potential camp models that could serve as a springboard for discussion among your fellow campers. One key factor to keep in mind while considering camp location is the *urban-rural gradient.* Urban centers are financial and cultural hubs. The trade-off to dwelling in cities has always been one of cost of living and access to high-paying jobs with that of the space and privacy afforded by rural dwelling. Wherever you are, the closer to city centers, the greater the cost but also the greater the institutional and cultural access for your camp. Therefore, let's take a look at four (non-exhaustive) templates from which to build your camp: *Urban Colocation* (city-neighbors), *Urban Cohabitation* (city-roommates), *Rural Colocation* (country-neighbors), and *Rural Cohabitation* (country roommates). At the core of each camp model is an inherent trade-off between cohabitation (living within the same dwelling) and colocation (living within the same local community). The first order of business you'll want to discuss with your camp is how you'll want to negotiate and balance between the assets and liabilities of each to best fit the needs of your camp.

The Vertical Neighborhood—An Urban Colocation Model

Eric Tauro (whose story is featured in an article by the Brick Underground, "What Co-living Is Like: An Insider Describes His Life in an Adult Dorm") is a twenty-eight-year-old architect who grew up in Los Angeles and moved to New York City after finishing graduate school. After some time deliberating his living options, he came across a coliving venture* in Long Island City. He lives there with a roommate (a friend that agreed to undertake the venture with him) and enjoys a short commute to his job at an engineering firm in Midtown. Tauro gets a lot for his monthly flat fee: a furnished apartment that comes with high-speed Wi-Fi, premium cable, and weekly cleaning. Additional perks include the use of the building's gym, pool, lounges, and rooftop deck. But what surprised him the most was the social community programming: "A lot of the events they have are events that are people doing activities together. We have multiple cooking classes. Those are really fun to do. You go with a friend and you cook together. You cook together and eat together. . . . [We do] beer tasting, we've had holiday events, they had an art show. They're having a holiday market in our building. We go to shows, too."

In fact, the social aspect of the living environment almost immediately insulated him from the typical isolation an individual can find themselves in when moving to an entirely new city: "I was able to meet a lot of people from my building. It's basically given me these new friends all of a sudden and it's made things so much easier to live in the city

* Alta+ by Ollie. One of a dozen or so companies investing and building coliving facilities throughout New York.

by having different people you can share experiences with and attend events with. . . . I feel like we're like a family already."[16] Tauro is a single man who, as a transplant, was looking for community in an urban landscape. Now imagine if he already had an honor group prepped to move to New York together in a similar colocation scenario.

Even without a company facilitating communal programing and spaces, most urban condos are equipped with shared spaces such as gyms, recreation rooms, and large cooking and dining spaces for groups. For example, San Jose is building the largest cohousing building in the world under the theme of "vertical neighborhoods." Similar ventures are popping up throughout the economically developed world, from Silicon Valley (i.e., WeLive) to London, England (i.e., The Collective). Yet, honor groups need not live in the same condo, as a camp, by our loose definition, is within walking distance—several city blocks could fall within these parameters. An advantage of urban colocation as a model is the greater HTH ratio, with each member possessing their own hearth, thus affording more privacy. An obvious disadvantage is the total space, relative to cost, in possession of the camp. In Toronto, for example, the price per square foot (at the time of the writing) is a whopping $1,000, and the average apartment is about 850 sq. ft. The sum acreage of a camp of four huts may approximate 0.06 acres. What is lost in space is arguably gained in access to high-paying professional jobs and cultural resources. If camps are strategically located near city parks, they can be ideal spots to hold camp meetings and bring children together to play for a little alloparenting and group bonding.

Mount Dennis—An Urban Cohabitation Model

Rebecca and Matt Daley are a married couple in their early thirties, with a newborn named Sullivan. Matt works as an assistant controller, and Rebecca, once she returns from maternity leave, will continue working as a registered nurse. Emily and Adam Truax are a married couple with a sixteen-month-old son, Lennon, and four-year-old daughter, Blakely. Adam is a church pastor and Emily the co-owner of True Pickles, a gourmet pickle company. Both families live in Toronto, Ontario, in a neighborhood called Mount Dennis, under the same roof. Having met at the same church, the friends started having monthly board game dinner parties. Shortly after the COVID-19 pandemic hit, with the news of a baby on the way, the Daleys moved into a two-bed plus den apartment for $2,250 a month, located in the same building as the Truaxes. In effect, the families were now collocated. This was an unexpected blessing, because the lockdown was hitting Toronto with severe restrictions, leading to acute social isolation for many.

Advantaged by colocation, the families could still hang out together. A topic they continually found themselves discussing was the fact that they yearned to live in the city, but neither could afford a place big enough to accommodate their growing families. This led Matt and Adam to speculate about what the possibilities could be if they actually pooled together resources. Several months and viewings later, they sealed the deal on a city home for $900,000, and now they are learning to live together:

> Adam, Emily, Blakely, and Lennon sleep in the bedrooms on the second floor and use the kitchen on the main floor. Blakely gets her own room, while Lennon stays with his parents. Adam uses the mudroom off the kitchen as an office and painting studio. Matt, Rebecca, and Sullivan sleep in a room on the top floor and use the second-floor kitchen. Matt uses the loft space next to the bedroom for an office. A student from George Brown who Adam knew from the church ended up moving into the basement, paying $1,100 a month. The families put that money toward the mortgage. . . . They all sit down for dinner at around 6 P.M. every day, bringing their own food. Blakely and Lennon keep Sullivan entertained, which makes things a bit easier on all the parents. Each couple helps out with babysitting. Rebecca and Matt took care of all the kids upstairs when Emily and Adam celebrated their anniversary in January, allowing them to have dinner alone. While there are some issues with noise traveling through the house, especially when Matt is working or the baby is down for a nap, the families are adapting.[17]

The Truax and Daley families are, on a small scale, engaged in an urban cohabitating intentional community. They also clearly illustrate the cohabitation-colocation trade-off. They can exploit all the wonderful things about city life because they are pooling together resources, but with the direct trade-off of reduced personal privacy. Rebecca admits on her blog how having an introverted personality makes coliving challenging at times: "As an introvert, I recharge by being alone; I like the quiet; I'm observant and self-aware (or at least I try to be). So, it's safe to say that a full house does not exactly fit that profile. . . . I strive for balance . . . [but] achieving balance, how and when I want, is not always possible anymore." Fortunately, they purchased a single house that also has multiple levels, with two kitchen-hearths (providing a better HTH ratio of 0.5) and outside space to retreat to when things get too overwhelming. It is also worthy of note that the families are bolstered by sharing religious tribal signaling, leveraging their evolved colation cognition that likely helps them weather the social challenges of coliving.

Bestie Row—A Rural Colocation Model

Google search "Bestie Row Texas" and you'll find a litany of stories celebrating a gang of lifelong friends that had settled in Austin for nearly twenty years and decided to build a little slice of intentional living overlooking the Llano River near Castell, Texas. Fred Zipp and his wife, Jodi, driven by conservation and environmental issues facing the region, joined three other couples who were inspired by the little house movement that champions smaller living spaces. Together, they partnered with Austin-based architect Matt Garcia for a campcrafting project that features small, energy-efficient cabins, each about 350 square feet and costing about $40,000 per house. They considered several alternatives in terms of designing their compound, including one large house, but they opted for each pair-bond to have their own personal home.

However, they were set on designing a communal space and Garcia obliged. The centerpiece of the compound is a 1,500-square-foot common building serving the purpose of binding the group together for social dining and entertainment. It's got a commercial-size oven and oversized refrigerator primed for dinner parties. According to one couple, Jodi "makes a mean pork and hominy posole," and the space provides ample room for cook-a-thons. The small camp, although not living in Castell full time, has plans that project into the long term: "We're going to be gray-haired friends." They playfully call the compound the "Llano Exit Strategy." Whether they knew it or not, they were engaging in the creation of a small-scale rural intentional community. The entire enterprise was relatively affordable once cost sharing was priced in. Finally, in terms of group purpose, they benefit from sharing an environmental ideology that informs their communal enterprise, which further enhances their cohesion as a group. In many ways this model, with the small-sized huts and a single, shared hearth, feels like a modern hunter-gatherer group plopped itself on a river in Texas.

The Spanish Sisters—A Rural Cohabitation Model

Minuca, Rene, and Maria Angeles—three sisters—have had a dream of living together since they were children. "I always said on this piece of land we would build a house here in the future. . . . I don't know why, but we always had this dream and . . . we decided to build the dream." Their dream was architecturally noteworthy enough to be featured on the television series *The World's Most Extraordinary Homes.* The end product became a 4,582-square-foot conjoined tri-functional construct that provides practical (yet stylish) individual and private dwellings interwoven at a common shared point and reflects the family's interpersonal relationships.

They inherited the property from their father and took the opportunity to build a legacy home for the shared price of $444,000—only $143,000 per household—well below the median price (which in 2020 was $284,600) for a home in the United States. To protect each other's privacy, the tapering structures extend outward to the landscape in opposing directions and open views far out to the horizon so that the families' sight lines do not overlap. The beautiful, natural-looking structure is composed of pressure-treated pine planks covering the exterior façades with a contextual appropriateness among the grapevines and olive trees, unifying the shared huts. Each home is literally a secluded social island for the family that lives in it, with only a few walls that connect the three into an architectural star-like shape if looked upon from a bird's-eye view. Thus, despite sharing the same roof, they have a high HTH ratio at one, given each family has their own kitchen space. This helps give the feeling that each sister, their mate, and their children have their own space. Yet, the dream is truly a communal one. Speaking more to the motivation that got them to build their dream, one sister said: "We spent our summer holidays very close to here where our grandparents were . . . and we were in a big house with about twenty cousins, and we always said we wanted the same for our kids." The communal center—the social jewel of the intentional community—is a courtyard that is shared by everyone. Weather permitting, they spend evenings outside dining, reveling in each other's company.

IN SUMMARY, THESE FOUR MODELS are in no way meant to be exhaustive or restrictive; we discussed them with the aim of opening our minds to the possibilities afforded by an honor group that has the determination, patience, and grit to build a camp but doesn't know where to start. Any number of hybrid combinations of colocation and cohabitation are possible. From turning a rural McMansion into a cohabitation space, uniting a few families into a single property with both house and carriage house in a more urban location, or simply moving into the same neighborhood together, there is a lot of flexibility inherent to any campcrafting project.

It also helps knowing that you're not alone; the past several years have seen the technological facilitation of the rise of new living arrangements that bring chosen family together. On the website CoAbode, a single mother can find other single mothers interested in coliving. For young, big-city-dwelling singles, a real estate development company called Common operates cohousing communities in several urban landscapes. Common has teamed up with a company called Kin to provide the same service, only for young parents. For those who are about to retire and seek some measure of community in their twilight years, there are currently 165 multigenerational cohousing communities (with 145 currently in development) such as Silver Sage Village Cohousing popping up across the country.

What all these efforts have in common is the idea that each family has its own living quarters, but the facilities also have shared play spaces to help coordinate childcare services and family-oriented activities. At the end of the day, any group of people that lives with some measure of proximity and meets under the banner of some type of shared identity and purpose has the heart of what it means to be coliving in a camp.

But once you've made some measure of progress in terms of the physical location of your dwelling, how do you actually figure out how to live together? Can applied evolutionary anthropology help small-group governance?

Prosocial—How to Create Social Norms and Govern Your Camp

Now you belong to an honor group with a strong, shared identity united under a common "living purpose" and you are dwelling with intentional community. The final piece is governance. Of primary concern is, How do we get things done? Fortunately, governance comes pretty naturally to humans in small groups—and cooperation is a natural outcropping of a well-functioning group. But challenges arise, and it's helpful to use the latest science to overcome them. To this end, a comprehensive guide was written in 2019 to help fledgling groups of all sizes align interest, support cooperation, and achieve shared goals. A consequential compendium for group governance* was written by Paul Atkins, David Sloan Wilson, and Steven Hayes called *Prosocial: Using Evolutionary Science to Build Productive, Equitable, and Collaborative Groups.* In the words of clinical psychologist Richard Ryan, who has over forty years of experience working with groups:

> [The Prosocial program] support[s] . . . the autonomy, competence, and relatedness of both individuals and the groups with which they identify. . . . The capacity to be autonomous—to endorse one's own

* The best I have come across and has inspired much of the Part II of this book. My camp has three well-worn copies of *Prosocial.* A powerful group exercise, used often in my camp, and described in the book is the Prosocial Matrix, designed to help groups collectively define values and overcome challenges to enact those values in the real world.

> actions—is critical. Prosocial emphasizes the importance of autonomy in its core design . . . (fair and inclusive decision making) as well as the collective autonomy of groups in its core design . . . (authority to self-govern). It thus helps groups, both large and small, to find their own inner compasses, and then to chart their own destiny.[18]

The prosocial method is based on design principles originally conceived by Elinor Ostrom, who was awarded a Nobel Prize for challenging basic economic assumptions about how and why people cooperate. The core design principles are the following:

1. Shared Identity and Purpose
2. Equitable Distribution of Contributions and Benefits
3. Fair and Inclusive Decision Making
4. Monitoring Agreed Behaviors
5. Graduate Responding to Helpful and Unhelpful Behavior
6. Fast and Fair Conflict Resolution

WE'VE TAKEN MUCH OF THIS chapter to address the importance of shared identity and purpose; now let's move through each of the other core design principles. Next on the list is how to figure out who contributes and benefits from efforts among the group. To do this, one of the first things we have to do is figure out roles.

What are the roles and responsibilities of the individuals in your camp? Ever since our first hominin ancestors began dwelling together in camps, *roles were the linchpin of survival.* The very name "hunter-gatherer" is a nod to this ancient lifeway and the power of roles. Unlike solitary animals that are all-in-one calorie-crunching machines, the delineation of roles allowed the first humans to develop and master skills they could contribute to the group level. We discussed in chapter six the power of social connection creating human capital and social capital. Humans have been doing this for 1.8 million years, and it's one of the things that got us here. It also turns out we crave it. When we execute a role that is clearly defined for us, and we excel at accomplishing it, it increases overall group cohesion. Thus, in another positive feedback loop, it can be said that role acceptance increases cohesiveness and, in turn, cohesion increases role acceptance. Specifically, in the context of teamwork, research has demonstrated that role involvement and cohesion are strongly tethered to each other. Role clarity (how well defined the individual's role is) and role acceptance (how much someone enjoys identifying with the role) predicts role performance.* On all facets, role ambiguity decreases cohesion. Therefore, in the roles that are envisioned for the project of intentional community, groups will be more cohesive if the role is well defined and enjoyed by the person performing it.[19]

As a quick, but important, aside, let's talk sex and gender. By definition, males are lifeforms that produce small, mobile gametes we usually call sperm, whereas females are lifeforms that produce large, typically immobile gametes we usually call eggs.[20] This most simple of distinctions—what gametes an animal produces—can be seen in all non-asexually reproducing animal and plant life. It also correlates with neuroendocrine systems that shape brains, hormones, and bodies, and on this basis alone, sex correlates significantly with behavior.[21] Gender differs from sex as it is a social role that can be independent of biological sex.[22] For example, the human sexually dimorphic trait of height difference between males and females is a consequence of sexual selection, while the "gender difference" typically seen in the North American cultures for hair length, where women on average have longer hair, is not a factor of biological sex but predicted by gender identity.[23]

However, some societies have historically acknowledged and even honored people who fulfill a gender role that exists more in the middle of the continuum between the feminine and masculine, which commonly is referred to as a third gender. For example, the Ojibwe *ikwekaazo,* "men who choose to function as women," or *ininiikaazo,* "women who function as men" and the Hawaiian *māhū,* who occupy "a place in the middle" between male and female.[24] Contemporary Native Americans who fulfill these traditional roles in their communities may also participate in the modern two-spirit community, and some proxy

* Another Dungeons & Dragons plug, tabletop RPGs are nothing if not an exercise in team roles.

for third genders that occupy a space between the masculine-feminine polarity is observed the world over.[25*]

Anthropological musings aside, in the context of groups overcoming challenges, gender takes on a powerful function—to help the group adapt and survive.[26] The two primary genders' roles are *extreme risk caretaking* and *moral courage*. Statistically, as recorded by the Carnegie Hero Fund Commission, men perform the majority of bystander rescues—90 percent—with women, children, and the elderly being the most commonly rescued. This form of extreme risk–caretaking is supported physiologically by the male disposition for "impulsive sensation seeking" and coupled with physical strength. However, women are significantly more likely to display moral courage. Moral courage is acting heroically within one's own moral universe, regardless of whether or not it is publicly known. For example, women are more likely to donate organs to non-relatives, and more women than men risked their lives helping Jews during the Holocaust.[27] A "hero" can be defined as one who risks their life for non-kin facing mortal danger. Fascinatingly, the gender difference in how heroism is expressed is so vital to group survival that it is replicated in same-sex groups during life-or-death challenges.

One incredible example occurred in 1958 in the Springhill mine in Nova Scotia, Canada.[28] Springhill, one of the deepest coal mines in the world, experienced a collapse that instantly killed 74 of the 174 men in the mine. Two miles below, sitting in absolute darkness, 19 men were trapped. Canadian psychologists interviewed those who survived from this group and discovered two key roles spontaneously emerged that were critical to their survival: physical risk-taking leadership and psychological and moral leadership.[29] Right after the disaster occurred, men with keen physical abilities (that far exceeded their verbal ability) took actions that saved several lives. They explored the reaches of their pitch-black enclosure, found precious lifesaving liters of water, and with their lack of empathy and characteristic emotional

* The *hijras* of India and Pakistan are often referenced as third gender. Another example may be the *muxe* found in the state of Oaxaca, in southern Mexico, and the Bugis people of Sulawesi, Indonesia.

control, they pushed the other men to eat coal and drink their own urine. But once the initial moment of action passed, these men became withdrawn and silent, whereas leaders emerged exhibiting another kind of courage. With their intellects and emotional sensitivity to others' moods, they helped reassure men who were starting to give up hope. As described by Sebastian Junger:

> Without exception, men who were leaders during one period were almost completely inactive during the other; no one, it seemed, was suited to both roles. These two kinds of leaders more or less correspond to the male and female roles that emerge spontaneously in open society during catastrophes such as earthquakes or the Blitz. They reflect an ancient duality that is masked by the ease and safety of modern life but that becomes immediately apparent when disaster strikes. If women aren't present to provide the empathic leadership that every group needs, certain men will do it. If men aren't present to take immediate action in an emergency, women will step in. To some degree the sexes are interchangeable—meaning they can easily be substituted for one another—but gender roles aren't. Both are necessary for the healthy functioning of society, and those roles will always be filled regardless of whether both sexes are available to do it.[30]

The key point is that whether your honor group is all female, all male, transsexual (and/or a different gender), or some mix thereof, your group won't be able to escape the inherent need for roles. Gender, which is not a fixed thing, will likely play some important and dynamically fluid part in this task, no matter the sex or gender identity of the members of your camp. When empathic leaders are needed, certain people with those capacities will step up, and when decisive, physically dangerous, or risk-assessment problems crop up, others will show their aptitude to take on the challenge. The roles that you and your camp come up with will be varied, and likely goal and task dependent. It's also important to keep in mind that it's not necessarily the person that is to be held accountable, but the *role itself*. For example, if someone fails at a task, instead of berating them specifically, collectively highlight the importance of the role during camp meetings; in this way, the person is reminded of the importance of their contributing to the camp, while saving face. Focusing on the role removes the potential for self-defensive stances to be taken on how tasks are being accomplished.[31] Finally, "care labor" should be appreciated publicly and acknowledged. From organizing someone's birthday to consoling someone when they are down or angry—these roles are primarily from the "moral courage" dimension and can often be unsung. Each role is crucial to a prosocial group and should be equally celebrated!

IF A GROUP IS TO achieve any of its goals and create anything of value, its members need to efficiently and effectively make decisions. When it comes to the core design principle of fair and inclusive decision making, one of the golden rules is—*if it affects someone, then they need to be included in the decision*. Thus, a critical decision-making venue is the camp meeting. Fortunately, technological communication apps such as WhatsApp, Signal, and Threema make private group communication easier than ever. Most of these apps also have polling and voting options for decision making. These are great for day-to-day check-ins on tasks, but cannot replace a formal space in which to meet and discuss camp priorities and goals. To this end, biweekly or monthly* camp meetings can facilitate most important group decisions for goal setting and reflection on progress. A quarterly meeting is effective at establishing several higher-order, more strategic objectives that link the work of the entire team and push forward the long-term purpose of the group. Every six to twelve months, big strategic meetings can be used to update and reflect upon group purpose, which can ultimately guide the next iteration of goals to be set for the group.[32]

Effective groups are goal-setting groups. Studies have shown that the average performance of a group that sets goals is far greater than that of

* Our camp, with the benefit of cohabitation, has a weekly Sunday dinner, after which we hold a camp meeting.

groups that don't.[33] The principles of goal setting have been formalized by sports psychologists as the following:[34]

1. Set long-term goals first, then set short-term goals.
2. Develop strategies and establish clear paths to reach the long-term goals by setting several short-term goals.
3. Involve members in the group goal-setting process.*
4. Monitor progress and provide regular feedback concerning group goals; display goals and team statistics in highly visible locations such as group public spaces.
5. Provide public praise for group progress rather than or in combination with individual member incentives.
6. Foster a sense of group confidence and collective efficacy toward shared goals (some goals should be easy, to show we can do it, others should develop realistic expectations for outcomes that are challenging).

If campcrafters share a home that happens to be multilevel, one example of executing this strategy would be to visibly set goals on the staircase so that the camp not only sees what the goals are (thus elevating them to each person's attention) but also gets to celebrate achieving the goals by "climbing the staircase together." Critically, as goals are achieved you may notice that the role structure provided by groups can have a profound effect on the personality change of an individual; this will especially be the case as you develop group norms. Individuals will change their behavior, and hence their personalities, based on the ideas in their social environment that emit rewards (positive reinforcement) and punishments (negative reinforcement). The power of group cohesion, shared identity and purpose, and social roles is particularly important; when an individual gets positive feedback for accomplishing a task that aligns with the group's shared identity and purpose using his or her group role's special skill set, group cohesion increases.[35]

* We'll detail more of this process in the next section when we discuss optimal governance strategies for small groups. For those with a penchant for statistics, the difference was massive as goal groups outperformed non-goal groups by a full standard deviation!

One final consideration should be taken with respect to decision making that stems from the guiding principle of including people in decisions that affect them. First, empower those members with the urgency to take initiative to make and actualize proposals. Second, if members are willing to take on the authority, those individuals that possess the best information on the topic should be enacting the decisions. These tips can go a long way to helping with the procedural fairness of your camp in the decision-making process.

YOUR CAMP WILL CREATE A CULTURE.

Much of that culture will be based upon another core design principle of the prosocial process—monitoring of agreed behaviors. The term *monitoring* is used to describe observation with the intent to coordinate helpful behavior. Although monitoring has a Big Brother feel to it, it really just means that as a group we're being transparent with our behavior and taking part in a type of shared consciousness.

Recall, from Part I, that the "primal template of culture" is the *tight-loose continuum*. To reiterate, this is the primary mechanism by which cultures shape the norms that the people who live within them monitor, observe, adhere, obey, or break. Michele Gelfand describes it as follows: "Social norms are the glue that holds groups together; they give us our identity, and they help us coordinate in unprecedented ways. Yet cultures vary in the *strength* [emphasis mine] of their social glue."[36] The general statement to be made about tight versus loose cultures is that the former have little tolerance for deviance, whereas the latter have weaker, more permissive social stances. Your camp will *create its own culture*—and understanding how this mechanism can be applied to strengthen your camp will be an essential tool in your campcrafting kit. The reason why this is so important is because you are creating a camp culture with the same tight-loose logic that explains differences across nations and states and also explains conflict or cooperation in the classroom, boardroom, bedroom, and the negotiation table of your camp. Using this knowledge wisely will do much to reduce conflict and increase cooperation.

For example, in 1998, Daimler-Benz and Chrysler Corporation underwent a highly heralded merger to the joint title of DaimlerChrysler.

The American Chrysler would finally crack into the European market, and the German Mercedes could sell more affordable cars; it appeared to be a match made in fiscal heaven—until the tight-loose cultures of the two chaotically collided. The Germans were a tight culture, keeping hands out of pockets during professional interactions and only using formal titles; whereas the loose Americans would rap out first names casually and treated meetings like unstructured spitballing sessions, much to the Germans' chagrin. Daimler was top-down, heavily managed, undergirded by rigid bureaucracy; Chrysler was looser from an operational standpoint, with a more egalitarian business culture and less red tape. After nine years of dysfunctionality between the two, their stock plummeted, and the pair divorced in 2007.

How, then, do we keep our own camp from suffering a divorce? First, each camp member should take the twenty-question *Tight or Loose Mindset Quiz* (www.michelegelfand.com/tl-quiz) that is used by Gelfand's lab and adapted for their research. This will allow you to position your own mindset on the tight-loose continuum. Knowing how tight or loose you are can help you understand yourself and others better to make a stronger camp. The value you will get will have you fall on a 1–100 scale, where you can be classified into one of four categories: Very loose (1–25), Moderately loose (26–50), Moderately tight (51–75), and Very tight (76–100). A loose mindset is typically less attentive to social norms, more impulsive, more comfortable with disorder and ambiguity, and therefore more prone to risk taking; a loose mindset also leans into change and welcomes novel situations. A tight mindset, in contrast, exhibits strong impulse control and a preference for structure and order, often created by routinizing their behavior; tight mindsets also are keenly sensitive to signs of disorder and thus pay acute attention to evidence of transgression of social norms. The average and the range of your group scores, which should be considered together, will tell you where your tight-loose score is as a group. To be clear, neither is a weakness as both mindsets are adaptive traits.

While the average can indicate which category describes your group, a wide range may portend conflict over tight-loose norm consideration. For your group though, we're looking for a kind of Goldilocks zone of tight-loose ambidexterity. In other words, both excessive freedom and excessive constraint can damage the well-being of your camp.[37] For individuals who are on the polar extremes, careful consideration should be taken to ensure open and clear communication about how to establish and enforce norms. Especially important for cohabitating camps, this occurs by negotiating clear boundaries around tight domains, and clear contexts when looseness is desired. Domains can even be associated with spaces, where the more public the space, a living room or kitchen, the more structured tightness may be beneficial. Bedrooms, on the other hand, are places where individuals can behave in their most unguarded manner and most norms are relaxed.

Taking from the broader literature on tight-loose cultures, tightness is a lever you want to pull up when the camp is threatened. Cultures tighten when under duress, and this allows you to have the wherewithal to integrate more structured tightness into a loose camp when the going gets tough.* Figure out where a conflict is happening—whether over how food is stocked in the refrigerator, messy dishes, or the toys that are left out in the living room—and discuss ways to tighten that specific domain to reduce overall conflict. Yet, know that this is a trade-off, and if misapplied with too many domains of tightness and control, nobody will ever have any fun and loving moments of creativity that tight domains tend to repress.

THE GUIDING CONCEPT FOR THE monitoring of agreed behaviors is accountability. One obvious way to make behaviors transparent is by sharing their progress with the group; this is easily facilitated with project management tools such as Trello and Asana. A visual record, shared by all members, is a great way to implement effective monitoring. There are several benefits to monitoring, as people act more prosocially when they know people are watching. Monitoring is known as a reputation-protective behavior that enhances motivation and increases shared identity, improves coordination, and decreases cheating.

* Again, check out the link in the appendix to see examples of campcrafting tight versus loose negotiations. In my own camp, tight versus loose conflict became an existential threat that we overcame with a better understanding of this important mechanism.

However, monitoring can go wrong when it deviates from the purpose of the group and is done coercively. To ensure this doesn't happen, it's always good to frame monitoring in the light of support—not control.

Imagine someone saying they will take out the recycling bin at the end of the month, but several months go by without this task getting accomplished. Instead of direct confrontation for failure, a more powerful yet indirect tactic would be to go to the member and *ask if they need any help* with the task. This way, they are reminded of the importance of the task, but instead of feeling policed, they feel supported. Typically, that's enough to ensure they double down on the task, and if they do ask for your help in good faith, they likely needed it. Meetings where monitored behavior is discussed should feel like opportunities for mutual learning and improvement. A norm like this, over time, will reduce surveillance and increase transparency. Ultimately, monitoring is not about exerting power over others, but sharing information that aligns behavior with the group purpose and identity.

Which brings us to the final prosocial design principle—graduated responding to conflict. The first line of defense when conflict arises is negotiation. Gelfand, besides being an expert on culture norms, also specializes in negotiation research and training.* The key to negotiation, when conflict arises, is framing it as a collaborative, win-win process, and not a zero-sum, winner-take-all game.

- **Focus on interests, not positions.** Your position is something you have decided on, whereas your interests are what caused you to endorse that position. For example, say two camp members' positions were to attain a big piece of an orange. If they understand each other's underlying interests, they might realize that one wants the pulp to eat and the other the peel to make a cocktail. By understanding each of their underlying interests, instead of focusing on positions, they can better craft win-win agreements.
- **Prioritize your interests.** Negotiators can't have everything, but they can likely get something that's important to them. To this end,

* A fantastic set of evidence-based tips for effective negotiation can be found on her eponymous website.

examine issues simultaneously, not sequentially. Negotiators who focus on single issues, one at a time, do much worse in achieving high-quality agreements. It's better to propose package deals that involve multiple issues; this allows people to barter with high- and low-priority issues.

- **Always have a way for negotiators to "save face."** Think of ways that the agreement helps both people to "win" and save face in the eyes of their camp members. One way forward is to frame the negotiation outcome as something to be gained, not lost. Also, adding a little chance (like drawing straws, die rolls, or rock-paper-scissors) to low-level negotiations saves face for both parties, because neither participant is seen as responsible for the outcome.*

In practice, before and during camp meetings, making it clear what domain each member prioritizes the most and being willing to trade and barter the lower domains in exchange for a strict norm adherence to higher domains is crucial. This list is not exhaustive, but some domains in a cohabitation context could be: rules of the kitchen and stove, dish-cleaning policy, food storing and sharing policy, living room and TV use, video game playing, gym equipment storage, acceptable noise levels for time of day, tool use and storage, office space use for work and videoconferencing, etc. For one member, a dirty sink could mean very little but a well-organized tool bin could feel like a life-or-death issue. If each member gets a single tight domain to apply, then bartering should be productive. Likely, several domains will have overlap, and cooperation will simply emerge through discussion of each member's priorities. Moreover, if conflict over domains seems entrenched, it may be helpful for the group to reorient toward the north star of purpose. Ask yourselves, what outcome best aligns with the group purpose?

What if we have created our camp's norms, and members continue to

---

* Rock-paper-scissors is the favored low-level conflict mechanism in our camp. One debate spawned over whether or not to have a carpet under the dining room table. Three members were agnostic, but two had firm opinions. Instead of a protracted verbal battle, we collectively determined rock-paper-scissors would decide. The result being the carpet is under the dining room table, and everyone involved had no loss of face except for laughter at the absurdity of randomness in determining the ultimate resting place for the carpet.

break them? Ostrom discovered, in her research, one of the most effective ways to combat this is by the use of sanctions. Critically, sanctions work best when they are a *graded escalation* and consistently applied with other core design principles. Initial sanctions for rule violations, in the most effective groups, start out very light. Nobody wants to live in a group where the smallest violation results in strong public shaming or banishment. In broad strokes, there are three levels of sanctions:

1. **Light sanctions**: the primary political tool here is gossip. When people talk, sooner or later, it gets back to the person whom the gossip is about. Well-placed jokes among the group that highlight the broken norm in a tactful but playful way can be just as effective. Light sanctions give violators an easy way out to reform.
2. **Medium sanctions**: escalating to this level can result in addressing the norm violation in official group settings. Typically, this is most effective if a pattern of norm violation is observed and evidence can be presented to the violator. Again, framing this in prosocial ways is most effective, and asking the violator *why* they continue to break the norm is a good place to start.
3. **Heavy sanctions**: the heaviest sanctions involve loss of group roles, responsibilities, and benefits. The ultimate expression of this is losing membership to the group.

On the topic of losing membership in one's camp, it's helpful on the outset to calibrate realistic expectations. In the quest to build a greater intentional community, you may extend and push the boundaries of your honor group by bringing in individuals that may not ultimately be compatible with your group.* The challenges may fracture some ties and create new ones with social and environmental challenges. As noted by Christakis: "Friendships begin and end. But the overall social organization of the species stays the same. Some turnover in social ties within groups may even be necessary for the networks to endure. It's

* My blood brother and his pair-bond separated after almost a year of cohabitation within our first camp. The strain in the relationship largely stemmed from differing expectations of tightness and looseness in household spaces and domains.

like replacing planks in a boat. This is required to keep the boat seaworthy. But the plan of the boat, like the topology of the social network, remains fixed, even if all the individual boards are eventually replaced."[38] In other words, the structure of the social networks arising from all the social ties is a fundamental human feature. It is likely why the quest for intentional community is one of the oldest and most continued in our species' history; from the first camps to the cohousing communities of the twenty-first century, humans will never stop trying to live together. Especially as we deviate in more pronounced and unexpected ways from this ancestral pattern, we will always run the risk of falling to the precarious vulnerabilities of mismatch.

Finally, we can end the section on governance with the topic of leadership. First, leaders embody the sacred values of their group. The psychologist Howard Gardner, in his book *Leading Minds*, argues that the universal commonality of the twentieth century's greatest leaders was their ability to tell stories about group identity, and embody those stories in their lives: "They told stories—in so many words—about themselves and their groups, about where they were coming from and where they were headed, about what was to be feared, struggled against, and dreamed about. . . . Leaders . . . convey their stories by the kinds of lives they themselves lead and, through example, seek to inspire in their followers."[39] One well-known example is that of Gandhi, who took great care to embody the identity story of Indian independence. He wore a simple garment—a spun fabric loincloth—which was a powerful symbolic juxtaposition dramatizing his connection to India's poor. This spurred a movement where Indians boycotted the British-controlled fabric industry by spinning their own clothing. In fact, the spinning wheel became a tribal symbol of the independence movement Gandhi was fighting for.

Leadership in hunter-gatherers is not coercive.[40] Recall, foragers vote with their feet, and if they feel like they are being treated as a lesser member on a strict hierarchy, they will simply leave for another camp. Hunter-gatherer groups possess an egalitarian ethic, and cohesion is the prized characteristic of functioning groups. This is not to say these groups are not hierarchical—they are explicitly hierarchical, but this type of *mild hierarchy* leans toward prestige as the principal value sought after by individuals in leadership positions. Christopher Boehm, in his discussion in *Moral Origins* of how power is distributed among hunter-gatherers, observes:

> A wise individual may be accorded the status of temporary or permanent band leader. However, that person is expected to behave with humility, for the accepted leadership style permits nothing more assertive than carefully listening to everyone else's opinion and then gently helping to implement a consensus. . . . Such decisions may involve a band's next move or the group's taking action against a serious deviant, but such leaders by themselves cannot settle on an outcome; this is a decision for the entire group.[41]

In fact, these types of leaders are generally appointed as such because they have a consistent track record of generosity in distributing their own resources, such as when they forage for food or hunt animals and share the spoils in displays of largesse.

Remarkably—and counterintuitive to the leadership styles heralded in twentieth-century corporate America—the most effective leaders are ones that sacrifice the most. One example is the potlatch ceremony, held by the Northwest indigenous groups of the Tlingit, Tsimshian, Haida, and Chinook communities. A potlatch is characterized by a ceremony in which possessions are given away in acts of generosity to others in their community.[42] Those who give the most not only adhere to the principles of central place provisioning but also accrue social prestige that helps them gain political traction during moments when group leadership is needed. This sentiment is also described in the book *Leaders Eat Last* by Simon Sinek.[43] One of the most effective ways leaders gain trust and earn the respect and loyalty of members is by making the most sacrifices to the group. In effect, trust is earned, not given.

Moreover, effective leaders don't force, but simply guide. In General McChrystal's *Team of Teams*, he describes leadership as being like a gardener.[44] The temptation to try and control and micromanage each group decision leads to disaster. Instead, each move of the organization must give way to an approach as a gardener does, enabling healthy growth rather than directing. Another source of leadership wisdom can be derived from the Navy SEALs Jocko Willink and Leif Babin, who describe true leadership in their book *Extreme Ownership*.[45] The most important rule of effective leadership is not to avoid, but to seize accountability. If something goes wrong, as a leader, you don't want to blame team members but figure out what you can do to help correct the course of your group. And for aspiring leaders that truly want to help their groups succeed, take note of studies that show that followers seek three qualities in their leaders: availability, motivation, and enthusiasm.[46] The groups that have leaders with these qualities will have the best chance at achieving cohesiveness and success. Ultimately, effective leaders manage social identity by helping the group articulate "who we are." Think of the role of countless shamans in the ancestral hunter-gatherer tribes as simply guides who cultivated shared identity through storytelling. We literally survived, and thrived, as a species because some leaders of our past told good stories about what made *us* special.

Sodality Forging—How to Level Up Your Camp to a Society

A sodality, described in detail in chapter three, is one of the most tribal social tools innovated by our species. The core function of a sodality is to embed a human social network with an added layer of social insurance, just in case times get existentially tough. This is why the instinct, instilled in us by the Tribe Drive, to form larger, community-wide coalitions was a key part of the social architecture that kept our species sheltered, fed, and alive.

A society can exist without secrets. Not-for-profits, charities, and scientific and professional societies dot the modern landscape of our culture at large. But these are usually open, transparent, and lacking soul. They don't adhere to the Tribe Drive's binding principles, or leverage tribal signals in a way that make them compelling enough for people to commit to their identity. For that you need secrets, exclusivity, and a strong in-group mentality. All this can be accomplished within an ethical framework—a topic we'll discuss in the next chapter. Secrets function as nothing more than trust-building mechanisms by which honest signals of loyalty to the group are hard to fake. Secrets are currency of commitment—those that have given most to the group know its innermost secrets because they have demonstrated the prerequisite trust to attain them. These members are honestly tribal signaling their love and devotion to the group, and so they alone gain access to its inner sanctum. For those readers that find the prospects of building your own sodality daunting, but are still attracted to the idea, societies already exist the world over that may be a nice fit. Several orders are active in the United States* and have been homegrown or inspired and transplanted from all over the planet. Many have checkered histories of controversy, but many also have been a net good to society.

But for those with the penchant to build their own society, listen closely to this instinct. It may be worth the effort.† With respect to this drive, Sasha Sagan put it eloquently when she wrote: "I crave a community of women. This is something I might have had if I were religious, a perk of being active at temple or mosque or church. So many religious sects have women's groups, for prayer or study, fund raising or charity, or just to join together to do whatever it is that is expected of us. It's one of the hardest parts of being secular: you have to work to congregate."[47] That's why she decided to craft her own sodality.

So how do you go about making your own sodality? The instructions are easily articulated, but difficult in application because the primary ingredients in a sodality are the time and energy put into the system. The good news is that it tends to give back more than gets put in, with dividends. For an organic example, let's follow Sagan's lead as to how she built her sodality.[48] First, she created a ritual: a monthly dinner at a cocktail bar, centrally located among her members. Then she created an identity for her sodality: the Ladies Dining Society. She sent out an invitation for membership to eighteen of "the coolest women I knew from childhood, college, work, and life. Six showed up. Including me." But with the initial investment in energy in the sodality, it flourished:

---

* Among them, the Freemasons, the Benevolent and Protective Order of the Elks, the Independent Order of Odd Fellows—typically male, but some have sister societies for women as well.

† Recall, sodality is still within the community, face-to-face levels of human social organization at the *band* level. This is even closer to your social self than the tribal, symbolic level.

> More ladies came the next month. Some of them were my lifelong besties from Ithaca, some I had met only recently. Filmmakers, teachers, writers, therapists, designers, executives, actresses, waitresses, resource managers, and stay-at-home moms. Among them, they have been all over the planet and experienced so many ways to be a woman on Earth.

Whether she knew it or not, the Tribe Drive manifested in the instinct to create a gendered network of widely versed and experienced *Homo sapien* females within which to navigate their environment. The social insurance provided by such an experienced group is obvious, as this network will not only benefit Sasha, but with the right tribal signaling, the sodality could persist intergenerationally; as is the case with previous sodalities, the benefits can also follow to her daughters. The monthly ritual soon evolved into a real tradition. As time moved forward, deep friendships bloomed from the inaugural invitation. Sasha's—and subsequently the group's purpose—was set: "I had wanted to create a community," and she realized that she had that when members began coalescing in ways independent of her. In large part, this occurred because she assumed the mantle of leadership. She had a fastidious and dogmatic approach to ensuring that the events were attended and experienced in the same fashion throughout the budding enterprise. She had a spreadsheet to keep track of the events that sprawled, over time, across the years; she had a roster of records detailing the attendees. She made the events uniform and consistent, to give each enough of the power of ritual to act as an adhesive to the female sodality. After five years of steadfast efforts, the community missed only a handful of monthly meetings. Her dream of community became a reality.

This was made clearest when her husband, like so many other people being constantly displaced in the quest for career, was offered a job in Boston, which resulted in her family moving there from New York. Yet, despite her being transplanted, the sodality had strong enough bones to withstand the displacement of their founder. There is no better measure of effective leadership than a test of whether what they built can thrive in their absence. In fact, a new chapter had been born of members who had transplanted from the East to West Coast. Today, the Ladies Dining

Society has taken up chapters in places including Los Angeles, Oklahoma City, and New York. Many of the principles we have explored in Part II of this book, in forging friendships, pursuing pair-bonds, and crafting camps, all can work in principle to help you build your own sodality. Use them well in this special quest.

Just as Sasha Sagan craves a community of women, I crave a community of men. In fact, looking back, I see that my participation in my male sodality has served as the moral lodestar of my life. For us, for our brotherhood, it was truly organic and a natural outcropping of a handful of pubescent boys calling each other by special names, and instinctually identifying with and creating secrecy as a currency by which that nascent identity was cultivated and protected. It was during that critical time in development that the sacred value of being a part of a brotherhood was instilled in me. Time and time again, when life presented seemingly insurmountable challenges, my brotherhood—literally and figuratively—was there. Literally, in that as a poor kid from Indiana finding his way in life, there were moments when I had nothing but the clothes on my back; they took me in, sheltered me, and fed me. When my parents divorced, and the bonds of biology felt shaky and uncertain, they were there. In times of emotional turmoil, when pair-bonds dissolved and I felt spiritually broken, they were there. The relationships that were forged when I was twelve years old guided my life's future in-group interactions and they persist to this day—no better proof that the evolution of friendship is a monument to human behavior. If it weren't for them—and the immense life-affirming support they gave for decades—this book would not exist, because in many ways this book is about them. It is about our instinct as a species to transcend the tyranny of biology and walk the path of life in our own way. A forever trust in who we are, and the knowledge that nothing else matters.*

A Ritual Performed, a Prayer Answered

A little over a year later, the gravity in tone of Benjamin Maxmillion Charles's words echoes in my mind as I begin the ritual. My honor

* *ERS* ~ Voronwë, otornassë, oialë

group, now camp, is assembled around the lone tree in our front yard. The tree stands as a spiritual sentinel, guarding the entrance to our newly cohabitating camp. It is after a Sunday feast and one of our first camp meetings and—after years of planning, sacrificing, and strategizing over thousands of miles overcome by videoconference—we are assembled together. The preceding moments were ones of surprise and joy, as two of our mated members have just announced the news that our camp, within nine months or so, would be blessed with an additional member. They are pregnant. As cells divide in her womb, with a magic as ancient as life itself, our camp literally is growing. But now, the mood is somber and serious. We are in a state the Yup'ik would call *piliriqatigiingniq*—meaning to work together for a common cause. I feel like for the first time in my life I truly understand its meaning. The scholar David Bollier argues that it is better to treat the word *commons* as a verb instead of a noun: "Commoning" rather than "the commons." *Commoning* is the care the community feels for its people. We are all, in that moment, *communing.*

I am inspired from rituals I have studied and, as a field anthropologist, participated in. Elements borrowed here and there, from the Huichol, the Yup'ik, the Hadza, indigenous Scandinavian traditions, and Christian symbology, cocrafted with my fellow camp mates, are integrated to perform our first group ceremony honoring the new home that protects us. Each camp member holds in their hands a token. It serves as a symbol of something of themselves to be sacrificed to this new land, this new camp. Deep words are spoken. Sacred actions are made. Oaths are uttered. And as I light the candles of each beloved of my honor group, they kneel down by the tree and bury their sacrifice—a small, but significant piece of themselves—under the earth and connected to the roots that feed and nourish it. A final prayer is made and the candles remain there until their light wanes.

As the honor group filters inside to begin their own, personal nightly rituals before sleep, my pair-bond (now wife) and I meditate together by the candlelit tree. My mind's eye drifts through space and time but then settles upon one thought. I recall the internal prayer I made when Charles, of the lineage of bear shamans of his people, taught me the sacred value of the ritual coming-of-age ceremony. So much effort and work had passed since that moment, but I realized my prayer—by some great ancestral and universal power—had been answered. I will gift my daughters and sons the *communing.* I found people to channel strength for my offspring, so that they may know the way. And when they transform from a girl to a woman, or from a boy to a man (or whatever role they adopt to bolster the lives of the people they love), they will know and be protected, supported, and guided by its power. As we continue with our final explorations of the Tribe Drive, it is my hope that you and your honor group will learn this power too.

10

THE TRIBE VIRUS

I can be loyal at one and the same time to several identities–to my family, my village, my profession, my country, and also to my planet and the whole human species. It is true that sometimes different loyalties might collide, and then it is not easy to decide what to do.

–Yuval Noah Harari, 2018

Primary identity is precious. It's the identity that supersedes all others. To build anything great–a company, a currency, a civilization–an affiliation must beat out the rest of the identity stack to become someone's primary identity.

–Balaji Srinivasan, 2022

Cyrus the Great may have been the first geopolitical genius in the world; he was to ancient politics what Einstein was to physics. Cyrus's greatest theorem—one might say his "$E = mc^2$"—was the discovery of *the principle of super-tribalism. Super-tribalism* is the symbolic appeal to a higher-order identity than the fundamental tribal unit. As a leader he took an upstart tribe of no-name Persians and cultivated the greatest empire the world had ever seen.[1] This radical theory was so powerful, Alexander the Great literally *copied* the super-tribalism playbook in the subsequent defeat of Cyrus's descendants hundreds of years after his brilliant innovation. Under Cyrus's rule, the empire embraced all of the previous civilized states—with the myriad tribes therein*—of the ancient Near

* Just to give a sense of what an accomplishment this was, historian William Durant described the Near East during this epoch, which would have "seemed like an ocean in which vast swarms of human beings moved about in turmoil, forming and dissolving groups, enslaving and being enslaved, eating and being eaten, killing and

East. From there he exported his super-tribal brand to the vast regions of Western and Central Asia.[2]

Before him, all would-be conquerors used a single, ruthless technique: completely erase and eradicate the intersubjective belief networks—tribal identities—of the groups you subjugate. Take, for example, the previous Assyrian Empire. For a millennium, the Assyrian kings committed symbolic genocide of the peoples they militarily defeated. Any time they conquered a people, aristocratic lineages were extirpated in family homicides, temples were ransacked, and literal gods (the physical statues thereof) and any key cultural symbols were either erased, stolen, or carted back to provincial capitals. The trouble is, it's tough to erase identities. They are sticky. To eliminate an identity you have to either eliminate the people wholesale, or give them the choice to voluntarily enter into an alternative. Because tribal roots are symbolic, you can't coerce anybody to hold them. For an identity to truly take hold, it must be taken on of one's own volition. Identities with true staying power are adopted by way of collaborative inquiry with others, not the tip end of a sword.

This was Cyrus's genius insight. The Assyrian kings didn't have the super-tribal playbook, and the karmic result was the same horror to the Assyrian tribal polity they had given to others—near-complete annihilation. Statues with the faces of Assyrian kings were toppled. The names of the kings were effaced from murals. The royal lineage killed down to the last soul. Their fall from grace, in a public relations nightmare with lasting consequences even to today, was detailed in the Bible as God's will. Cyrus the Great, however, got lots of props in the Bible for restoring the Jewish temple in Jerusalem. Whatever the cost of rebuilding the temple was, its propaganda is still paying dividends. Cyrus wrote the playbook of how to—with the least bloodshed—most efficiently create and apply the principles of empire. To be clear, there was violence, but it

getting killed, endlessly. Behind and around the great empires—Egypt, Babylonia, Assyria and Persia—flowered this medley of half-nomad, half-settled tribes: Cimmerians, Cilicians, Cappadocians, Bithynians, Ashkanians, Mysians, Mæonians, Carians, Lycians, Pamphylians, Pisidians, Lycaonians, Philistines, Amorites, Canaanites, Edomites, Ammonites, Moabites and a hundred other peoples each of which felt itself the center of geography and history, and would have marveled at the ignorant prejudice of an historian who would reduce them to a paragraph."

was typically applied to the battlefield. War is a type of group-level force projection that demonstrates that one group's economic (read, military) power is keener than its competitors'. Once his power was projected, however, he was notoriously tolerant of the people who fell within his newly redrawn economic sphere of influence.

He gave full license for the people within his empire to savor their tribal identities. Gods, temples, and symbols remained, on the whole, intact. He even co-opted the tribal memes of the day by proclaiming that his coming was the will of the locals' gods. For example, he brought back the physical gods to the Babylonians (the previous king had absconded into a desert oasis with them) after the defeat of their military, and without a single civilian loss of life restored them to their place within the city. His arrival into the city, and many other places he conquered militarily, was heralded. In the words of modern cipher punks,* Cyrus controlled the memes of production! In his application of this novel political theorem, he saved countless lives. The Persian state then became the super-identity of the day by which all tribal units co-existed as long as they paid taxes and defended the super-tribe when called upon. By the standards of his day, his application of geopolitical strategy would be given a Nobel Peace Prize.

Tribes had been innovated three hundred thousand years beforehand, and a human being in the fourth century BC mastered the secrets of the Tribe Drive to become the first (in actuality) self-titled *King of Kings*. What we need to do now as a species, with the emergent technological tools at our disposal, is pull off a Cyrus the Great moment in the twenty-first century. Nothing less than the fate of humanity is at stake.

ONE OF THE AIMS OF this book has been to outline the trajectory of our species' evolutionary innovations to the Trust Paradox. Until the present era, there have been three solutions to the equation of trust, which can be expressed in levels of complexity: (i) *kinship level*: trust genetically similar organisms (kin); (ii) *friendship level:* trust people you can

* gm @punk6529.

rely on when you are in trouble; and (iii) *tribal level*: trust people who believe in the same symbolic stuff as you.

The evolution of tribalism, then, was a radical update for *Homo sapiens*—its primary purpose was to bootstrap trust among strangers. But embedded deep within it was a malicious virus. The *tribe virus* can be lethal to mental immune systems. Thanks to the budding science of cognitive immunology and scholars such as Andy Norman, who are actively working in the field, we now are coming to understand mind viruses as not just metaphors but real, *measurable* things that adversely affect human wellness.[3] The mind *actually* has an immune system that protects against parasitic, bad ideas, and certain types of thinking that are disruptors to healthy mental well-being. The Tribe Drive launched our species to unparalleled success; this ascendency has even been dubbed the "Social Conquest of Earth."[4] But even through this conquest and in the process of its spread, there has been embedded within it a tribal virus. The result is a global pandemic of tribalism that threatens the very existence of our species.

We need a vaccine for this tribal virus and a novel solution to the Trust Paradox.

And just like a biological virus, the key to creating a tribe vaccine is by finding weakened parts of the virus—tribal antigens, if you will—that trigger a mental immune response that strengthens the mind against its more pernicious forms. In these final chapters, we have been granted emergency authorization to undertake a tribe vaccine development. Creating vaccines under a time crunch is no small feat, but as we have seen in recent years, it is possible. Fortunately, there is hope.

The human scale that exists today would be completely incomprehensible to our Paleolithic ancestors who, long ago, dwelt only in small, close-knit camps numbering in the hundreds. Today, we have myriad tribes that range from the thousands to more than a billion in number! We've tried our best to adapt; however, the best we've come up with at this scale of humanity is more and more centralized sources of authority and power to help coerce and enforce cooperation among members of coalitions. The chains of centralization, with elite autocrats and tyrants commanding bureaucratic thralls that hold the keys to our shackles, has been a favored stratagem; around four billion humans currently live under authoritarian regimes.[5] The result has been a propaganda barrage of mind control leading to a devastating loss of life that has pitted tribe against tribe, prompting dozens of civil wars and two devastating world wars in the last century that laid waste to millions of human souls.

In our finer moments there have been blips of inspiration and collective action, where some groups have attempted to experiment with novel kinds of government and economic systems to improve the human condition. Patchworks of something resembling democracy, enmeshed with Enlightenment values and mixed with social and free-market economies, have miraculously improved the average human life—all while respecting the authority of people to dwell within the Social Suite.[6] But this ad hoc system of checks and balances appears to be under existential strain. The Tribe Drive, once a magnificent adaptation for its time, now seems antiquated and to have outlasted its initial inherent value. The benefits no longer outweigh the cost, as the tribe virus has too many hosts to multiply within, causing a great deal of human misery. In this crucial moment of human history, when *Homo sapiens* are on the precipice of becoming a multiplanet species, we need to keep some of the beneficial characteristics of the tribe while continuing to think about how we can move beyond it.

We need a *new* answer to the question "In whom do we put our trust?"

The Costly Expanding Circle

Everything I've said so far about how important our in-groups—our social selves—are to our well-being leads to an obvious ethical conundrum. A skeptic* can counter: "If everybody was to preference their group, then doesn't that inevitably lead groups to tribal conflict?" Isn't that what we in fact see today, with groups, states, countries, and societies nepotistically privileging only themselves?

* With an understanding of Immanuel Kant's Principle of Universalizability. This is one of Kant's categorical imperatives in which one should "act only in accordance with that maxim through which you can at the same time will that it become a universal law." In other words, if everyone does it, does it contradict itself? If contradictions emerge, then the behavior is immoral.

In 1981, the philosopher Peter Singer postulated a truly revolutionary concept into the philosophy of ethics called the *expanding circle*. Singer argued that the primary moral dictum for our species is the *principle of equal consideration*. This principle demands that we give equal weight to the interests of individuals, and that reason is enough to guide the human species to an "expanding circle" of morality. Specifically, he argues that the philosophical idea of "the principle of impartial consideration of interest" is the sole rationale for ethics. This concept hinges on the idea that one's own interests are one among many sets of interests and that any rational being—sentient enough to have interests themselves—can observe that the interests of any given individual are no more important than the similar interests of others. At its core, this is a consequentialist argument that holds that the consequences of one's conduct are the ultimate basis for any judgment about the rightness or wrongness of that conduct. Furthermore, Singer argues, if his principle is the ultimate basis for morality, then we are left with the task of furthering the interests of all sentient life—including non-human life. Thus, conflicts over differing moral ideals should be treated like any subjective preference, by assessing them impartially and, all things being equal, satisfying the most possible preferences.

Singer poignantly describes the problem as such: "Obviously there are actions one can defend in a manner that is acceptable within one's own society, but unacceptable to members of other societies. Tribal moralities often take exactly this form. Obligations are limited to members of the tribe; strangers have very limited rights or no rights at all. Killing a member of the tribe is wrong and will be punished, but killing a member of another tribe whose path you happen to cross is laudable." In other words, if you can't trust them, they are better off dead. Furthermore, the arc of history bears evidence of the expanding circle in action, like Cyrus the Great's ingenious expansion of the moral circle of concern to a super-tribal level. Another, grander extension occurred millennia later, with the inclusion of a "brotherhood of man" among the ideals alongside liberty and equality espoused by the leaders of the French Revolution, a message that was considered radical at the time.

It appears that there has indeed been an expansion in the circle of moral concern over time. The core question posed by Singer is "Why should our capacity to reason require anything more than *disinterestedness* within one's own group?" In his view, public thought should prohibit justifications that give the interest of my group more weight than the interests of other groups; the key to an ethical future is to limit or remove altogether the capacity to privilege your own group in any circumstance.

The postulate of the expanding circle, and the ideals it promotes, are laudable. Philosophically, I *want* to believe deductive reasoning is enough to compel our species to climb peaks on the moral landscape,* but as an evolutionary biologist, I believe it is time for an honest acknowledgment of the limitations of our species. The past forty years of evolutionary science that have followed the publication of Singer's thesis shed new light on our cognitive limitations and the litany of inbred biases—and intersubjective warped realities—we possess as a species. It is no use to begrudge these limitations, as every species has them. But do we actually have the capacity to rationally take broader points of view, from which our interests are no more valid than the interests of others? The science explored in the first part of this book details many of these limitations. If this were in fact possible, in our species' current form, the Trust Paradox could be solved by the neocortex alone; but we also have a rooted nervous system that has other, competing answers to moral concerns. The limitations square on a type of *evolved preciousness*—that is, *irreplaceability*—of the human beings that fall within the channel capacity of your sympathy group. Put simply, thinking impartially among all is energetically costly.

Morality Is Not an Ethical Crisis—It's an Energy Crisis

In practice, moral decision making can be wildly context dependent—and the strongest predictor of moral behavior is proximity itself; the closer in proximity the moral agents act, the more the behavior is moral.[7]

* To borrow a phrase from neuroscientist and philosopher Sam Harris.

In addition, the human capacity for perspective-taking is limited. Imagine a series of concentric circles. Dunbar's number is the furthest circle outward—the channel capacity stretched to its furthest possible extensions. It's this outer circle where the computational ability of the human mind fails at being able to calculate the intricacies of social networks.[8] Now imagine the circles closer to ourselves; the sympathy concentric circle is much smaller and subsequently more robust. Recall that the average size for most individuals is twelve people.[9] Now think of your honor group concentric circle, which is smaller still. Critically, the new science that has emerged since the publication of the expanding circle demonstrates how important *irreplaceability* is in human relationships. In fact, friends are evolved to be irreplaceable—that's what makes them real friends—and the science to support this is only now coming to full fruition.

As observed by Christakis considering the role of friendship in the evolution of morality: "Friendship is a fundamental category . . . yet scientists have tended to neglect the role friends have played in the life of our species. An extreme focus by our species on kinship and marriage has obscured the more numerous relationships people have with unrelated friends. These friends are also, after all, the primary members of the social groups we form and live within." The evolutionary psychology that underpins friendship is a kind of evolved preciousness.

Attachment theory demonstrates that evolution has programmed us to single out a few specific individuals in our lives and make them irreplaceable.[10] This is a type of bred dependence, without which our species would have never survived. In fact, this attachment penetrates our psyches so thoroughly that there are times we extend it beyond the confines of our species. For example, one study found that, if given the choice between saving the life of their dog or a stranger, 46 percent of women* would save their dog over a foreign tourist.[11] This does not appear to bode well for the expanding circle.

Human need can sometimes be desperate. Large institutions, in general, are notoriously horrible at providing for that kind of need at the most desperate of times. This phenomenon is called "the Banker's Par-

* Men chose this too, just not as much as women.

adox," the catch-22 of help and risk that helps explain the importance of irreplaceability of friends as a cornerstone to the evolution of human altruism.[12] Imagine you are in desperate need of a loan. Of course, the bank doesn't want to give you one because, in your current state of desperation, you are obviously a bad credit risk. But when you are in good shape financially and you don't really need a loan, then the bank is more than happy to give you one. Frustrating, isn't it? It is not a paradox for the banker, but for the person who is in desperate need. In a world governed by reciprocation alone, how can one get help when the chips are stacked against them? Friendship is the answer. In the lives of our Paleolithic ancestors, the need was likely a more serious conundrum than a bank loan—such as severe illness, injury, or privation—which created a selection pressure for a deeper relationship. These relationships centered on specific properties, like honor, courage, sacrifice, and loyalty. These sacred values among friends serve as a form of insurance against times of desperate need.

Thus, baked into the human concept of friendship is irreplaceability. When you are in desperate need and you are a bad "credit risk," your slow-to-act reciprocation partners (modern-day health insurance companies, or ineffective government agencies) will be better off abandoning you. This is true only to the extent that the benefits they can normally obtain from you are available from others; in other words, to the extent you are replaceable. By becoming irreplaceable to a small number of others, those tried-and-true friends now have an existential stake in your well-being and thriving, thus making it beneficial for them to help pull you through the times when you are in desperate need. Embedded within the friendship construct is the fact that you both have a *shared fate* and your collective fates are intertwined in a type *of mutual fate control*.[13] The fact that these individuals are among the few in the world with a stake in your well-being gives you a stake in theirs, leading to a continual deepening of the engagement that makes friends truly irreplaceable.

This dependence extends beyond kin in our species because sometimes families lack resources, skills, or connections that can be obtained by expanding a social network into the realm of non-kin.[14] You may be completely unremarkable to strangers, yet you are irreplaceable to your friends because they know you. Groups are small in conjunction with the channel capacity (the ability to get to know people well takes time) because they need to be small to be individually appreciated. This is why larger institutions and massive market economies make the individual feel alienated and inconsequential. The size, scale, and scope of these alienating institutions drive a large part of the maligned forces we discussed in chapter two on the topic of social mismatch. The reason why we suffer and feel naked without friends is because without them, and the irreplaceability of the relationships they represent, we are naked.

Not all humans can be friends. Humans have finite energy to distribute among social networks (this is the channel capacity we investigated in chapter four). Thus, to "love all equally" is *reductio ad absurdum*—a premise proved false because it stands in contradiction to itself. The logic is contradictory because loving all is to love none at all. Why is this so? One is compelled by positive injunction (a popular technique of centralized government propaganda) to consider everyone's needs equally—no one is supposed to be above anybody else—but this doesn't mean it's easy to enact.

Let's ground this with a thought experiment. Take a moment to imagine your most beloved active relationship. It could be someone in your sympathy or honor group. It could be your best friend, your lover, your mother and father, or son or daughter. Dwell on the image of your closest human companion. Consider the love and support they've given you over the years. The good and bad times you have shared and conquered together. Now, imagine you hear the news of this person's death. Do not focus on the circumstances of the passing, but mentally highlight the feeling evoked by the person's loss. Now take that feeling and amplify it 150,000 times. Why 150,000? That is the number of people that die each day—of war, famine, murder, disease, and random chance or old age.[15] Imagine feeling that amplified pain every day for the rest of your life. The result of such distress would likely end in a psychological catatonic state—you would be a shell of your former self, stewing in suffering and loss. In line with Kant's categorical imperative, we can even universalize this exercise. What would happen if everyone did this? Would not all of human civilization come to a screeching halt? The moral landscape would descend into a valley of despair.

This is the kind of cognitive cost you would need to incur if you were to literally adopt the precept of the expanding circle. Empathy is incredibly expensive. Several studies demonstrate that the cognitive "costs" of empathizing with someone distant are shown by increasing people's cognitive load (i.e., making their frontal cortex work harder by forcing it to override a habitual behavior); they become less helpful to strangers but not to family members.[16] "Empathy fatigue" can thus be viewed as a literal demonstration that the frontal cortex is challenged more and exhausted more readily when people are in a state of repeated exposure to out-group pain. This same cost, however, is not incurred among people considering their in-group. Human empathy, at times, appears feeble. Anyone with humanistic sensibility that "cares about you" truly does feel your pain, but to expand this same feeling to everyone is impossible because it's too finite and expensive a resource to spread itself across billions of strangers.

Selfing is the psychological phenomenon where the in-group becomes so self-identified that it is no longer psychologically separate from the self. But does selfing automatically lead to us seeing those who are outside the circle of our social self as less than human? In other words, does it explicitly lead to dehumanization? The psychologist James Coan's answer is informative:

> What you see as out-group dehumanization is usually a reframing of an in-group preference. When you pit groups against each other and then introduce a sense of danger, people *don't* [emphasis his] tend to devote resources to going after the out-group. *What they actually tend to do is to devote more resources to their in-group* [emphasis mine]. So they become more groupish, but it is more inwardly focused than outwardly focused.

To Coan there is a distinction between *dehumanization* and *deselfing*: it isn't that the person under threat is less human than you; they just don't belong to your identity in the same way that someone from your in-group does. In essence, when you forge an in-group, you are not actively denying people in out-groups their humanity. You are actively identifying a few select people to be given preference, as they are neurologically imprinted as less indistinguishable to yourself. The difference is subtle, but the implications are important. Ultimately, it takes hard work for humans to dehumanize. The default isn't to dehumanize the out-group, but to only see a select number of humans as we see ourselves.

This also passes Kant's categorical imperative of universalizing this ethic to the rest of the world. Before worrying about society, focus on the health of your group. If we lived in healthy groups, society would flourish. This is not a radical idea: only you have the power to ensure your own health and well-being. You can't be the best member of a society unless you are a healthy individual. Extend this logic to the social self and your in-group. Only your group has the power to ensure its own health and well-being. How can your group be a healthy part of a larger community if your group is unhealthy? Healthy groups are the foundation, and the unit of change, of healthy societies.

Empathy is a valuable and energy-expensive resource. If we apply it to everyone equally at all times, we are soon depleted and become inoperable. Thus, whether we recognize it or not, we all have what can be called a kind of moral gearshift.[17]

Joshua Greene uses the metaphor of a camera's dual modes in his book *Moral Tribes.*[18] The human brain is like a dual-mode camera with both *automatic* and *manual settings*. A camera's automatic setting is optimized for efficiency and can be used quickly with the click of a single button. The user of the manual mode can configure the aperture, exposure, and ISO by hand, but the trade-off is that it takes time and learned expertise. The automatic mode is great for considerations of "me and us" morality (that evolved over 1.8 million years). It is the more ancient limbic system working to enhance social cohesion. Whereas the manual mode (which requires massive recruitment of the prefrontal cortex) is much less efficient (more recently evolved) but more flexible (and utilitarian) when considering "Us versus Them" morality. In essence, the moral gearshift is the difference between "point and shoot" and "take it slow to line up your shot." Because empathy is cognitively expensive, Greene points out that it is a mistake to use the same type of fast-twitch, gut moral determinations when beyond face-to-face "me versus us" considerations of morality.

How do we know when to turn off the empathy valves that efficiently, quickly, and relatively reliably fuel our "me versus us" moral decisions? According to Greene, as soon as we see an issue is *controversial.* Typically, the controversy means that different moral tribes are using the norms generated by *their* face-to-face considerations beyond what they were designed for and at an intertribal level. This doesn't end well, which is why we need to shift to reason-morality mode. In other words, when you feel controversy, you are being challenged by the Trust Paradox.

The Concentric Circle—Ethical Energy Conservation

Given that empathy is a resource, it should be applied in maximum only to those within the moral circle of concern; above the level of the community it is neither plausible nor feasible for us to share true empathy with others. *Reason,* on the other hand, is to be applied to the area of the moral circle of concern that deals with numbers of individuals beyond the capability for the evolved human brain to process. These are our moral gearshifts—empathy and reason. The former you reserve for your innermost concentric circles and the latter for all others. In fact, we can make this an official tribal ethic and golden rule for group-to-group interaction. To this end, I offer a concise moral precept that is compatible with reason, Enlightenment values, *and* evolutionary biology. I call it the *Concentric Circle,* and it is a postulate that refracts the concept of Singer's "limited principle of equal consideration" through a tribal, evolutionary lens. The Concentric Circle is in essence: *limited equal consideration among groups.* Simply stated: *privilege your in-group unless to do so causes pain to an out-group.**

In-group and out-group need to be treated equally when the outcome of doing otherwise will ultimately be harmful to your in-group. Thus, this updated version of the limited principle of equal consideration of interest demands that we give equal weight to the interests of in-group and out-group only when the loss of human thriving is a

* When reviewing my original book proposal, Josh DeYoung commented and paraphrased the core essence of the principle of equal consideration among groups with the following: "Favor your boys, just don't be a dick about it."

consequence of interest to both groups. In other words, when comparing similar interests among groups results in the mutual reduction of well-being, the action is immoral; whereas the intuitive (i.e., evolved) and automatic preference for one's group is moral up until the point where actions taken on behalf of the group result in direct or indirect harm to the out-groups. In practice, the concentric circle needs a recognition of three postulates:

1. All humans have the capacity to feel pleasure or pain as a sensory state in the brain and should be included in an expanded circle of altruism.
2. The recognition of this expanded circle does not require all humans to be treated equally in all circumstances.
3. It is only in the instance when we directly compare similar interest—using the consequence of pain as the ultimate arbiter—where there is a demand of equal weight to the interests of all individuals.

Jack Donovan is the author of the alt-right-leaning cult hits *The Way of Men* and *Becoming a Barbarian*.[19] He is a modern tribalist who advocates a counternarrative against the institutions of globalist monoculture. He dreams of a future with meaning and purpose restored where men are "scattered across the Earth in a million virile, competing cultures, tribes, and gangs." His primary thesis is that civilization, or "The Empire of Nothing," weakens specifically men with universalist thinking: "This concern for the feelings of others is an intertribal perversion of his intratribal moral sensibility . . . the Empire isn't my tribe. The government isn't my people. . . . *Not my people, not my problem* [emphasis his]." If articulated by way of a moral precept, Donovan's would look something like: *privilege your in-group in all circumstances.*

I find Donovan's premise is both shortsighted and unethical. He speaks from a place of rare privilege in our species' history, living within the confines of a state that protects his property rights, has public goods, and mitigates intercommunity violence and the effects of natural disaster. Moreover, through innovation, the medical sciences buffer him and his loved ones against disease, and the military defends his borders from invasion. An effective counter to his tribalist ethos is John Rawls's *veil of ignorance*.[20] This is a method of determining the morality of issues by asking a decision maker to make a choice about a social or moral issue, all while assuming they have enough information to know the intended consequences. The "ignorance" part of the equation (the veil) refers to the fact that the persons on the receiving end of the decision would be unknown to the decision maker, so no bias could affect the decision itself. The theory contends that not knowing one's ultimate position in society would lead to the creation of a just system, as the decision maker would not want to make decisions that benefit a certain group at the expense of another, because the decision maker could theoretically end up in either group.

Let's make a minor tweak and call this thought experiment the *Tribal Veil of Ignorance*. Imagine the Donovanian world where only tribes exist, and no intervening super-tribal systems serve as an intermediary. This is a world (much akin to the post-agricultural ancient worlds of emerging cilizations) without nations and where tribes of 1,500 people compete with one another. Now imagine that you are randomly assigned a tribe at birth. The question is, Would a modern tribalist roll the dice for the chance to live within the makeup of a purely tribal world? Would Donovan take the chance of being randomly reassigned to a different tribe than the one he has the privilege of identifying with today? Maybe he would. Or would he prefer just to play at being a tribalist within the confines of a free society with institutions that foster his and his in-group's well-being? It's easy to write and proselytize about the masculine virtue of purely tribal life within the safety and comfort of the most privileged society of human beings to ever live.

In fact, some level of networking between groups raises the tide for all ships. As observed by Sarah Cavanagh, "We may not be able to escape our ultrasociality or our tendency to form in-groups, but we can train ourselves to be more sensitive to whom we invite into our circles. . . . Think of humanity as a giant hive of cooperating honeybees rather than an amalgam of warring clans and allow the reappraisal to include all human beings in your in-group. You can do this without relieving the solidarity you share with your in-groups. Being great at in-groups doesn't have to translate automatically into being hostile to out-groups."[21] I want to live in a world where I have both irreplaceable relationships in my local honor group, camp, and community *and* where those irreplaceable relationships also receive the benefits of a super-tribal society that has roads, universities, hospitals, and police as an added layer of insurance for my in-group.

The fact is that the benefits of living in societies with basic protective institutions outweigh the costs—even to the most tribalist among us. What are the chances that in a small group of bonded individuals there resides someone with any one of the specialized skill sets that improve our lot? For example, does your tribe have an oncologist on call when your loved one is diagnosed with cancer? Paying taxes directly redistributes resources from within your in-group to neighboring out-groups but returns the collective benefit of access to enriching resources that one lone tribe simply cannot facilitate. When your tribe lacks a needed resource due to changes in the environment or bad luck, are you willing to war with a neighboring community to gain that resource? In the game of society, it need not be zero-sum winner take all. If the society in which you dwell can both foster the freedom of the Social Suite on the community level and at the same time support space-faring institutions like NASA or SpaceX, then we can say it's a pretty lucky roll of the die in playing the Tribal Viel of Ignorance game. This is the potential power of adherence to the Concentric Circle.

There is no escape in cultural or ethical relativism, either. In other words, you can't say that one culture can treat their tribe as superior to another because *they deem it so, and who are we to say otherwise?* Cultural relativism is the idea that there are no moral absolutes or truths but that each culture can determine for itself, without value judgments being made by outside and different cultures. Can we see the flaw in this kind of thinking? Value, including moral value, does not depend on one's own opinion and a relatively arbitrary set of preferences—it can be reasoned out. Sam Harris posits in his book *The Moral Landscape* the following argument: The medical sciences build on an explicitly nonarbitrary premise of value—to be healthy is better than to be sick. The same applies for secular ethics: mutual and shared well-being is better than mutual and shared suffering. The greater the shared well-being, the greater the moral peak a society experiences. Inversely, the greater the shared misery a society experiences, the closer it is to a valley on the moral landscape.[22] Cultural relativism has been a major source of error in thinking for thousands of years and was even identified as such by ancient thinkers like Plato.[23] It is a mind parasite that erodes the cognitive immune system and needs to be eradicated from its human hosts.[24]

Extreme tribalist philosophy succumbs to the naturalistic fallacy—just because in-group bias is natural does not mean it is good in all circumstances. The potential for those who adhere to such philosophies for mutual reduction of well-being—and the massive distribution for pain among every participant for both in-group and out-group—is perilously costly. The Concentric Circle protects us from this while keeping in place the emotional primacy of your sympathy group. It does so by acknowledging that the theory behind the expanding circle is sound, but only for a species without energy capacity limits on empathy. Perhaps the biotechnological beings we could *become* may have less energy constraints and thus a greater capacity to expand such moral circles of concern. Until that day though, understanding that tribalism is an energy crisis and not a moral crisis is step one to prolonging the lifespan of our fledgling species to the point where we evolve—or technologically innovate—cognitions with greater energy efficiency.

By adhering to the Concentric Circle, one can privilege one's in-group while not undercutting other out-groups. With this in mind, the Concentric Circle may also solve the religion problem. Popular anti-religionist thinkers such as Richard Dawkins, Sam Harris, Daniel Dennett, and Christopher Hitchens—together known as "the four horsemen"—have brought atheism to the mainstream. The general line of attack is that religions corrupt cultures by making them more prone to bloody, winner-take-all, intergroup conflicts over their respective mythical, imagined orders. The rationale, then, is "Why not just get rid of religion?" I'm generally sympathetic toward this reasoning; having grown up in a fundamentalist Christian cult, I've witnessed firsthand the corrosive potential of religion to shatter human wellness. In reality, isn't the religion problem actually a *tribalist problem*? Religion is a tool within the suite of the Tribe Drives arsenal. Yet, tools aren't inherently bad or good. Can we not envision a religion that adheres to the Concentric Circle's moral precept, where individuals cherish the symbols that bind their community together and stay mindful to never let these symbols stray in function from tools that bind people together to tools that cut people apart? An atheist may counter: How then does one go about ensuring their religion remains inwardly binding, and not outwardly harming? The final chapter of the book will explore this idea further.

The philosopher Philippa Foot uses the metaphor of a plant to help verify what is good.[25] That which is good helps the plant function and thrive—an environment that is rich in minerals, water, and sunlight. The same is true for people dwelling within a socially rich perimeter of irreplaceable relationships that help them thrive. It is certainly possible to arrange multiple plants in such a way that each plant gets what it needs to thrive while not posing an existential threat to other plants. It is even better to imagine an interconnected ecosystem where mutually dependent thriving occurs by way of nutrient exchanges. Thus, the primary philosophical—and original—contribution of the Concentric Circle is a way forward for our species to embrace its evolutionary heritage, while remaining within a moral framework. We humans can be potted like plants in a way that ensures we all thrive together. We have done this before, and in many ways it's our default setting.

Are you a tribalist or an individual with tribalist tendencies? The truth is, many of us are without even knowing it. Tribalist methods are employed by both the extreme right (for example, racism) and the extreme left (for example, doxxing and canceling free speech) as weapons in the culture wars. This is the sneakiness of the Tribe Drive.

It begs the question: Have *you* been infected with the tribe virus? Let's consider the problem of racism from the tribalist perspective. Racism is a simple, nasty subcategory of tribalism. If you are a racist, you are a tribalist who overinflates the importance of skin color as a signal of coalitionary alliance. The greater sin is being a tribalist. Being a tribalist means that you will break the Concentric Circle and "favor your in-group in all circumstances." Skin color is only one of hundreds of tribalist signals people use to assign group membership. Note that today when people use the term *racist* they often mean *tribalist.* The stakes for being able to measure our own tribalist thinking are high because if we can solve tribalism, we can also solve racism. Consider this next section a rapid antigen test to determine whether virus levels in your mind are at critical thresholds. It also provides some cognitive immune inoculants to help in case the virus risks compromising your mental health.

How Tribal Are You? Measuring Our Tribe Drive

In *The Power of Us,* Jay Von Bavel and Dominic Packer argue that "understanding how identity works provides a special type of wisdom: the ability to see, make sense of, and (sometimes) resist the social forces that influence you. It also gives you the tools to influence the groups you belong to. Among other things, you can learn how to provide effective leadership, avoid groupthink, promote cooperation, and fight discrimination." Learning this wisdom is the point of assessing your own "my side" bias—a core bug of the Tribe Drive. "My side" bias, first described in 1989, was discovered by asking subjects to list thoughts that occur to them when they consider controversial issues. This initial research uncovered that people can be easily prompted for additional arguments on the other side, although prompting for further arguments on their favored side is less effective; in essence, this means that the failure to think of counter-arguments is not that you don't know the arguments, but that you don't identify with the groups who hold those arguments as values.[26] This lack of identification leads to blind spots in our thinking.

In the words of Steven Pinker: "The discovery that political Tribalism is the most insidious form of irrationality today is still fresh and mostly unknown."[27] That's the way blind spots work; you don't know they exist until they are brought to your attention. The fact that people remain blissfully unaware of their own tribalism is a product of the fact that our understanding of these cognitive biases are recent scientific discoveries. That was the lone solace I had when I came face-to-face with my own compromised cognitive immune system. Only recently, I miserably failed all three of the diagnostic tests I am about to share with you. Numerous have been my tribal sins.

My confession needs to go a bit further. As the son of a former fundamentalist minister, I witnessed firsthand how religion affected the minds of those I cared about the most. Although many of the functions of growing up in the Worldwide Church of God were either benign or outright prosocial, some were dangerous. Although my father and mother were not hard-liners as a young adolescent forming my nascent identity, I still came to resent the legalistic and regimented life that I, by chance, had been born into. Because of this experience, as I matured, I traded one tribe for another: I became deeply, politically tribal. In direct violation of the precept of the Concentric Circle, I was willing to cause pain to an out-group to benefit my in-group. Worst of all, I was driven to this end because of the moral superiority I felt was the foundation of my cause. That's how the tribe virus works—in a kind of nasty biomimicry—you feel you are something you are not. The reality was that I was ethically compromised because my cognitive immune system had been overtaken with a mental immune disorder.

As so often happens, converts to a new tribe can be the most ardent members. In stark contrast to the *Conservative Christian Coalition* that I had known from my early life, I was now a warrior for what Amy Chua, legal scholar and author of *Political Tribes,* calls the "American elites":

> American elites often like to think of themselves as the exact opposite of tribal, as "citizens of the world" who celebrate universal humanity and embrace global, cosmopolitan values. But what these elites don't see is how tribal their cosmopolitanism is in reality. For well-educated, well-traveled Americans, cosmopolitanism is its own highly exclusionary clan with clear out-group members and bogeymen—in this case, the flag-waving bumpkins. . . . There is nothing more tribal than disdain for the provincial, the plebeian, the patriotic.[28]

In retrospect, I can see how it all happened. Having left my home in Indiana, and striking out on my own into academia, I was socially isolated and vulnerable to the need to signal tribal allegiance to something, *anything* that could give me the subliminal solace of group membership. I became a secular warrior. A progressive proselytizer. I signaled my virtues constantly. I trained every day. Read voraciously. Debated vigorously. Learned how to effectively wield my tribal weapons. I was becoming an evolution missionary to convert the religious to atheism and a liberal democrat to channel my secularism into activism. As Pinker notes: "Professing a belief in evolution is not a gift of scientific literacy, but an affirmation of loyalty to a liberal secular subculture as opposed to a conservative religious one." Tapping into, perhaps, the skill set of rhetorical ability I inherited from my father, I was enmeshing myself in this tribal ideology, in hopes of securing my place of membership among its ranks and giving meaning and purpose to my life.

As my education furthered, I became well versed in evolutionary theory but also the other talking points that correlate with membership in this tribe, and as a result I was becoming a near-perfect liberal secular stereotype. There had been a critical loophole in my academic training of "critical thinking," and my newfound ideology was exploiting it. Now I know psychologists have a term for this: "motivated reasoning."[29] As a motivated reasoner, I could "think critically" about views I disliked, but unconsciously and selectively not apply the same critique to my own comparable views. Some people who think they are trained in critical thinking don't become more fair-minded; they just become more skilled propagandists. In this sense, some are merely ideologues who are simply highly motivated and leverage their critical faculties as instruments to lobby for some preconceived agenda.[30] The worst part of being told *you* have been given the skill of critical analysis is that one's education, instead of enhancing the cognitive immune system, can actually disrupt it.

Many academics use critical thinking as a sledgehammer against ideological opponents. They accuse others of not thinking critically when what they really mean is that their opponents simply do not agree with them.

I learned the above later in life, but as a young man, I was utterly oblivious to the fact that sometimes education doesn't make people less tribal, but simply more equipped to rationalize and justify their tribal behavior. I was unconscious to the ways in which my Tribe Drive was coercing my behavior. In retrospect, my most acute regrets were the family relationships I strained and friendships I lost, not based on an individual's character, but on the difference of tribal ideologies. Fortunately, there is a way out of the ideological quagmire. I have applied it to my own life and have substantially changed it for the better.

SOCIAL PSYCHOLOGISTS ASK A NUMBER of questions to figure out one's group identity. The answers to these probing questions give critical clues as to the fundamental nature of who you are.

There are four principles of identity that are important predictors of people's responses to their own and other groups. First, groups of belonging define our sense of selves. Second, common experiences and shared characteristics (even if totally arbitrary) have a remarkable power to produce collective solidarity. Third, priming a particular social identity has powerful influences over one's motivation and behavior. Fourth, an active identity—that is, an identity salient in the moment a behavior is made—will determine the norms to be adhered to that are associated with that identity, even at great personal sacrifice. These four principles of identity are universal and the product of human evolution. We are governed, cajoled, and influenced by them in ways subtle and profound. Now, let's figure out how to avoid being manipulated by them.

Tribal Diagnostic 1: The Moral Mirror Game (or, the Moral Equivalency Test)

We'll define political tribalism as the phenomenon where loyalty to the political tribe is more important than loyalty to other aspects of your identity. That is, if your identity as a member of a political party is weighted greater, and as a result supersedes, other identities. This correlates with a suite of cursed human behaviors. It predicts tribe members going to any lengths to defend their tribal leaders—be it Donald Trump or Joe Biden—from any criticisms or wrongdoings.

One self-diagnostic tool is to play the Moral Mirror Game (aka, Moral Equivalency Test). Take any recent accusation of wrongdoing leveled at a political leader and imagine two instances—one where the leader of the "other" party is being accused and a second where the leader of "your" party is being accused. Let's use a simple example of tax evasion. Imagine Trump being accused of tax evasion. Not too far of a stretch. Now, reverse it with the Moral Mirror Game, and imagine Biden being accused of tax evasion under the same circumstances. Does the narrative you play in your mind change at all? Is there a shifting in moral equivalency? Do you go from wanting to pillory one to viscerally defending the other? If so, again, you fail the moral mirror game.

Now, to get a more formal measure of moral equivalency, you can go to the appendix and follow the QR code to a workshop on measuring and assessing your coalition cognition. The basic idea is based on a peer-reviewed paper titled "Bridging Political Divides by Correcting the Basic Morality Bias." With this survey tool, we can actually measure the strength of our morality biases by quantifying the *actual* versus *perceived* number of immoral acts that people expected the outgroup political tribe to approve. This research demonstrates that people vastly overestimate the immorality of those of the other political party.[31] There are subtle differences between party (for example, Republicans are more likely to approve of tax fraud, whereas Democrats are more likely to approve of cheating on a spouse), but they are both statistically very close to massive disapproval of moral wrongs. The realization that your politically tribal opponent knows the difference between right and wrong turns out to be a powerful intervention to restoring a common sense of humanity of the other side.

Political tribalism is a loyalty to the ideology of your political team irrespective of facts. In this framework, tribalists never concede an inch to other tribes. It is an "Us versus Them" zero-sum mentality, where "they" are seen as morally suspect and morally dangerous while "we" are morally righteous by comparison. This topic is explored masterfully and at length by the psychologist Jonathan Haidt in his book *The Righteous Mind.* In addition, given what we know about the Tribe Drive, we also understand that, in reality, political tribalism is about identities and the deeply rooted emotions that protect them from harm. Even lies themselves are not exempt. One study showed that even when you *know* your team is lying, it's OK, because it's for the moral good. "Your team's lies are a sign of moral decay, but my team's lies are morally justified."[32] The Moral Mirror Game exposes these in-group–warped mental hypocrisies.

The linguistic and rhetorical subcategory of the Mirror Game is known as the Russell Conjugation (named after Bertrand Russell, who coined the term).[33] In essence, it is a psychological trick where we assess subliminal cues we take from communication that prime our conclusions. The key is what adjective is used and whether or not it taps into empathy. Russell gave the following example: I am firm (positive empathy). You are obstinate (neutral to mildly negative empathy). He/she/it is pigheaded (very negative empathy). Note this the next time you read a news article. Opinion journalists use this all the time as a form of (perhaps unconscious) manipulation of their readership.

One more example of Mirror Game psychology. Michelle Dresbold, a handwriting expert who has been trained and worked with the Secret Service, analyzed Donald Trump's signature. In her analysis, she described his handwriting as "bold, condensed, angular signature shows someone who is rough, tough, aggressive, competitive, can never relax and is not nurturing."[34] She further expounds that his angular writing style with minimal curves shows up in the signatures of some sharp-minded and competitive workaholics that are prone to anger, hostility, and fear. Whether there is truth to her analysis is beside the point. The insight is how people *interpreted* her observation about Trump's signature. Dresbold was shocked by the public response, as conservative audiences (corporations, business groups, and entrepreneurs) reacted positively to her analysis of what she considered mostly negative attributes. However, liberals in the academy, at universities, and other progressive hot spots had mirror-opposite responses, using the identical data as proof of Trump's corruption. Dresbold recalls: "When I say something like 'his check-mark-like stroke, called a tick-mark [in the bottom left-hand corner of the *D* in Donald], it indicates that Trump has explosive anger and a very bad temper,' the conservative interpretation is, 'Of course, he's angry about what's happened to America.' The liberal interpretation is, 'Yes, he's a very angry man with childlike temper tantrums.'" Continuing on theme, if Dresbold remarks that his signature is unreadable, which indicates that he keeps his feelings hidden from the public, the liberals' interpretation is that he's sneaky and untrustworthy, whereas the conservative interpretation is that he is intelligent because he doesn't want our enemies to know what is in his mind. Same data, same analysis, two different universes of interpretation.

There is nothing more human than believing in collective narratives. These tribal narratives were the glue that kept our camps, bands, and tribal signals and symbols coordinated in collective action and resilient to a host of existential challenges throughout human evolution. In the context of Hoffman's Fitness-Beats-Truth theorem, *faith* is the ultimate fitness coalitionary alliance signal. It says to your compatriots, "It doesn't matter how improbable our collective story about reality is—I believe it because you believe it. We are one." The best summation of the meaning can be found in the German word *Weltanschauung*, which translates to our "worldview," but includes much more—our entire psychological profile as to how we interpret the world we inhabit. The example of how two very different political tribes interpret Trump's signature brings to bear an important idea; our interpretation of the world is filtered through tribal multiverses to fit the collective narrative of *our people*.* Like an optical illusion, group beliefs have this fantastical, strange, perception-warping property. That's because throughout human evolution, there was a naked urgency to belong to a coalition—if you did not belong to one, you were at the mercy of everyone else's. Thus, in a state where people feel isolated, which is now endemic, people are vulnerable and willing to cling to any identity; unfortunately, political teams, and the ideologies that are transmitted by them, are low-hanging fruit for people hungry for group membership.[35]

Erwin Schrödinger proposed (on the basis of his equations that described several different histories of the universe) that these were not alternative histories but, in fact, were happening simultaneously, in a phenomenon he dubbed "superposition." Whatever they have been called ("alternate universes," "quantum universes," "parallel universes," or "interpenetrating dimensions") this is a socially real phenomenon. They happen every time two tribes look at the same evidence brought

* In an interview, an army officer deployed in the post-9/11 "war on terror" recounted how she was shocked when she found out that the terrorists believed they were the Rebels in *Star Wars*, and that the United States was the Galactic Empire.

before them for explanation. In his book *Win Bigly*, published right off the crest of the Trump election wave, Scott Adams describes events like this as follows: it is as though two tribes seeing the same data results in "ripping a hole in the fabric of the universe." He argues that for the liberal establishment to make sense of the transition of Trump going from "not a chance of being our president" to president of the United States in one blurry-eyed moment, they in effect had to rewrite the previous movies playing in their head that narrated the story. It was the moment that "your entire worldview dissolves in front of your eyes, and you have to rebuild it from scratch." This quote aptly describes what I, as a 2016 card-carrying Democrat, experienced that election night.*

Tribal Diagnostic 2: The Moral Valence Test (or, How Virulent Is Your Tribalism?)

Imagine a flag unfurling in the wind. Consider the images that choreograph the constellation of symbols that is embedded in line and color within the rectangular piece of cloth in your mind.

Now carefully picture the mosaic of a red-and-blue donkey with three stars along its torso. Do any emotions manifest to the surface of consciousness? For Americans (and perhaps internationally), feelings will vary in predictable ways, depending on your political tribal allegiance. They could range from a sense of pride and moral certitude, to contempt and moral depravity. Next, picture the same shapes and colors in the form of an elephant. What emotions emerge? You may feel the same emotions, possibly in the inverse order of what you experienced previously. It is noteworthy that these images control for everything (color-coding, number of stars, design) except for the simple shape of the animal the flag adorns. The association between colors and objects is a biocultural phenomenon. Associations are forged over eons by evolution, over centuries by culture, and over decades by personal experience, and as the science of color and texture association

* Among the several factors that prompted the research for this book was the night Trump won the presidency. It was a moment that demanded epistemic humility.

matures, we are beginning to understand increasingly more powerful ways to tribal signal. It's just the outline of an animal that differs, and coded in that outline is either a team you feel is closer to or further from your identity. Moreover, as a matter of experience, the feelings that manifest when you see these identity symbols can serve as a warning for whether or not you are unconsciously committing a real sin against humanity—dehumanization.

To assess this important factor, let's ask a follow-up question: In either instance of imaging the identity symbols of the other side, do you feel *contempt*? Even more importantly, in either instance do you feel *disgust*? If so, you have unfortunately failed a very important test (one we'll come back to later in the chapter) whose outcome predicts the success or failure of nations. Consider this exercise a diagnostic tool to analyze the *intensity* of not just your political tribalism, but also the extent to which you dehumanize the out-group. Again, you can measure this more formally as well, by way of survey (see the QR code in the appendix). Another visual way to do this is to bring to mind the classic image of the human evolution scale.

On the left side is a silhouette of a primitive, gibbon-looking primate, and as we move to the right, we see forms that look sequentially more and more human. In the middle is a chimpanzee-looking image, and then some type of archaic human and neanderthal, then on the right we see a fully upright human being. If you were to draw these images out on a paper, and then number the far left with a 0 and the far right with a 100, bringing the political out-group to mind, where on the scale would you say the "average representative of that group" falls. This scale is literally called "the dehumanization scale," and it's an explicit test of your moral valence. If you rank an out-group on this scale as less than human, you've dehumanized them.

Tribal Diagnostic Test 3: What Is My Identity Stack?

Last, let's perform what may be the most important of the diagnostics: an identity-ranking exercise. The goal is to see if we can't reweigh some of these identities in a way that augments, not impairs, human wellness. Take out a piece of paper. The first exercise is to write down in a stream of consciousness, without any particular order, all the identities you hold. It could be your job title, your local bowling or softball team, your family name, the nation you are a citizen of, the ethnicity your ancestors identified with, your political party, or even your species. The following table is not exhaustive, but simply a series of examples that could help get the identity juices flowing:

SOCIAL IDENTITY CATEGORIES	EXAMPLES
Age	Adolescent, Child, Young adult, Elder
Biological sex	Female, Intersex, Male
Clan/Family origin	Surname
Club/Society member/Sodality	Freemason, Gaming club, Scottish Rite, American Legion, Musical fan club
National origin	Barbados, Canada, France, Japan, United States
Political affiliation	Conservative, Democrat, Liberal, Republican
Profession	Actor, Data analyst, Consultant, Teacher, Server, Construction worker
Race	Asian/Pacific Islander, Biracial, Black, Indigenous, Middle Eastern, White
Religion or spiritual affiliation	Agnostic, Athiest, Baha'i Christian, Hindu, Jewish, Muslim
Sexual orientation	Bisexual, Fluid, Gay, Heterosexual, Lesbian, Queer, Questioning
Socieconomic class	Owning class, Upper class, Middle class, Working class, Poor
Team association	Sports team fan, Sports team participant

You just have to have some significant part of *your* identity as that category. Take your time and be deliberate. Now, once that is done, get out a fresh piece of paper, and rank them by order of importance to you. To what extent do these categories truly capture your identity? If you are having trouble, it may be worthwhile to frame the question similarly to social psychologists attempting to measure identity:

What groups are you proud to belong to? What group memberships do you find yourself thinking about a lot? Which ones affect how you are treated by other people? With which groups do you feel solidarity?

To figure out how strong these identities are, let's assign some metrics to them. Consider each group you thought up. How strong would you say your attachment is to it? Would you say your attachment is not strong at all (1), slightly strong (2), somewhat strong (3), or very strong (4)? Now list them, weighted, from the strongest to weakest identity attachment. What you have just done, in effect, is quantify your tribalism. This identity stack is a powerful tool.

Let's analyze the results. To be clear, there is a line to demarcate between two types of identities. There are tribal and sub-tribal face-to-face identities. Recall, tribes are groups of strangers nested by symbols. Thus, any identity that has strangers in it is by definition a tribal identity. If your identity category has real, tangible, face-to-face relationships, then they are sub-tribal identities (at the level of bands and camps). For example, people in your camp can all identify as Libertarians, Catholics, or Yankees fans, and thus be part of a tribe you identify with. But not all Libertarians, Catholics, or Yankees fans can be part of your camp. One is a tribal identity, the other—*much more important in determining your well-being*—is a sub-tribal identity.

Now consider your *identity stack*; this is a list of identities weighted by significance. If you rank your levels of identity, and you find that your ideological tribal identities are listed higher than your non-tribal identities then *you are a tribalist.* You are being tribal, and letting your tribalism seep into your identity in ways that can reduce the well-being of you and your loved ones. It's also wasteful. An individual has little power to influence their tribe. It takes massive amounts of energy to change tribal networks, whereas energy spent focused on your non-tribal—face-to-face—identities can have huge impacts on you and your social network's well-being. Imagine, for example, the influence you have over your local sodalities (such as the Parent Teacher Association, clubs of various types, churches, etc.) versus the influence you have over who wins national elections. This exercise is thinking about how to best channel energy in ways that maximize its impact. As we will explore in the next chapter, the real unit of societal change that impacts people's daily lives for better or worse is the community level. Thus, by prioritizing your tribal identities over your face-to-face relationships, you are committing the sin of modern-day tribalism.

Let me use myself as an example. The young, adolescent, cultural warrior version of myself would have listed the following identities in order of most to least important:

1. Democrat
2. Scientist
3. American
4. Sodality
5. Samson clan
6. Quebecois

In effect, I allowed my political tribal identity to be near the top of my identity stack and disproportionally influence my behavior in ways that hindered, not helped, my face-to-face networks.* The top of the stack is one of the most precious cognitive resources in our possession, because it determines much of the moral reasoning that occurs unconsciously when we face ethical challenges. Since then, I have completely reweighted my rankings to fall within the parameters of the Social Suite—that is, decentralized from the ground up. My closest, most important identities are my best friends and pair-bond. Then there are the commitments I honor to my camp and greatest friends within my sodality (fraternal guild). Then there is the identity I share with my extended family. All these are *real* relationships within real social networks of varying levels of relatedness. By the time I get to the tribal line between tribal identity and sub-tribal identity (see below), we are making the moral gearshift from sympathy morality up to reason morality. You may note that once that happens, seeing our species as a single

* As noted by Balaji Srinivasan in *The Network State,* primary identity is a precious resource leveraged by all successful groups. To see a great visual of the identity stack, visit: https://thenetworkstate.com/sociopolitical-axes.

tribe is one of the most important moral moves we can make (we will explore this topic in greater depth in the next chapter).

Sub-tribal **(Face-to-Face) Networks**

1. Blood brother / Pair-bond
2. Honor group
3. Sodality
4. Samson clan

Tribal **(Beyond Face-to-Face) Networks**

1. Human
2. Scientist
3. American
4. Quebecois

This exercise is valuable, because once you identify the ways you are ranking your identities, then you can become aware of the negative influence they can have on your day-to-day, real-life, sub-tribal relationships that truly impact you for the better or worse. In a democracy, you have one vote among millions to influence your society. One study demonstrated that those who are most politically engaged are the most corrupted by their motivational and "my side" biases.[36] One conclusion the authors came up with was: "All of this has an ironic and unfortunate upshot: the people who seem the most capable of political objectivity are the least likely to participate in politics." Another study, which was later replicated, found that "political commitments can scramble your brain and impair your ability to reason—especially if you're smart."[37] In summary, deeply consider whether or not it's worth identifying with a political party. You may be doing your country an ironic disservice by pledging allegiance to one over the other.

I'm not suggesting here to politically disengage, but simply to reorient and reweigh how you think about your identity in light of political tribalism. In the sympathy and honor groups you are networked within, your influence is vastly greater to change lives for the better. Rank these groups—and identify with them—accordingly.

You have now taken three political tribalism diagnostic exams. If you felt contempt in any of them when considering the rival political tribe, you failed the ideological test. You are a tribalist ideologue. And for critical reasons, we will soon establish that being an ideologue compromises your mental immunity to mind parasites. That is, in the words of the philosopher Andy Norman, your cognitive immune system will be vulnerable to bad ideas. More on this later. However, there is a more severe failure to be had. Consider your reaction to the identity symbols of the out-group, if you felt *disgust* then you may have committed a grave sin. If you hate the sinner as well as what you deem to be the person's sin then you are on the cognitive path to dehumanizing another group of people. This cognitive classification, it turns out, is the precursor to the most immoral of social behaviors—genocide.

Treating Chronic Political Tribalism

Political tribalism is not an acute societal disease. It eats away at the health of a system slowly over time. To reverse it, it takes systemic, positive changes over the life of a society to better its health. Political tribalism feeds off of tribal propaganda, fake news, and "alternative facts"—that is, all the lies that lead to the metastasizing of ideologies.

If you are a "political animal" in the United States, then let's examine your brain as it considers "Us versus Them." Republicans and Democrats are the Eagles and Rattlers of our contemporary political landscape. On the eve of the 2016 election, for the first time in modern political history, they moralistically ranked each other in *majority terms* as having an extremely unfavorable view of the political out-group. We can literally see this at work in the brain when we hook up members of each political party to brain scanners. When they were asked to think about their in-group, the ventral medial prefrontal cortex lit up. In contrast, when queried to contemplate their out-group, the dorsal prefrontal cortex lit up. This demonstrates that liberal and conservative tribes activate *the same neurons* when thinking about their own tribe and different neurons to think about a different tribe. In essence, there is different neural real estate for us and them. In the words of Walt Kelly, "We have met the enemy and he is us." The enemy being our brains. This way of thinking also leads to what Norman calls "the Way of the Culture Warrior."

According to Norman, "A culture warrior understands that ideas can be dangerous and works to combat the ones she or he deems evil. The aim is to defeat those ideas and make the world safe for beliefs that are good and true. Culture warriors typically locate the evil ideas in others and the good ones in members of their own *tribe* [emphasis mine]."[38] Culture warriors can be either secular or religious, liberal or conservative. You can tell them by their methods—abusive and disabusive approaches to combating ideas *they think are dangerous.* In other words, they use language as weapons, like verbal daggers and swords meant to harm others. They are ideologues and tribalists, and their cognitive immune systems have been compromised. Their obsession blinds them to the fact that their strategies have the opposite effect they intended.

Although "righting wrongs" is the purported purpose of donning the mantle of a culture warrior, the alternative function can materialize in fantastic gains of status. Will Storr, in his book *The Status Game: On Human Life and How to Play It,* describes how moral panics are especially attractive to ideologues—as they serve as a gold rush of virtue currency to their tribe—they provide a way of usurping the current status hierarchy.[39] As we recall from the chapter on trust signals, the weirder and more extreme the beliefs purported, the more virtue currency an adherent can rack up and use to attain greater tribal status.

That these partisan passions end up alienating potential allies who may otherwise (with a little humility and openness to inquiry) befriend is not the point. The point is to achieve status within their in-group. Along the way, they damage relationships, sow division, and push otherwise good people into defensive camps. They are a large part of the reason why so many people in our society have contempt toward each other and—in its worst expression—dehumanizing disgust for other members within it.

Truly bad people are statistical rounding errors in the population. They exist, but at the margins. Psychopaths are individuals with persistently impaired empathy and remorse and disinhibited egotistical traits that produce antisocial behavior; they account for, at most, one percentage point of the total population.[40] But from the current data on polarization in the United States, we would assume half the population falls into this category: "truly bad people." In truth, the members of other political tribes are human beings with different life experiences, genetics, social connectivity, and cultural backgrounds that are tight or loose and many emphasize honor over dignity, or dignity over honor. Both have their place in our society. First, it is important to acknowledge fear's role in all of this. In fearful environments, bias and misinformation reigns. To our evolved brains, truth matters little when we feel threatened.

When we feel threatened, we are vulnerable to the extreme forms of political tribalism on both sides, as groups begin defending tribal identities with extreme actions. This is why "identity politics" is one of the most pernicious dangers to prosocial collective action as a society. It is a case of mistaken identity, where people *identify with their beliefs*. Its perniciousness as an idea is compounded when identity is linked with skin pigmentation. As discussed at length in Part I of the book, skin color is not necessarily relevant when it comes to any particular tribe. Humans are not their beliefs, nor are they their skin color, just in the same way that mindfulness meditation identifies that you are not your thoughts, feelings, emotions, and senses. When people make this mistake, it often results in contempt toward "the other," and contempt dissolves unions, disgust dehumanizes the other, culture warriors turn into actual warriors, and sometimes, blood is spilled. Ideologies flourished unabated in the twentieth century and are on an accelerated pace in the twenty-first century so far. The death toll is astonishing, as ideologically fueled conflicts have killed more than a hundred million humans.

Evolutionarily, we are extremely sensitive to the threat of loss; be it the death of members of our tribe, or resources, such as a water source, or even something that may appear insignificant on the surface, such as the loss of a tooth. A loss of any one of these elements could mean death to the individual and potentially the extirpation of a small group of people living off the land. To unleash preference for in-group bias takes very little stimulus—the hint of threat can activate the protective systems housed in our limbic brain. Stevan Hobfoll, a behavioral scientist, formerly an army officer, and author of *Tribalism: The Evolutionary*

Origins of Fear Politics, calls this the *defend and aggress system*. Hobfoll notes that its activation—and subsequent response on individual and group levels—is proportionate to the perceived strength of the environmental signal of threat to loss of personal, social, and material resources that we value. If we do not effectively address our ability to deal with potential loss, then we severely hamstring our ability to face the common enemies of our species: technological change (AI), climate change, and nuclear proliferation—all of which will require collective, multinational, and super-tribal cooperation.

Christopher Boehm argues in *Moral Origins* that both the political left and right are guided by evolutionarily honed moral instincts.[41] The hunter-gatherer groups of the Paleolithic selected camp and band members that were neither overly tyrannical, stingy with collective resources, nor prone to freeloading. Any of these would pose a grave threat to the stability of the camp. With a tyrant, you have a single human (typically male) attempting to monopolize mates and resources through coercion, resulting in an antisocial agent that threatens the well-being of everyone else in the group. Foragers are central place provisioners and therefore share calories they find in their environments in a kind of group calorie insurance policy. With freeloaders, you are put in just as much risk. If you have individuals in your camp that take resources without contributing to the overall welfare of the group, it is obviously a detriment. Both present serious, even existential problems to the survivability of the group. Thus, two strains of righteousness evolved, where humans were deemed immoral if they took more than they gave, didn't share their winnings, *or* tried to use physical dominance to serve their own interests. The answer to this immorality, according to Boehm, was capital punishment or excommunication from the group.

The mark of this moral evolution remains in both the left and the right today—with the politics of the left being focused more on legislation against monopolies and redistributing resources, and the politics of the right being more focused on anti-freeloading strategies that manifest in the form of fiscal conservatism. The politics of libertarianism are intrinsically anti-tyrannical, pushing against any government overreach. All politics then is a weird bleed-through of our evolved instincts as camp-dwelling creatures. This realization is a powerful mental inoculant that provides mental antibodies against the worst forms of political tribalism. Both parties, in their hearts, serve a moral end that exists *because*, not in spite of, being a prosocial species. Consider this the next time you look upon someone who identifies with another political tribe. Perhaps there is some ancient wisdom embedded in their policy preferences.

Moreover, each political tribe shares the same instinct to hold sacred, but differing values, encircled by identity-protective cognition. One of the most effective ways to begin reversing our political tribalism is by removing contempt. Contempt is one of the most predictive expressions of behavior that foreshadows divorce in married couples.[42] It is used by political parties to declare the other side unworthy of the benefits of public good. It is used by governments to provide rhetorical cover for abuse or torture. Contempt is the dissolver of unions. Be mindful when this emotion crests to the surface of your consciousness, because it is priming you to eliminate a relationship.

By acknowledging the other tribe's sacred value,* you can retrace the steps that led you to contempt's corrosive bonds. The act of acknowledging sacred values suppresses the worst symptoms of the tribe virus, allowing individuals to actively fight off infection long enough to act and think rationally. The American-French anthropologist Scott Atran, along with Robert Axelrod and Richard Davis, published a paper in *Science* that considered the power of what they called "sacred values" in conflict resolution.[43] Sacred values, as we've discussed, are identity defining and thus are defended far out of proportion to their actual importance. In other words, the quickest path to contempt from my group to yours is by dishonoring that which my group holds sacred.

For example, Atran and colleagues analyzed the Israeli—Palestinian conflict. They discovered that sacred values are at the very heart of the conflict, and it is likely that until this fact is addressed, bringing peace to Israel and Palestine will not revolve solely around the concrete, practical specifics of borders and jurisdictions. Just as important for reconciliation is the symbolic recognition that each tribe has the capacity to feel

* As long as it does not break the moral precept of the Concentric Circle.

pain, thereby restoring an aspect of their humanity. In the case of the Palestinian-Israeli conflict, this included reparations to Palestinians for the homes and lands they lost almost seventy years ago, but also an apology for what the Palestinians call "the 1948 tragedy." As for the Israelis, Benjamin Netanyahu cited not only instrumental issues of security but also how the Palestinians must "change their textbooks and anti-Semitic characterizations." Rationally, an apology and characterizations in a textbook should not decide the fate of nations, but in this instance were clearly standing in the way of peace. The reason is that in recognizing the enemy's sacred symbols, you are also recognizing their right to be proud, their right to unity, and connection to their identity—all basic human needs.[44]

The science of identity shows there are really poor ways for groups to politically engage. Don't use labels, don't use divisive symbols, and don't use moralizing words; unfortunately, the words that get the most retweets are laden with such emotional content: *attack*, *bad*, *blame*, *care*, *destroy*, *fight*, *hate*, *kill*, *murder*, *peace*, *safe*, *shame*, *terrorism*, *war*, and *wrong*. Examples exist from the annals of history's greatest leaders of the opposite of such polarizing tactics. In a powerful use of tribal symbols, Nelson Mandela united Black and white South Africans by co-opting a symbol of colonial oppression—the Springbocks white-only Rugby team—and wearing the jersey of the team during the Rugby World Cup. This brilliant, super-tribal stroke of genius helps dissolve the narrow, race-identity politics coalitions by uniting a greater "us." Previous presidential candidate Andrew Yang, now the leader of the Forward Party aiming to bust up the U.S. political duopoly, proposes measures that could help reduce political tribalism; namely, ranked-choice voting and open primaries to help enable the promotion and adoption of nonpartisan, centrist policies.

We must reframe the political process not as a zero-sum game between two teams, but as a type of *adversarial collaboration*, where camps are committed to good-faith arguments backed up by a process of "steel-manning" their opponents' arguments.[45] "The Steel Man" is a form of discourse opposite of "straw-manning" the other side's argument. With the Steel Man you help others articulate the strongest form of their argument, mainly by asking clarifying questions of their positions, and in doing so, you come to an understanding that permits compromises where each side gets something, but not everything of what they want. Under the guiding question, "What would it take for you to change your mind?" representatives with differing perspectives could craft empirical tests that they publicly agree would settle an argument, then they both save face and share the credit for contributing to society irrespective of the productive outcome.

David Huff has argued that our nation's obsession with the political process has rendered us dangerously enamored with the false hope of political salvation: "The finances, energy, media attention, and zealous devotion heaped upon candidates for high office at times reaches messianic proportions. They provide further evidence that what was once a valid political process now borders on idolatry."[46] Political obsession is a kind of tribal idolatry; absent a strong, communal identity, people are heaping disproportionate levels of energy into a system that will almost always fail to channel that energy into something useful. Huff calls this phenomenon a drift toward a *Pyramid Society*, which "is a culture in which a majority of the people spend most of their time transforming the civil government to the near exclusion of themselves, their families, churches, schools, and businesses." By changing the centralized institutions, we are led to believe, the trickle down will blossom into a better society—but this ignores one important fact that we will explore in the next chapter: the unit of change is at the community level. Importantly, there is an emerging technology of trust that will enable the community (band) level of societies to flourish in ways previously unimaginable.

IF WE CAN SOLVE THE Trust Paradox, then we can freely inoculate humanity with a tribe vaccine without losing our bearing as a species and jettison the most pernicious forms of tribalism to the wastebin of evolutionary history. We have already taken the first steps, tackling a long-standing challenge in the philosophy of ethics by answering the question "Is it possible to simultaneously privilege your family, your honor group, and your camp and still be a moral person?" The Concentric Circle says yes. With this in mind, we can now move to the final steps—exploring the most important innovation in the technology of trust ever seen in our species and manufacturing, at scale, a vaccine for the tribe virus.

11

OUR TRIBAL FUTURE

How Tribalism Can Save the World

If we can solve tribalism, all problems admitting a human solution would be answered.

–Sam Harris, 2022

As a humanist, I think it's important to think of humanity as one big tribe. And if we need a common enemy, let bad ideas be that enemy, and mental immunity be our shield against that enemy.

–Andy Norman, 2021

From time immemorial, decentralized human existence was the default. Ramón Medina Silva, shaman to the indigenous Huichol of Mexico, captured this sentiment when he said: "The Huichol lives freely. . . . Out in the wind, everywhere. We work as we wish, we go as we please. . . . And the Spaniard? He cannot do this. . . . The Spaniard must do as the government says. The government has him in this way."[1] One way to think about the devastatingly harmful outcomes of social mismatch (described in chapter two) is to consider it as a function of humanity being pulled further and further away from decentralized communities to centralized regimes. Within this framework, a society is mismatched to the degree with which centralization occurs. For millions of years, humans dwelled in decentralized social networks of about 30 to 150 people, where everybody knew everybody, and the greatest extensions of one's primary tribal identity spanned a few thousand individuals at most. That made it hard, and very costly, to do bad things. The evolution of tribal affiliation was the first nascent outreach toward a symbolic centralization as a way to cope with an increasing number of strangers. The Tribe Drive was the go-to social norm generator, and these cues imbued the people within those tribes that adopted them with unprecedented cooperative powers that jump-started civilizational progress. But with agriculture coming online, a boom cycle occurred, and we lost our shared index of social memory, which was the result of the rise of anonymity. At this point in the story, a dark, difficult chapter—Ann Druyan calls it "post-agricultural syndrome"—plays out in the human narrative. Sendentism became the norm, and populations were bursting at the seams. With every seeming advancement, so, too, followed a perilous derivitative of that innovation. This threatened social order. Strong men arose to restore that order. The strong men wielded the tribal weapons of top-down authoritarian abuse in a command-and-control system. And history shows, they can't help but overreach.[2]

Decentralization and the Technology of Trust—A New Solution to the Trust Paradox

When totalitarian regimes disintegrate, the same pattern plays out—a variety of local assemblies effectively administer societal affairs better than the central authorities and bureaucrats. The author Kirkpatrick Sale wrote of this pattern: "It is striking to reread history with eyes opened to the persistence of this tradition, because at once you begin to see the existence of the anti-authoritarian, independent, self-regulating, local community is every bit as basic to the human record as the existence of the centralized, imperial, hierarchical state, and far more ancient, more durable, and more widespread."[3] What Sale was tapping into was that evolution had shaped an effective default to social order—the constellation of camps within a band.

It's why the French social theorist Alex de Tocqueville, upon visiting America, remarked: "The village or township is the only association which is so perfectly natural that, wherever a number of men are collected, it seems to constitute itself."[4*] A recognition of this by the founding fathers was part of the genius of their enterprise. Decentralist philosophy was a driving force in America's nascent years, and leaders such as Thomas Jefferson were acutely sensitive to centralist encroachment. Jefferson saw in decentralization the spirit of free and sovereign units of society, where each individual would participate in government on a local community level. As one of the architects of a new country, he wrote of people living in free association with such independence from central authority that the individual would "let the heart be torn out of his body sooner than his power wrested from him by a Caesar or a Bonaparte."[5] Thinkers like Jefferson were always skeptical of the capacity of centralized government to serve as a panacea for the ills of the individual.

There is data to support Jefferson's hypothesis. A recent study led by Harvard economist Raj Chetty analyzed two California neighborhoods: Compton and Watts. They are demographically similar and geographically close to each other, yet in Watts 44 percent of the men who grew up there ended up incarcerated compared to only 6.2 percent in Compton. Moreover, measures of social mobility were lower in Watts.[6] Why are Compton kids getting better chances in life, despite both communities being disadvantaged? The answer, according to David Brooks, is community infrastructure and the authority of those community groups to enact change. Only the people on the ground know what challenges they face, which brings us to the important conclusion—*band-level community is the unit of change*. In addition, they can enact change much more effectively when authority resides, not in some outside, federal jurisdiction, but in the community itself. This is good news because your community is something you can not only choose, but unlike national or international politics, you can actually influence. This is another reason why sodalities are important. A prosocial sodality is a strong social network of trust with the power to effect change at the community level. And it's worth influencing, because your camp and band is the most powerful

* Intriguingly, several states have articles of incorporation for townships that literally use the value of 150 people as the minimum to incorporation. That number—Dunbar's number—should be very familiar by now.

hedge against bad luck you can have.[7] Again, community is the social unit of effective change.

The challenge to the actualization of Jeffersonian ideals was that the technology of trust had not yet matured to the point where an early United States could survive in global competition without the militarized fist of centralization to pound out their agenda. Importantly, something had happened with the help of the scientific method only a few hundred years earlier that, for the first time in the ten-thousand-year reign of tyrants, was to slowly stem the tide of their power. In fact, one of the most important characteristics of the scientific method is that it is decentralized. Good societies are scientific societies because they regulate beliefs with reasons, not centralized epistemic authority that reinforces prevailing loyalty and obedience to the state. The scientific method, by enabling the measure of reality, spurred on a new decentralizing technology.[8] A mere five hundred years ago was the first drop of a rainfall of innovations destined to reshape the world. It was the invention of the printing press.

People-based trust systems are capped by the channel capacity and Dunbar's number, but *technology scales.* It has no memory limits. As observed by James Dale Davidson and Lord William Rees-Mogg:

> Printing rapidly undermined the church's monopoly on the word of God, even as it created a new market for heresy. Ideas inimical to the closed feudal society spread rapidly as 10 million books were published by the final decade of the 15th century. Because the church attempted to suppress the printing press, most of the new volumes were published in those areas of Europe where the writ of the established authority were the weakest.[9]

Freely printing ideas and distributing them was the first challenge to the theocratic state's stronghold on human society.

From Printing Press to Proof-of-Work—Trust Tech Evolves

The printing press was the progenitor—and spreading vector—of even more decentralization technologies. The internet, which many thought would be a passing fad, has been the prerequisite technological innovation* for pretty much everything twenty-first century. One fascinating, and underappreciated innovation is that of new money in the form of cryptocurrency. Between 2008 and 2012, a programmer (or programmers) going by the pseudonym Satoshi Nakamoto shared with the world their solution to the Trust Paradox. Embedded within lines of code was a vision for our species where stranger interacts with strangers *knowing* beyond any doubt that the interaction is trustworthy. This technology is called *proof-of-work*. It was published open access (freely given to all), in what is called "The Original White Paper."[10] Nakamoto modestly wrote in chains of emails to the development team working on the Trust Paradox, "I'm better with code than with words," but it is worth hearing from the source: "The proof-of-work chain is the solution . . . to knowing what the globally shared view is without having to trust anyone." Of all the Trust Paradox solutions that natural selection has given Earthlings, this one is uniquely human. The technology—like the first human who flint knapped a handaxe from bone and rock to open an animal carcass—was created by an engineer elegantly solving a problem.

Proof-of-work scales trust.

The profound nature of this innovation is described by the Austrian economist Saifedean Ammous: "This . . . process . . . produces a ledger of ownership and transactions that is beyond dispute, without having to rely on the trustworthiness of a single third party. [It] is built on 100 percent verification and 0 percent trust." Most important, this solution renders obsolete the need for tribal cheat sheets to bootstrap the odds of cooperation among strangers because a transaction can be—for the first time in human history—100 percent verified.

TRUST PARADOX SOLUTION	ANCESTRAL CONDITION	TIMELINE
Kinship	First organisms with intra- and extracellular interdependencies	500 million years ago

* As Srinivasan likes to analogize, like the "tech tree" of a video game like *Civilization*: shorturl.at/bgKP9.

TRUST PARADOX SOLUTION	ANCESTRAL CONDITION	TIMELINE
Friendship	First social species with large brains and social networks	30 million years ago
Tribalism	First social species with capacity for symbolic manipulation at scale	300,000 years ago
Proof-of-work	First species with mastery of mathematics at the level of cryptographically protected decentralized ledgers of chains of custody	Starting in 2008?

All combined, these technologies are what Johann Gevers calls "the technology of trust."[11] With these radical innovations—powered by proof-of-work—trust *infinitely* scales past Dunbar's number without having to rely on easy-to-bias tribal signals! Imagine, as a quick science fiction example, a merchant on Mars recording a transaction with an Earthling with a cryptographically protected transaction that needs no government currency or third-party banking system to serve as the centralized broker of trust. This is a radical challenge to the status quo. Let's assess how this seemingly amazing claim could be possible, with an interrogation of what Gevers terms "the four pillars of a decentralized society."

The Four Pillars of a Decentralized Society

The first pillar is *decentralized communication.* The core technology propping this pillar up is the internet. In what's been dubbed the "Great Reset," the decentralized communication revolution got boosted by the COVID-19 pandemic. As more people worked remotely, leaders in the space of decentralized work emerged to show best practices for companies and institutions struggling to adapt. Speaking on the topic of the future of work, Matt Mullenweg, a successful CEO of 1,200 employees at Automatic, says: "We are going to have to reexamine every aspect of our society that is overly reliant on colocation, because that makes us . . . particularly vulnerable . . . for something [a disease, or disruption] that could be a lot more deadly."[12] This appears to be a perfect opportunity to reframe the way we look at the role of communication to facilitate remote work.

In the spirit of the times, a period where terms like the "Great Resignation," the "Great Reset," and the "Great Migration" are commonplace, honor groups of people looking to put into practice the principle of intentional proximity by building a community have more power than ever. Camps in the beginning stages of conceiving the locations they want to live can use data to figure out the potential communities they would dwell within.

If realized, it will allow people to work 100 percent remotely, living within the perimeter of their camps and bands, connected to a larger empathy group of community. It also furthers free speech and privacy protection. Creating a pseudonym protects not only your data (being used by corporations to target ads or for other more nefarious deeds) but also your real identity. Imagine living in a society where reputation mattered but you couldn't be canceled for voicing countervailing opinions. By using your real name everywhere online, you are vulnerable to personal and physical attacks at the craze of social media mobs by doxxing. This society would advance considerably relative to societies that are more restrictive because of the good-idea-generating properties of free speech. Finally, and perhaps most importantly, the pseudonymous economy undercuts racism. If you do not know the true sex, sexual orientation, skin pigmentation, height, weight, and level of attractiveness—and any other tribal affiliations—of your coworker, there is nothing to discriminate and to consciously or unconsciously bias for or against. In essence, rip the "my side" bias out by the root. If a society wants to guarantee the destruction of institutional, systemic racism—adopt a pseudonymous economy.

The second pillar is *decentralized law*. With good communication, we can then come to agreements on the laws that govern our cooperation. For decentralized law to flourish, we need choice of law, choice of adjudicator, and choice of enforcer. In other words, *we choose* the law that represents how we draw up legal contracts. I imagine a hunter-gatherer, who is used to cocrafting social norms would find as the most satisfying and obvious solution to binding agreements between two parties.

If both parties agree to the law in full understanding, they are acting within an ethical framework without national boundaries but between people. Two people could choose their agreement to be based on English law, American law, international law, or simply make up the law as they see fit and fair. Two people can also choose who hears each side of a theoretical dispute if the attempted cooperation comes to conflict. Finally, the two can choose who enforces the ruling of such a theoretical dispute. These three principles sound utterly radical, but they are actually the evolutionary cornerstone of modern law; they are the very essence of the hunter-gatherer legal system (which we could alternatively call forager-level legal agreements)—peer-to-peer communication that facilitates adherence to social norms. Also, with greater investment in the local, a new, more humane legal system can take place called *restorative justice*. The current centralized approach to criminal justice causes suffering for both criminals and victims. Restorative justice can be thought of as a community of justice, where the aim of justice is not only to punish but also *repair* the harm that was done. The method is to have the affected parties decide together, which leads to fundamental changes in people, relationships, and communities. Where enacted, the application of restorative justice has had some remarkable prosocial results for the community at large.[13]

Moreover, with the advent of Web3 (a decentralized internet where digital property is safeguarded through cryptography), decentralized autonomous organizations (DAO)* are going to be able to connect online communities in ways previously unimaginable. DAOs could be a game changer in the way groups organize. What's required is the staking of funds, creating a building block and token of commerce that aligns behaviors seamlessly with group incentives. Groups, clubs, and sodalities can also leverage this emerging tech to bolster their identity

* A DAO is essentially a programmable organization of people that coalesce around a shared mission and fosters an emergent online community. The governance of DAOs are hyper efficient as their operations are written in smart contracts, consisting of automated if-then statements, making them transparent and auditable. The group jointly control a crypto multisignature wallet, ensuring that its objectives are met with incredible efficiency.

coalitions in a new digital landscape. The beauty of this societal shift is that while we move toward decentralized work, it also gives us a unique opportunity to use technology to socially fortify our real-world lives.

The third and fourth pillars are *decentralized production* and *decentralized finance*. In the same way we've bypassed censorship in communication, we need to be able to do so with materials production and energy production. One example is with 3-D printing, where goods are produced beyond national boundaries. Second, decentralized energy production will be produced cheaply at home, or collectively with camps and band-level communities that have their own grids. The core principles that underlie decentralized finance are the creation of decentralized contracting systems and decentralized currency. There are several cryptographic technologies that facilitate these new channels of currency; most notable is the innovation called Bitcoin and Ethereum. Today's financial system is highly centralized, yet these decentralized technologies disrupt the centralized systems. In sum, these pillars will force the old monopolies to change and radically transform governments that decide to go the route of decentralization as opposed to the centralized regimes of the world. In the decentralized states, there will be a radical realignment of incentives, which could unleash a new era of human flourishing the likes of which we have not seen since the Enlightenment.

According to Gevers, the goal is "a decentralized, networked society where people once again have the power in their own hands to live lives of their own choosing in peace, freedom, and prosperity."[14] Humans are now free to interact with each other without needing tribal signals as tools to promote biased cooperation, because we can use proof-of-work to validate interactions with strangers. As of 2008, proof-of-work was a new tool at humanity's disposal. We're only now beginning to feel the consequences of this new engineering solution to the Trust Paradox as it grows from root to shoot. It brings us one step closer to a prosocial future, where incentives—rather than driving strangers to competition—influence human behavior more toward cooperation.

Decentralization, Then Recentralization—The Four Pillars Propping Up the Network State

Twentieth-century nationalism appealed to Cyrus the Great's playbook of super-tribal theory, where tribes live under a nation-state with one super-tribal creed. For the United States, the imagined order or mythology is the Constitution. It worked for its day, but with the emerging blockchain technologies, we may be on the cusp of alternatives that better align with human wellness.

Balaji Srinivasan envisions a future where the nation-state* takes a back seat to a novel form of government: the *Network State*. Defined by Srinivasan:

> A network state is a social network with a moral innovation, a sense of national consciousness, a recognized founder, a capacity for collective action, an in-person level of civility, an integrated cryptocurrency, a consensual government limited by a social smart contract, an archipelago of crowdfunded physical territories, a virtual capital, and an on-chain census that proves a large enough population, income, and real-estate footprint to attain a measure of diplomatic recognition.[15]

If the twentieth century was defined as the conflict between capitalism and communism, this vision of the future sees the twenty-first century as a testing of two radically different models for society—decentralized versus centralized zones of influences.

One of the key units of change in the decentralized model that moves toward the realization—and recentralization—of the Network State is that of the startup society. Whereas the centralized future will dwell in establishment global powers, emblemized in the Chinese Communist Party, decentralized zones will be ripe for a kind of natural selection for societies. These startup societies—which serve as the units of selection—can compete with each other as kinds of "societies subscrip-

* For a fantastic summary of the definition of both nations and states, see Srinivasan's https://thenetworkstate.com/on-nation-states. The book, which is considered more a live, working document by him and his team, is freely accessible online.

tions." The key is having the ability to opt out of a society at any time. In this model, you get competition between governments to *attract,* not control, their "netizens." The number of network citizens that opt in (which can be monitored via cryptographically verified dashboards) is the empirical fitness value of the society in question. Here is where moral innovation comes into play, moving away from stale, *low-trust* nation-states to a *high-trust society*. If societies run in parallel with their current nation-states, they can use the software of the legacy system while targeting one moral innovation by which all its members identify, to help leverage the Tribe Drive to create a shared consciousness. If Srinivasan is right, and the future of governance for our species looks something like a Network State, then I believe the societies that adhere closest to the Social Suite (which is the blueprint of the evolutionary origins of a good society) will attract the most citizens and will achieve the greatest societal fitness.

There is even an argument to use trust technologies in order to reduce global conflict. MIT Space Force Fellow Jason Lowery has developed a thesis that claims Bitcoin has properties of a "powerful war-deterrent protocol."[16] The idea is that it is the violability of property rights that encourages state-level violence and war as "proof-of-work" to determine ownership. Bitcoin, Lowery claims, makes property rights inviolable (i.e., cryptographically protected) and thus reduces the incentives for war. Goods, such as gold, art, land, livestock, facilities, ports, companies, natural resources, and people, are all spoils of war. The core tenet of this thought-provoking theory is that a decentralized ledger of record serves as the digital surrogate to war for defending and establishing zero-trust access to monetary property. Bitcoin is the first property in human history that you cannot take with violence.

Perhaps humanity is on the cusp of a novel solution to the Trust Paradox problem. We don't yet know, as only time will tell. But if we are, it is indeed an exciting time to be alive. What proof-of-work could allow us is the possibility of a future where cooperation can exist without the appeal to tribal yokes that corrupt one identity group over another. It's a way around the pernicious, nasty, winner-take-all zero-sum of identity politics. Yet, and perhaps most critically, we need one more thing to help humanity transcend the most pernicious forms of tribalism. As noted by Srinivasan, we need a moral innovation to inoculate our future societies. Proof-of-work gives us the technological quantum leap forward to open up the possibility, but it's just a prerequisite to perhaps the ultimate goal. Just as we eradicated smallpox from the human species forever, we need to extract tribal antibodies to produce a vaccine to inoculate it from our species forever.

Developing the Tribe Vaccine: Using Identity-Protective Cognition to Strengthen Mental Immunity

So far, the twenty-first century has been waylaid by ideological derangement. Conspiracy thinking, hyper-partisan politics, and extremist culture warriors dot the landscape in a winner-take-all attack on each of their enemies. Terror bombings (spurred by religious tribalism), hate crimes (spurred by racist tribalism), and mass shootings (spurred by mismatched lonely humans in social isolation) are endemic; one could go so far as to state tribalism is at the very center of the modern epidemic of bad ideas. How do we bridge the gap from a dysfunctional tribal now, to get to a post-tribal world?

Within tribalism lies the key to the ultimate prosocial enterprise—species-level cooperation. In the same way that we "needed" smallpox to develop a weakened form of the virus so we could use it to inoculate it into oblivion, we can use the existing evolutionary virulent tribalism in this historical moment. I believe it is possible for the tribal biases embedded in the human code to be used as the very catalyst for human cooperation on a scale never before witnessed. This could be what David Sloan Wilson calls a "major evolutionary transition" for *Homo sapiens*. A concept to be fully fleshed out in the epilogue, a major evolutionary transition has happened a rare number of times in history. Major evolutionary transitions happen when species "choose" cooperation over competition as the predominant strategy. If it were to occur for humanity in this moment, it makes a world where living with the ethic of the Concentric Circle is more feasible to enact. We can be free as a species to shower our sympathy groups and camps and friends with love and affection, without fear of betrayal that may happen from outsider groups and strangers. We can be sure, through mathematical certainty, that the fruits of both our physical and social labor will come to fruition. Perhaps most important is decentralizing away from inept, and often corrupt, institutions and recentralizing into *real living human communities* interconnected by a Network State. I believe we can be free of ideological tribalism that wracks the societies in conflict. But how?

For us to understand how this is possible, we need to deep dive the concept of mental immunity. Andy Norman's timely 2021 book *Mental Immunity* gives us the equipment needed for this quest.[17] According to Norman and the new science of cognitive immunology, the mental immune system is the mind's capacity to filter and shed bad ideas. The new approach centers on an unnerving concept: bad ideas act like mind parasites—literal pathogens that infect our mind and its ability to sort through both positive and negative ideas. This is a serious and substantive issue with real-life consequences because you do possess cognitive infrastructure to eliminate bad ideas. Mental immunity, when operating correctly in the human psyche, has three areas of self-reflection, areas of inquiry, that help us to (i) test ideas, (ii) harbor reservations, and (iii) revise opinions. When an entire society has compromised mental immune systems, it results in a *cultural immune disorder*. Memeplexes—interlocking webs of ideas—can be either adaptive (good) or maladaptive (bad) overall to human wellness. Just like we can measure health, we can also measure wellness, and empirical indices can thus note whether a society has wellness-enhancing memeplexes by its scores on indices of wellness.

Are all ideologies bad? Perhaps not, but they *are* biased. I define ideologies as memeplexes that have been integrated into group-level identity; in other words, they have embedded with and are defended by identity-protective cognition. Good ideologies* can enhance wellness. But a problematic ideology, as Norman notes, can be "a system of ideas . . . that is infectious, dysfunctional, harmful, manipulative, or stubbornly resistant to rational revision." Society-wide outbreaks of bad ideologies have disastrous effects, typically ending in tragedy for the civilization in question.

---

* Like the sci-fi system of ideas that governs Starfleet's United Federation of Planets in Gene Roddenberry's *Star Trek*.

What is a bad idea? Simple—a bad idea measurably reduces wellness in its host. In the same way a virus harms its host's physical health, a mind parasite compromises its host's ability to clearly reason. Bad ideas use other minds as vectors to spread, because minds *host* them. Ultimately, a bad idea hinders the capacity for humans to thrive on both the individual and social level. How does this happen?

An idea can benefit its host by way of inspiring it to action or providing comfort—yet still be false. An idea can have short-term benefits, but long-term harmful consequences. An idea can benefit its host while harming others. All these are properties of real viruses. Virus-like replication occurs when they spread, and just like the flu spreading infection with a sneeze, they can induce their hosts to infect other minds. When an analogy fits so well as to be indistinguishable from its proxy, it no longer can be said to be an analogy but a real phenomenon. If mental immunity, mind parasites, and cultural immune disorders are real, then so, too, are the inoculants to help restore fully functioning mental immunity. As recounted by Norman:

> Imagine a world where cognitive immunologists design interventions—"immunotherapies"—that restore mental immune health. Where mental immune "boosters" and "mind vaccines" prevent epidemics of partisan thinking. Where people think more clearly, reason more collaboratively, and change their minds when reasons show them to be in the wrong. Imagine reason-giving dialogue flourishing and breaking down ideological barriers. Imagine "fixed" mindsets unfurling, like flowers freed from frost.

Imagine, then, a world freed from the most virulent forms of the tribe virus. To get there, we need to identify the core compromiser or "disruptor" of mental immune systems. It turns out the science of cognitive immunology has identified it. That *thing* that drives our vulnerability to ideology is *willful unreason*. Also known as "willful belief," it corrupts cognitive immune systems by degrading the mechanical linkage between critical thinking and belief revision. One example illustrating this point emerged from a debate (now watched by over eight million viewers on YouTube) in 2014 between Bill Nye "the Science

Guy" and creationist Ken Ham. The moderator asked the key question of the night, targeting the beliefs of each of the debaters: "What would change your mind?" Nye displayed the openness to reformation of his belief system in light of evidence by responding with a single word: "Evidence." Ham responded with classic willful unreason: "I'm a Christian; nothing is going to convince me that the word of God is not true."[18] I recall watching this debate live, and I can still hear in my mind the outburst of applause for Ham with this clear and concise tribal virtue signal to his coalitionary alliance. This is the worst kind of immune-compromising disruption—*identity-based unwillingness to yield to evidence.* When your identity is at stake, there is no world where your position can be permitted to prove false. If willful unreason is the root cause of the spread of bad ideas, it should come as no surprise that its opposite is the key mental immunity booster that gives the greatest number of protections against them. Cognitive immunologists call this *metabelief*, which simply means "belief about beliefs."

Recently, Gordon Pennycook and his team[19] published peer-reviewed research that revealed a special kind of mental immune-enhancing metabelief. That work, titled "On the Belief That Beliefs Should Change According to Evidence: Implications for Conspiratorial, Moral, Paranormal, Political, Religious, and Science Beliefs" has found metabeliefs to be the key ingredient to resilient minds. This powerful force has been called many names. Lee McIntyre calls it the "scientific attitude," Pennycook and colleagues call it "actively open-minded thinking," and Norman calls it "reason's fulcrum."

When I use the term *metabelief*, I specifically mean the good, cognitive, immune-enhancing kind. Metabelief, therefore, is—*the precept that beliefs should change in response to evidence.* It is your mind's primary defense against mind parasites like fake news, misinformation, science denial, propaganda, fundamentalism, and extreme divisive ideologies. Metabelief is a key factor that distinguishes real critical thinking from what some academics claim is critical thinking, when what they really mean is "you are only thinking critically *if* you agree with me." It is perhaps the most important aspect of real critical thinking, and hedges against unconscious selectivity.

A common theme throughout this book has been that the things we commonly assume protect against tribal bias—education, expertise, intelligence, numeracy, or political affiliation—are not even close to being the primary factors that give us an inoculation to the tribe virus. Studies have uncovered a disturbing paradox to formal education. Some suggest that those who are better educated are more likely to craft convincing narratives to justify long-held ideas and thought patterns, making them even better in justifying their actions; people with specialized training become better at deceiving themselves and others when truth conflicts with their prior beliefs. Critical thinking (especially abused and miswielded by academics in the humanities, resulting in harmful postmodern ideologies that make illogical claims attacking the concept of truth itself*) is often a dog whistle for the specific ideology of the professors that spread them. The spread of these bad ideas has harmful consequences to society, as noted by Norman: "If students graduate thinking they're entitled to their political ideology's articles of faith, we've failed to create responsible citizens." Thus, if we are to achieve a tribe vaccine, metabelief needs to be a primary ingredient. But how do we get enough people to adopt metabelief? In other words, how do we as a species reach herd immunity to the tribe virus?

First, we need to update our definition of the tribe virus. The tribe virus is the combination of *willful unreason* that is pegged to identity. As discussed at length throughout Part I of the book, individuals that share group identity unconsciously bias nearly every category of information in the social world.[20] To restate the Fitness-Beats-Truth theorem, tribe trumps truth in value as a signal because coalitions, not truth itself, result in fitness benefits. It should be clear, then, that if we are going to have any shot at overcoming the tribe virus, we'll need to leverage identity-protective cognition to use group membership to counter its worst effects. Fortunately, science points the way on what type of construct is up to this seemingly impossible task. We will have to fight fire with fire, and use the power of *sacred values* to enshrine the tribe vaccine's distribution to humanity and stop the global epidemic of harmful ideas.

* A favorite and tragically flawed line of false reasoning by postmodernists is the *reductio ad absurdum* that "truth does not exist." This statement contradicts itself. If one posits under the guise of relativism that "truth does not exist," then the very statement itself cannot be true.

Recall, a sacred value is one that an individual observes as absolute and inviolable—even thinking of breaking a sacred value is a social taboo. Sacred values differ from material values in that individuals feel honor-bound to uphold them. Importantly, sacred values are powered by identity-protective cognition that comes with identifying with the community that originated that value. Norman cautions the use of sacred values in any instance, but also admits there may be one case in which an exception should be made—as a remedy: "It's always dangerous to treat something as sacred, especially if doing so places it beyond the reach of critical questioning. But the more carefully I examine 'Thou shalt yield to the better reason,' the more reverent my attitude toward it becomes. . . . For the purposes of building mental immunity—and resilient communities—it's hard to do better." Thus, to counteract the tribe virus with a tribe vaccine, we need to combine metabelief *with* identity-protective cognition. To vaccinate the world from the tribe virus, we need herd immunity of people who hold metabelief as their sacred value pegged to a single "community of inquiry" tribal identity.

This is what I call a *Metatribe.*

THE IDEA OF HUMANITY IDENTIFYING as one universal tribe is not new. One could make the argument that the holy grail of the humanist and spiritualist alike is a world where more people than not identify as the same tribe. The concept has been around since the first tribe allied with another tribe to cooperate within the framework of a super-tribe. This is and has been the recipe for the success of many civilizations. In unmatched prose, Carl Sagan and Ann Druyan outlined this recipe:

> Human history can be viewed as a slowly dawning awareness that we are members of a larger group. Initially our loyalties were to ourselves and our immediate family, next, to bands of wandering hunter-gatherers, then to tribes, small settlements, city-states, nations. We have broadened the circle of those we love. We have now organized what are modestly described as super-powers, which include groups of people from divergent ethnic and cultural backgrounds working in some sense together—surely a humanizing and character-building ex-

> perience. If we are to survive, our loyalties must be broadened further, to include the whole human community, the entire planet Earth. Many of those who run the nations will find this idea unpleasant. They will fear the loss of power. We will hear much about treason and disloyalty. Rich nation-states will have to share their wealth with poor ones. But the choice . . . is clearly the universe or nothing.[21]

Three hundred thousand years ago, the Tribe Drive virus infected our DNA. Now, we can take this infection and transform it into the most powerful amplifier for wellness the human species has ever witnessed. We can use the virus *against itself.*

The key lies in *identity-protective cognition protecting the spread of good ideas.*

The blind spots created by our Tribe Drive are not random. This book has outlined them in some detail. The available heuristic, negativity bias, motivated reasoning, biased evaluating—all these corruptions of human reason have been slaves to the Tribe Drive for three hundred thousand years. But by becoming aware of this fact, we are now alerted to the pressures that have brought us to this stage in human evolution. The hard data we are continuing to uncover can now be accessed in multiple fields and scienctific disciplines with myriad possibilities for the future. This allows us to take the science of tribalism and devise methods and norms to correct for their negative distortions upon human behavior. It will require deliberate mental effort. But as with all mental habits, once implemented, it can become second nature. Bootstrapped by the technologies of trust, we are now free to sacrifice all tribes, whether they be national, civil, ethnic, or any other mutant tribal existence fashioned by the evolutionary algorithm upon the altar of the one, single human tribe.

The Metatribe.

But how do we go about doing this? E. O. Wilson once wrote: "Human beings are consistent in their codes of honor but endlessly fickle with reference to whom the codes apply . . . the precise location of the dividing line is shifted back and forth with ease." Our goal is to use the science of the Tribe Drive to permanently affix the line. It is time for the tribal line of identification to come home to its final resting place. Norms change.

Overton windows (the current spectrum of ideas on policy considered socially acceptable by the general public) shift, and evolving zeitgeists turn the taboos of the day into celebrated freedoms of the present. But it's not as easy as simply changing the tribalism norm to an anti-tribalist norm. We need first to hack the tribal signaling system, to augment shared identity.

Once the Metatribe identity is elevated to a primary identity, we are then freed to follow the ethical precepts of the Concentric Circle to do no harm, but also to preference your in-group. This also leaves room for the peaceful coexistence of religions that adhere to both metabelief and the moral innovation of the Concentric Circle. Religions with these qualities could, in fact, be a profoundly human-wellness-enhancing tool in the *Homo sapiens* tool kit in the twenty-first century and beyond. What are the signaling tools at our disposal? Recall from chapter five the nested hierarchy of signals that are co-opted by our coalitionary alliance instinct to predict other social agents' group membership. Clothing, art, music, body modification, language and its idioms, and the slogans it can create—all these signals emit to the social world our group membership.

A tribe is essentially a signaling algorithm, a set of outputs that demonstrate social network fidelity. It's a password, a code, for an individual to be able to demonstrate the relationship of the self to the social groups competing within a complex landscape. Critically, due to our advanced cognition, we can learn multiple codes, multiple algorithms to demonstrate fidelity on many group levels. Humanity's way forward, then, is not to destroy our current identities (unless they are identities that harm other groups, breaking the moral precept of the Concentric Circle), but to adopt a unified series of multilevel coding and at the top, beyond face-to-face—one single team human identity, with a single sacred value. This is the Metatribe.

An example of this comes from one of science fiction's most revered narratives: *Star Trek*. In Roddenberry's utopic vision of the twenty-fourth-century future, humanity has not only overcome resource competition, discrimination, poverty, and war on planet Earth, but—in a future where the expanding circle of moral concern has burst beyond our solar system—is part of a unified interstellar community of sentient life. Starfleet is a kind of singular space-tribe with branding that unites not only humanity, but life across the galaxy. Membership in Starfleet is not compulsory, but if you sign up to explore the galaxy, you adopt a singular, color-coded uniform. As a Starfleet member you can bring with you a piece of your species' identity and interweave it into your Starfleet identity. One example of this is Lieutenant Warf, who is an alien Klingon and wears a "baldric" around his torso. He does this, though, as a sub-tribal identity; it is never overshadowed by the coalitionary alliance signaling to the Starfleet intergalactic Metatribe. No matter the life-form, if you are part of Starfleet, you adopt the uniform. It unifies their identity, which coalesces around their moral innovation, which is a type of space metatribal creed: the "Prime Directive." The Prime Directive serves as a kind of Hippocratic oath, to do no harm to evolving species. It's time for humanity to find our Prime Directive.

Fortunately, back on earth, brilliant minds have been working on the problem. Anti-tribalist politicians like Andrew Yang (who has remarkably managed to craft coalitions along the entire political spectrum) have captured the spirit of this idea in slogans like "Humanity First." Douglas Rushkoff has composed a twenty-first-century rallying cry for the spirit of human community in his manifesto: *Team Human*.[22] We need much more of this kind of thinking. My instinct is that just renouncing all tribes won't cut it. That would be akin to renouncing your interest in attaining and keeping status. It's too deeply evolved, too embedded in our subroutines . . . and is much easier said than done. To defeat the tribe virus we will need something much more directed, precise, and intentional. One factor in our advantage is that humans are superspecialists at creating new identities. This is one of the most powerful tools of the disenfranchised; perhaps, as we decolonize the symbols in our institutions (at the time of this writing, protesters around the world are dismantling symbols of oppressive history, and in the United States alone, almost one hundred Confederate monuments were removed in 2020), we must be careful to rebuild these with a shared identity, a creed, that appeals to and rallies the majority without alienating would-be allies. The branding of "Team Human" has to be so viscerally good that people *want* to adopt it. It needs to be a Nike-level brand, permeating both digital and metaspace alike. It's tempting to want to use the tools of the centralized state to enforce top-down singular identity, but it should be evidently clear by now that this will always fail.* I am not a marketer, but there exists the perfect combination of signals that will make everyone *feel* that they are party to a singular membership deep down in their bones. What we need is the greatest marketing campaign in the history of the world, to project a new model Metatribe: *Team Human.*

Imagine a world where this idea was crowdsourced, or the billionaires of the world funded the most important of all marketing campaigns, adopting the latest science of brand consumerism, to serve as a unifying force for the human species. In a way, this movement is and has been underway for some time, propelled by globalism. When people watch the same collective stories and consume the same products, they trade off a bit of the idiosyncrasies of their singular, ethnic, or genetic tribe for a super-tribe. But a super-tribe is not the Metatribe. Some anthropologists have speculated that by the year 3000 there will only be a single human language. This would radically transform our views of membership.[23] A global lingua franca is feasible in our lifetime, through the development of technology, given the capacity for what will eventually be a seamless universal translator. These are all unifying forces. However we get there, we need a human brand. One identity grounded in accuracy and prosocial, positive, good faith feedback.

Combining the Network State and Mental Immunity

The decentralized internet of the future will dramatically accelerate the extent to which we can share information and make decisions together in a social Network State that uses the "prosocial process" to foster polycentric governance.[24] As written by Wilson, Atkins, and Hayes: "We are literally at the beginning of an age when a global commons might be possible, where humanity might choose to create, test, and refine previously unimaginable new models of collective decision making that

* One could argue this was the ultimate end to the Nazi regime as envisioned by Adolf Hitler.

empower, include, and enable human beings across the globe." Novel institutions, such as the Human Energy Project, "envision a global human society that recognizes itself as a meaningful unity—a new kind of superorganism."[25] Inspired by Pierre Teilhard de Chardin's concept of the noosphere, these groups aptly want to create a third narrative—beyond the aging twin Leviathans of ancient myth and scientific reductionism—that combines the best of both to articulate a vision of global integration for the immense variety of human beings and communities that continually engage in its evolution. To do this, we need to be consciously aware of the uniting properties of the memeplexes we extol. There has never been a better moment for this unifying effort in the evolution of our species. In the words of the cyberpunk Punk6529 on Twitter: "memes > land." Those who control the memes of production will control the fate of nations.

It is worth now taking a moment to describe the two most critical ingredients that make up vaccines. The first element is the *immunogen*. An immunogen refers to a molecule that is capable of eliciting an immune response by an organism's immune system. By introducing weakened forms of the virus to the immune system, an antigen binds to the immune system and is indexed as "bad." The second element is an *adjuvant*. An adjuvant is used to prime the immune response with a set of instructions that result in a stronger, more effective immune response. The tribe vaccine's immunogen is the antigen of identity-protective cognition coupled with the adjuvant set of instructions of metabelief. To "administer" the tribe vaccine to people who choose to be inoculated from it, the key is identifying and practicing a Metatribal creed.

What is Team Human's creed? The recitation and internalization of a possible unifying creed is the single greatest prophylactic practice we can conceive to improve humanity's chances of survival and strengthen our cultural cognitive immune systems. The tribal vaccine uses the antigens of tribalism to train the body's immune system to identify the attacking agent. This cognitive immune booster serves as a mind inoculant against ideological thinking. Network societies that integrate this vaccine into the base code of their moral systems will be well adapted to confront the challenges of the future. We can share it with others and administer it to ourselves. The way to administer the vaccine to ourselves is to wake up in the morning and chant the only *tribal creed* that will matter in humanity's Metatribe of tomorrow, where all other tribes we may have been born in, or become a part of, all become less significant and our one metabelief becomes preeminent in our minds.

What is that long-awaited tribe vaccine?

I am a member of Team Human.

Our creed is that beliefs can change in light of evidence.

We are a community of inquiry where beliefs are deemed reasonable if they can withstand reasonable challenges to their veracity.

We are the Metatribe.

IF ENOUGH PEOPLE CAN BE convinced to participate in this mindset, then perhaps the entire human species can achieve herd immunity to individual tribal mind parasites. Skeptics may be thinking this vision is impractical, but Cyrus the Great showed us that—when the principles of the Tribe Drive are deftly applied—radical change can happen. Critically, major social norm changes have been experimentally shown to take a shockingly low threshold to overtake previous positions. One study quantifying the threshold by which a minority position overtakes a majority position stunningly revealed that it only takes 25 percent of a group to accept the new position for it to take root as the new norm.[26*] More importantly, if only a few Network State societies adopt them, they would be characterized by these wellness-enhancing properties, which could attract a threshold number of individuals to their cause. Like the diaspora of wellness-seeking humans coming from different parts of the world immigrating in waves to a young United States, people would flock to these decentralized refugia to stake a claim on their futures. If this happens, the chances of our species' survival throughout the coming millennia significantly increases.

Unfortunately, in the evolutionary "short run," xenophobia has been an adaptation for human survival; thus, it has been imprinted deep

* It's quite specific, as 20 percent adherence wasn't enough to shift the norm.

within us on a tribal level. But perhaps transcending beyond its boundaries is possible. If we intentionally nurture our species' capacity to do so, then maybe we can enter a new level of collective human consciousness. *Homo sapiens* may be at the dawn of a major evolutionary transition. The arc of the tribal universe is long, but it bends toward oneness.

Tribalism itself, in all its aspects, may yet be the key to saving the world.

To Reign with God as Kings and Priests over All the Earth

I know the tribe vaccine works. I have seen its protective effects with my own eyes.

My father was a shaman for his religious tribe. He was a fundamentalist minister for the Worldwide Church of God (WCG) for over twenty years, and as such, the world that was my childhood might be seen as unusual compared to a typical North American Protestant upbringing. In my younger years, up to the age of twelve, my life was strict and authoritarian. I never had a birthday or celebrated Christmas or Easter, as they were classified as pagan holidays. Thus, I distinguished myself from out-groups by way of abstaining from tribal signaling during more traditional cultural rituals. Our church was essentially a form of messianic Judaism, as we believed in Christ as the Divine Messiah and the most important figure in human history, but still adhered to all the Jewish Old Testament food laws (no pork or oceanic bottom feeders like clams or shrimp). To distinguish ourselves tribally from the "gentiles" (a term used by those who claim Jewish heritage to label others as out-group) we celebrated the Days of Unleavened Bread. This is a celebration of the journey of the children of Israel out of Egypt and through the wilderness, when following Passover and the Exodus, they ate unleavened bread for seven days. The church also kept the Feast of Tabernacles (the Jewish bible calls it Succoth) annually in various "feast sites" around the world. At larger sites thousands of people would gather together for eight days to worship and hear sermons about the coming millennium—when the Kingdom of God will reign on Earth and all people will be taught by God directly (Isaiah 2:3), "Come and let us go up to the mountain of the Lord, To the House of the God of Jacob and He will teach us His ways."

I recall the child-like pride I felt during the many voluntary acts of virtue signaling by my objecting to participate in a host of activities—from avoiding consuming any foods with yeast (during Unleavened Bread festival, which Orthodox Jews still do today), to voluntarily leaving the classroom if the exercise involved drawing or coloring a Santa Claus, or singing Christmas-themed songs. The most prohibitive—and costly—tribal signal of all was the weekly adherence to the Sabbath as a day of spiritual rest. That is, from sunset on Friday to sunset Saturday, all members were required to refrain from their regular work. This made it impossible for some to continue in their current employment, because those who became true believers in our particular biblical interpretation and converted could no longer continue to work on Saturdays. They had no choice if they wanted to become members of the church; they had to observe the Seventh Day of rest, they had "to obey God." As a tribally indoctrinated child, I, too, "kept the Sabbath." I did not perform any activity that was considered a distraction from spiritual reflection, which ranged from playing sports to watching television. The Sabbath was one of the Ten Commandments and was therefore viewed as the Test Commandment of obedience that God demanded. I look back now and see that this was the WCG's sacred value and its chief tribal signal. If you kept the Sabbath correctly, you were one of us.

By distancing ourselves from "worldly people," the women and men who joined our religious tribe were offered all the benefits that humans have coveted since they evolved the Paleolithic coalitionary instinct: for women, the group provided an inbuilt supportive community replete with emotional solace, and for men it promised a glorious future and gifted status as "one of God's chosen." It also provided discipline and a more predictable life. Men who had little power "in this world" could practice and signal the WCG tribal creed, which promised them all power in "the World Tomorrow" when we "would reign with God as Kings and Priests over all the Earth."* All other tribes would be subjugate to our one "true" tribe. This was certainly an exhilarating mindset for members of the church. We believed every word of the Bible quite literally.

* Revelation 5:10

With all the benefits of belonging to a community of faith, what we lost was something far greater, and more precious—a touchstone on objective reality and truth. This led to a very compromised mental immune system. Fundamentalism is the belief that one's holy book, whether the Jewish Torah, the Christian Greek scriptures, or the Muslim Qur'an is *literally true,* without any scholarly thinking applied to discover metaphors, or seek out historical context and meaning. All fundamentalists, whether Jewish, Christian, or Muslim, deny the theory of Darwinian evolution. This is classic identity-protective willful belief and is a mind-parasite.

Today, my father calls some of these negative traits of fundamentalism, "theological schizophrenia." Just as schizophrenia is a mental illness that disconnects people from the real world, theological schizophrenia is an anti-scientific mental virus. Despite benefits for the people within some fundamentalist Christian sects who are not "too legalistic," the ultimate costs can be immense. On a societal level, it ultimately pits tribe versus tribe in a zero-sum game and erodes people's mental immunity to bad, ultimately harmful ideas. Any tribal creed that ultimately diminishes trust among strangers is a negative value that needs to be addressed by those in power. This results in a breaking of the precept of the Concentric Circle—because it causes untold, immeasurable harm to others' groups at the expense of your in-group.

My father, like so many other human beings, relied on the Tribe Drive's three-hundred-thousand-year-old answer to the Trust Paradox. He joined a religion. Yet, after dedicating two decades to his religious tribe, he resigned from the church. How was this possible? How could he renounce the religious tribe that he had dedicated and sacrificed most of his adult life to?

My father retraces the chain of cause and effect in his life and has told me that he owes a lot to one high school English teacher during his freshman year who offered a course on Transcendentalist philosophy (the writings of Emerson and Thoreau), in which she inculcated in her students a love of the truth. This high school memory was triggered much later in life when my father was pastoring two French churches in New Brunswick, Canada, in the 1980s. After eight years serving there, the church transferred us to Lennoxville, Quebec, home of Bishop's

University in 1989. It was there, while scouring the library in an attempt to answer a geology student who challenged him on his literal reading of the Book of Genesis, that my father was first confronted with the concept of "theistic evolution." This is the idea that God could have used evolution to bring about all of creation over billions of years. Of course, this meant that parts of the Bible could not be taken literally.

My father soon learned that several mainstream Christian churches accepted the theory of theistic evolution as fact, even the Catholic church. In his studies he learned the Greek definition of truth, that "truth is always consistent within itself." He was also struck by something he read about Mahatma Gandhi's conception of truth itself. Gandhi once said, "When I was young I used to believe that God is truth, but when I became older and wiser, I came to understand that Truth is God." Every religion lays claim to the truth and they all say that "God is truth." But when we press further, we find that what they really mean is that "only my version of God is truth." Gandhi's reversal of this common expression to "Truth is God" makes all the difference in the world. Truth does not contradict itself. Thus, much of the evolution of thought that propelled my father forward was with the transition of his mental framework from "God is truth" to "Truth is God." The former is tribal, meant to empower one tribe over others; the latter is universal, meant to bind all tribes together in unity of reason under the *Epistemic Golden Rule*: "Observe the same standard of belief you would have others observe."[27] My father's teacher had inculcated a love of philosophy and this stayed with him, and was, perhaps, his first inoculation of metabelief.

Little did he know that when he was a teenager, he had been given a cognitive immune booster that produced just enough mental antibodies to eventually rid his mind of the tribe virus. This resulted in a rocky, tumultuous journey to the Worldwide Church of God.* He went from an indoctrinated Catholic by upbringing, to him and his brother joining the Jehovah's Witnesses after an encounter with missionaries. The Jehovah's Witnesses' doctrine proved to have too many inconsistencies within itself, so within one year, he left while his brother stayed. The per-

* It was said of my father, mockingly, by his brother's Jehovah's Witness friends, he "changes his religion as often as he changes his shirt."

son who demonstrated these inconsistencies to my father was a pastor of the Church of Christ. Subsequently, my father began attending this church for six months all the while continuing his study. But the inconsistencies within the Church of Christ doctrine soon led to him leaving them and beginning a two-year deep study of all the denominations of Christianity without committing to any of them. He eventually heard the charismatic radio voice of Herbert Armstrong, pastor general of the Worldwide Church of God. One thing in particular attracted him to Armstrong; the televangelist claimed to "believe in always being willing to reject error and march toward truth." This jibed with my father. When I asked him about this period of his life when he was seeking God, he responded: "If certain ideas I held could be proven to be in error, then I would change and update my beliefs." And update he did—many times with the new information he learned. He felt that each step he took to reject error and move toward the truth was a "step toward the Divine."

However, this germ of metabelief in Armstrong's otherwise fundamentalist ideology was to unknowingly set in motion events that were to come back to haunt half of the Worldwide Church of God after his death in 1986. My father resigned from the ministry in 1999 to write a book on theistic evolution. But unlike other pastors who left the church, my father kept good communication with the leaders of WCG. One such leader, John Halford, who had read my father's book, called him up in 2009 and asked him to send six copies to the church's governing body in Pasadena, California. To my father's astonishment, the WCG, now known as Grace Communion, decided that a member of the church could now in good conscience accept the theory of evolution and still be considered a member in good standing. This was historic: *it was the first fundamentalist denomination in U.S. history to publicly accept the theory of evolution.*

The Metabelief that Armstrong had often spoken—"to be willing to change when confronted with error"—became a reality in the church, but not without some severe consequences. The church's income, over two hundred million dollars per year, fell to around forty million after the Sabbath doctrine was changed and became no longer obligatory in the late 1990s. And the membership of 150,000 and another 100,000 "coworkers" who donated regularly to the church fell to around 65,000 as half of the ministry left and took members with them to "keep the faith once delivered"—in other words, they could not accept the Sabbath change as this had been a key tribal signal and sacred value for the church. The change on evolution cost Grace Communion a loss of four more churches. At the core of this tectonic shift away from a fundamentalist group's transition to stronger cultural mental immunity was Armstrong's mantra "Don't believe me, prove it for yourselves!" And so, Armstrong was "hoisted by his own petard" as the ancient Romans used to say.

Can metabelief really make a difference to our world?

Sometimes, it only takes one person who approaches another tribe and honestly begins a dialogue—a community of inquiry—for a false belief to be changed. My father had begun his journey by embarking on his quest to "disprove evolutionary theory once and for all." On this journey he had to confront the great scientists and intellectuals of the nineteenth and twentieth centuries. This quickly led to his realization that evolution was a fact. This exposure to the scientific method and natural history were booster shots of mental antibodies that eventually forced a critical mass. It led to full-blown immunity to the tribe virus. By the time of his resignation, he had embodied the creed of the Metatribe. What occurred was nothing short of a miraculous, radical transformation that has the potential to help those infected everywhere. But even more miraculous was what followed—the change that occurred to over fifty thousand members of the former Worldwide Church of God. My father was one of thousands that identified as a massive religious tribe who overcame errors in thinking and freed themselves from a mind-controlling parasite. This ultimately led to the first fundamentalist denomination in the history of the United States to obey the metabelief commandment: "Thou shalt yield to better reasons," and accept evolutionary theory.[28] Somewhere, deep within his mind, lay dormant antibodies of metabelief that "beliefs should change with the evidence." If that had not existed, it would have kept his cognitive immune system—and perhaps mine by exposure—compromised and unable to filter out bad ideas. He had everything to lose—his community, friends, source of income, station in life—

and every incentive to stay a shaman of his tribe. But he didn't, because he came to fully embrace his metabelief precept: "Truth is God."

The tribe vaccine works. We've seen it. We've experienced it. Will you help us, members of the Metatribe—to administer this vaccine to save humanity?

EPILOGUE

The Timeless Hero

Nothing vast enters the life of mortals without a curse.

–Sophocles

Each epoch of cosmic time has witnessed a great enemy. A great "evil." Something has arisen, again and again, to be a scourge on the land. It is a kind of metaphorical Sauron, that commands marauding (antisocial) hordes of orcs, goblins, and wraiths to bring devastation to the world of (prosocial) elf, dwarf, and man.* You may insert the villain of your choice if you like—be it Thanos, Agent Smith, or Lord Voldemort—but the theme remains the same. It is what Darwin's Bulldog, the biologist Thomas Henry Huxley, termed "survival of the fittest," that selfish, antisocial behavior that benefits the individual (or their collective) in opposition to you and your collective.

In the history of life, self-interest has appeared to be the dominant force. And why shouldn't it be? From Huxley's main proposition on Darwin's theory, cooperation should not exist. *Prima facie* it appears that life-forms that look out for their own self-interests should always be able to prey upon, take advantage of, or outwit those life-forms that

* If this reference fails to hit the mark of recognition for the reader, I highly encourage you to read what is arguably the greatest fantasy saga ever told, J. R. R. Tolkien's *Lord of the Rings*. I believe it is timeless in both theme and scope, because the "us" of *our* people are fighting a primordial, eldritch evil of *them*.

sacrifice any part of themselves for other units in an ultimate zero-sum game of evolution. Cooperation should not exist . . . but it does—otherwise you would not be reading these words. *"Tho' Nature, red in tooth and claw"* has its advantages, if it were the only law of organismal interaction, the human species would never have come into being.

You are an organic monument to several primordial alliances of molecules, membrane-protected replicators called cells, and amalgams of cells that we call organs when serving a higher-order purpose through collaboration. The sum is greater than its parts. That sum is you. In very real terms, you are a *nested group of us and them* that at some point, long ago and far away, overcame their differences. Perhaps it is possible that this iterative ancestral experience—of a group of *us* overcoming an existentially threatening *them,* traced all the way back to the very first sparks of organized molecules—may be the reason that humans are addicted to storytelling. The formula of protagonist and antagonist, good guys and bad guys, is a timeless tradition that knows no cultural boundaries. The story is baked into the universe. It self-assembled into *you.*

Overcoming differences? Forgoing separateness? What power could compel life to this end? Scientists have discovered the processes and we've called them major evolutionary transitions. The implications of this discovery are still being unpacked as a new generation approaches evolutionary theory with a different light and vantage point: *evolution can sometimes occur without the mutations of the individual and instead come from multiple individuals forming cooperative groups.* When did an apparently selfish Darwinian law begin to permit cooperation? How were the molecular, cancerous "Saurons" beaten back, restoring justice to the land? It is rare, but when the benefits of cooperation are strategically *elevated* to constitute an organism in its own right, then something incredible happens—there is a dawning major evolutionary transition. Major evolutionary transitions are metaphorically akin to major events in what Joseph Campbell dubbed "the hero's journey," only these are evolutionarily very real and on the grandest possible scale.

Here's how they work. Nearly all life is embedded in a web of groups. Yet, a single selfless actor in a group loses energy by being selfless toward the group. Typically, this results in single actors being selfish. In order to overcome selfishness, groups with more selfless actors become super competitive against their rivals. A scale exists, where success can be had at multiple levels, but the selfish-individual lower levels can only undermine the higher-group levels until groups elevate cooperation as the dominant strategy. When this happens, the magic of major evolutionary transition takes hold and those species begin to thrive. In other words, the only way to survive competition is by doubling down on cooperation.

Let's consider an example from ancient military history. The Roman Empire was a nest of cooperating ethnic groups that started as a melting pot of warring tribes and ended in a monolithic super-tribe. Using the lens of major evolutionary transition, we see that the Romans elevated themselves to the super level, moving from the previous stage of constant local warfare into one competing tribal identity among the many rival human tribes that surrounded them. They were an evolving group in a networked web of groups.

Rome was beset on nearly all fronts by other tribes with long-standing warrior traditions, which—for Rome's early fledgling existence—gleefully sacked their lands to the point that they nearly went extinct.* *Vae victus* or "woe to the vanquished" were the terms of the day, applied ruthlessly to Rome when it was occupied by the Gauls in the fourth century BC. In existential response, the Roman military innovated a hyper (within group) prosocial tactic: compulsory military service. This service to the collective was a massively costly endeavor for the men who were conscripted in the name of Rome. But not only did this service save the tribe, it turned it into the most powerful institution the world had ever witnessed: the Roman Empire.

The source of this power was the crucial innovation permitted by the application of novel military tactics: the Roman testudo. In the testudo (Latin for "tortoise formation"), soldiers aligned their shields to create a packed formation covered from top to front. The men in the first row would hold their shields perpendicular to the ground from about the height of their shins to their eyes, providing cover for the formation's front. The men in the back ranks would balance their shields on their helmets in an interlocking knit, protecting their heads. This innovation,

* An army of Gauls led by Brennus sacked Rome in 387 BC, capturing most of the city and occupying it for months.

which became the backbone of the Roman military for the next seven hundred years, redrew the lines of tribal identity (and gene pools, for that matter) across the known world.

It even had its own phenotype: typically, shield surfaces were first covered with canvas and then calf skin and painted the color that represented Mars, the god of war—a deep, bold, bloodred. When a testudo is fully articulated, each shield is like an armored plate on the body of a pangolin (a type of scaly anteater*). The extended group phenotypes innovated by the Roman military shaped and sculpted the population genetics of the day—stretching from Hadrian's wall to the sacred places of Jerusalem.

Major evolutionary transition theory provides a theoretically grounded explanation for why some species experienced astounding success in certain moments throughout the history of life on Earth. Among the set list of protagonist species, 3.5 billion years ago, heroic membrane-creating molecules self-assembled a perimeter—*the first perimeter*—protecting genetic precursors to life, as long as they could outwit the enemy of defector cannibal molecules that consume those inside the cell wall. In the time of *the evil mutation* some seven hundred million years ago, symbiotic mitochondria, nuclei, and other organelles defied the voracious hunger of cancers attempting to extinguish them. In the epic origin of eusocial invertebrates, the predecessors to today's ants formed colonies of hyper cooperators, destined to defend their nests successfully over rival colonies, parasites, and free riders within. In fact, they have been so successful there are an estimated twenty quadrillion individual ants across the globe, which together weigh more than all wild birds and mammals combined.[1] The ancestors of naked mole rats battled the great enemy of rival species competing for underground tunnels and turf, and defectors within their ranks, to become the world's first eusocial vertebrates twenty-five million years ago. And of course, our heroic ancestors, dwelling in East Africa two million years ago, fended off coercive dominant males and camp free riders in a process of self-domestication that has led to the most cooperative—and premeditatedly vicious—ape on the planet.

* Perhaps the testudo should have been called the Latin derivative of anteater instead of tortoise.

The overarching theme is that each epoch's heroes (prosocial species) has to overcome either an internal or external threat to conquer the "evil" of its time—and this evil is pernicious. It is a constant. The cold and the darkness are ever present. Entropy—the second law of thermodynamics that seeks to tear down complexity into simplicity—is a patient stalker with an excellent track record of hunting prey. Entropy always wins. It has time on its side. Entropy states that in isolated systems, absence of energy spontaneously evolves toward thermodynamic equilibrium; in other words, a state of maximum entropy. Humans put a word to this law long before we knew the mathematics of the phenomenon—we call it death.* From the draining off of heat produced by thermosynthetic hot springs of primordial earth to the blackness of space, entropy has always had the advantage. But "good" can win. It is an auspicious and rare event . . . but when it happens, the organisms in question use the forces of prosocial behavior to defeat antisocial behavior, and from the ashes arises a new world order. The question still remains: Will humans remain the heroes of our own narrative?

By positing group-level selection, Darwin deflected the blow of the paradox of how and why prosocial traits evolved: "Although a high standard of morality gives but a slight or no advantage to each individual man and his children over the other men of the same tribe . . . an advancement in the standard of morality will certainly give an immense advantage to one tribe over another."[2] Yet this argument has a glaring deficit: destructive conflict is not eliminated, but punted up the hierarchical rung to the next level of intergroup interfacing, where the cancerous conflict can metastasize once more and spread even greater devastation.

This is important. It explains why cooperation has not run roughshod over competition. For example, two ethnic groups may opt not to go to war with each other by forming an alliance, but it could be to serve the function of becoming a super-tribal coalition that can exact

---

* It's why I identify as an anti-entropist. Consider this the opposite of dangerous human identities, like the postmodern "deconstructionists" whose force in the world bolsters entropy. It's easy to tear stuff down. You are doing death a favor. I believe humanity is at its best not when it tears down "deconstructively," but when it builds constructively.

even more catastrophic violence upon another population. Violence, death, and lethal competition in this model are not extinguished but delayed, only to be applied later on a greater scale. When Rome dominated their rivals with the testudo, a final tenet was enacted. The scale of cooperation—and conflict—was elevated to a higher multitier level, where in order for other groups to survive, they, too, would need to elevate cooperation as their dominant strategy. Empires and civilizations were born in this primordial supertribal soup, and so, too, was a level of warfare never before seen by our species.

At some point between 2 and 1.8 million years ago, hope manifested as small groups of protohumans collectively embarked on a radical social experiment that eventually became a major evolutionary transition that would transform the world. This species-defining event occurred when more of our ancestral lineage began emphasizing the social self as the key unit of survival. This was likely a compounding model, where relationships in early groups oriented to increase the likelihood of win-win outcomes. The rewards compounded slowly, but over time they became immensely valuable. This event of social assembly was miraculous. Those early human groups that began leaning on each other—in unique, special, and never-before-practiced ways—sparked a type of social revolution that resulted in our species' ascendancy . . . but there was a cost. A dark, deeply embedded trade-off. The indelible mark of Cain* remains. Our ancestors overcame the enemy of their time and bestowed a great *blessing* upon their progeny. But in doing so, they unknowingly cast a terrible *curse* upon us. The forces of evil did not vanish, but crept into the shadows to reform, regroup, and return with more power than ever before. The great evil of our time has driven our species into two world wars, and to the brink of nuclear holocaust and the myriad existential threats we now stare dead in the eye.

Our fate is uncertain. This great evil may yet win. Never before has a species controlled the power of the atom. Tribal out-group psychology—as of now a three-hundred-thousand-year-old adaptation—simultaneously

* The biblical story of the curse of Cain (Genesis 4:11–16) was the tale of the first-born son of Adam and Eve. Cain, murdering his brother Abel and lying about the murder to God, was cast from the settled lands. It was said that when Abel's blood was let and taken into the ground, the earth became cursed.

drove the horrors of the attempted genocide of a Jewish tribe over years of meticulous and industrialized murder while it also snuffed out the lives of nearly half a million Japanese people at Hiroshima and Nagasaki in a momentary flash of energy and heat. The very element that gave us the capacity to unlock subatomic secrets, by way of incredible levels of cooperation for the sake of an in-group, also compelled us to unleash its power to devastating and deadly ends on an out-group. The mushroom-shaped plume that ensued is an iconic symbol that represents the worst of the curse hexed upon us at the dawning of our greatest Paleolithic achievement.

The blessing of our ancestors was that they and their descendants could, for the first time, possess the capacity to see a social self as one single protective identity. Yet, as they basked in the light of identifying as one group, united in the common cause of survival, the evil that lurked in the *shadows of forgotten ancestors* lingered and grew in power until the moment was right to strike. It is now manifest in its full power—feeding those who would wield its dark art to stoke the fires of political tribalism and strong-arm populism across the globe. Like all dark arts, the power is so surreptitiously embedded that it proves difficult to reveal; and so, until now, like some wizened Gríma Wormtongue it has whispered in the ears of antisocial, self-aggrandizing big men with the power of the atom at their fingertips. To understand how we got to be the way we are, we traveled back in time to the place where the blessing and curse was first bestowed. And that should give us hope; after all, we are the only Earthling species that has ever existed that knows what a major evolutionary transition is. Could this awareness be the source of our salvation? By retracing our steps, maybe we can reveal the secret talisman to fight the evil before it renders us slaves to its will. And answer—by the timeless heroes that lay embedded in us all—the greatest question of our time:

Can we lift the curse?

APPENDIX

Testing the Tribe Drive Hypothesis

I believe the implications of the existence of the Tribe Drive for both science and humanity are profound. Yet, this topic has only recently captured the attention of many researchers and there are many as of yet unexplored facets of the Tribe Drive theory. The theory stands on the shoulders of scientific giants who have outlined signaling theory in the animal kingdom,[1] group and multilevel selection theory,[2] in-group/out-group/minimal group psychology,[3] sexual and virtue signaling,[4] and other forms of human social selection,[5] including self-domestication.[6] To my knowledge, the anthropologist John Tooby was the first to term the *coalitional instinct.*[7] Here, we can operationalize the concept of coalitional instinct as a hypothesis that can be modeled in a new theoretical branch of science called *coalition cognition*. By integrating signaling theory into coalition cognition, we can begin by a testing of a novel idea I call *the Tribal Signaling Hypothesis* (TSH). The TSH is the postulate that natural selection shaped at least some of our distinctively human cognition to process signals of symbolic affiliation to tribal coalitions. Here are just a few predictions stemming from the TSH:

1. *Mate preference and selection*
   a) The signaling of tribal virtues should be favored in mate choice. Potential mates should weight aspects of individuals that are motivated

to tribal signal in socio-sexual situations, thereby testing the specific tribally compatible virtues of mates.

b) Individuals seeking long-term mating relationships should favor costly tribal signals. In contrast, individuals seeking short-term mating relationships should prefer cheaper tribal signals.

c) If tribal signals are valued in mate selection, same-sex rivals should derogate each other respective to the honesty of their coalitionary alliance signaling.

2. *Phenotypic features of tribal signaling*

a) If tribal signals are valued as phenotypes, they will be displayed with greater variance in males. As with many species with a degree of sexual dimorphism (hence extra-pair copulation resulting in polygyny), there should be higher male variance and skew in reproductive success, which will favor risk-seeking patterns of trait expression. Specifically, males are predicted to be more conspicuous tribal signalers than females.

b) Males lacking sexually attractive traits should favor alternative mating strategies by enhancing tribal signals. Males are predicted to enhance tribal signaling, and costlier signaling, to circumvent mate choice by the opposite sex and mate-guarding by same-sex rivals.

3. *Genetic features of tribal cognition*

a) If tribal in-group preferential behavior are "good genes indicators," they should prove to be positively heritable. This could be tested by way of genetic twin and adoption studies.

b) If out-group stereotyping behaviors are costly and evolved under natural or sexual selection, the underlying genes and behaviors should be expressed differentially throughout life history stages. Specifically, tribal signaling should be most pronounced after sexual maturity and be correlated with certain sex hormones (especially the status hormone testosterone). This should be expressed by higher heritability of in-group and out-group psychology in adults than in children. Alternatively, if tribal psychology is typically partner-quality signaling, then they should show a life-history pattern that shifts depending on sexual maturity and whether one is mated securely or unmated.

The proposed TSH model is intended to complement (not replace) other models of human social, moral, and psychological evolution. If empirical testing supports the TSH, it will dovetail alongside many other models of social selection theory, including kin selection, reciprocal altruism, commitment mechanisms, risk-sharing mechanisms, social norms and cultural evolution, group selection, equilibrium selection, among others. It is my hope that this line of research will be critical to overcoming the challenges that political tribalism presents to us on a societal level, all while identifying the features of the coalition instinct that can—on the small group level—enhance our health, well-being, and ability to flourish.

Part II of the book introduces many practical concepts best explicated outside *Our Tribal Future*. Therefore, I have crafted an external site where workshops, surveys, instruments, and other useful tools for those who are interested in taking on the challenge of campcrafting their own community or want to more formally quantify and assess their own coalition cognition. This site is a live, working series of articles that is a practical guide to those who want to know more about their Tribe Drive and how to leverage it, in practice, to enhance their lives.

ACKNOWLEDGMENTS

The Buddhist monk Thich Nhat Hanh once said: "In this food I see clearly the presence of the entire universe supporting my existence." Something similar could be said for a book. To the whole, beautiful web of my social network—from the moment gamete met gamete, son of a father and mother, brother to brother, friend to friend, pair-bond to pair-bond, campmate to campmate, and mentee to mentor—I owe everything to the tapestry of human beings that stitched me together.

In this book I see clearly the presence of an entire social universe supporting my existence.

To all those who inspired, encouraged, enabled, and contributed to *Our Tribal Future*, I am eternally grateful. To my extraordinary agent, Don Fehr: thank you for seeing this project's potential and guiding me from meme to memeplex. This book, written throughout the COVID-19 pandemic (a tumultuous time for the publishing industry), graced the editorial minds of three editors. Thank you, Pronoy Sarkar, for bringing me into St. Martin's with a shared vision of what could be. You inspired me to go big, be bold, and aim for epic. Your initial mark remains. Daniela Rapp, thanks for helping me cull an overwritten manuscript. And crucially, thank you, Kevin Reilly, for believing in my vision and guiding me to the book's final form.

Specifically, with respect to the writing of *Our Tribal Future*: to Mike Smith I owe the ultimate butterfly effect acknowledgment. I still recall our hiking with his dog, Sagan, and both of us playing out a "what if we wrote books about tribalism" thought experiment. Deep gratitude to my father, Dan Samson, who was a surrogate editor and philosophical interlocutor for the entire journey—not only for the book but for my life. Brandon Minton, for taking me into the Hall when it mattered most and for embodying the hero's journey with every breath; the ideas we discussed in those halcyon days are the applied bedrock of this book. August Costa, for keeping me on the path of discipline when I otherwise would have faltered. Patrick Heyes, for insights on the life of a tribal maven. Joshua DeYoung, for reading the early proposal and distilling the Concentric Circle into its most pure essence. Kyle Potts, for trailblazing the twenty-first-century chivalric path, PMing my life, and so much more.

Other thinkers have inspired, throughout my intellectual life, my view of the cosmos, and the writing of this book: Kevin Hunt, my academic father and mentor. Charles Nunn, a mensch and awe-inspiring evolutionary scientist. Sam Harris. Nicholas Christakis. Richard Wrangham. Balaji Srinivasan. David Sloan Wilson. Jane Goodall. Satoshi Nakamoto. Peter Singer. Yuval Noah Harari. Charlie Gehy. Andy Norman. Ann Druyan and the family of Carl Sagan: I'll treasure the night we met at Sasha's book signing and the moments afterward when I visited Carl's grave to make him a promise and an oath I hope I never break.

For reading and providing feedback on manuscript drafts: Allyssa Crittenden, Daniel Benyshek, Eric Shattuck, Heather Shattuck, Dan Pardi, Pierre Lienard, Leela McKinnon, Luke Louden, Sarah Van Tassel, Randi Griffin, Karisa Hoke (chapter six is dedicated to you). Fortuitous conversation with Tim Gosnel and Jason Howard—thanks for the orange pill, dudes. A special thanks to the Department of Anthropology at the University of Toronto, Mississauga, without whose support this project would not have been possible. I couldn't ask for more supportive colleagues. Of particular note, thank you, Esteban Parra, for wisdom, counsel, hands-on feedback, and inexhaustible support. Thank you, Tracey Galloway, Lauren Schroeder, Rasmus Larsen, Genevieve Dewar, Sherry Fukuzawa, Francis Cody, and Carolyn Loos, for expert opinion and support during the editing process. To those who provided a sounding board, encouragement, or other forms of aid:

Martin LaPointe, Yuri Kazimirov, Peter Lai, (Sister) Helen Pan, Sean Evans; and to the family and friends that took the Big Five for me: (Sister) Pearl Truong, Logan Meneely, Aaron Madison, Holly Green, Luke Louden, Leslie Hudson, Kay Yarnall, Kaleigh Reyes, Luke Casey, Kai-Lani Rutland, Ming Fei Li, Noor Abbas, Megan Portnoy, Nicolle Hodges, James Yu, Olivia Clavio, Ujas Patel, John Givans, Erica Kilius, Jane Weldon, Ben Swain, Christy Laguardia, Michael Dyer, Alex Riddhagni, Jonathan Miller, Lydsee Miller, Jenny Thompson, Brett Thompson, Kenny VanHouten, Kayla VanHouten, Krystal Stier, Mike Stier, Alyssa Riddhagni, Allison Shelby, Ryan Olson, Amanda, Barbara, Elana, Margarita, Mili.

ERS: Branders, Bretters, Chrisers, Danners, Drewers, Fanners, Jers, Jonners, Joshers, Kenners, Kyelers, Loggers, Louders, Madders, Matters, Mikers, Riggers, Smithers, Peters, Potters, Gussers, Gozners, Zachers. And POIs to come. In the SCA, the following Knights have proven invaluable role models while walking the path: Sir Cecil de Tueurleon, EikBrander Solgyafi, Savaric de Pardieu, Aleric le Fevre, and Gunnar RedBoar. To the camps I've lived with, fought alongside, and feasted with: the Brotherhood of Steel and Trotheim of the Middle Kingdom. The Atlanteans. I couldn't dream of a better camp to build a life with. The Samson clan, who protected me from seeking corruptive identities by giving me a strong one: *Pejus Letho Flagitium*.

And last but not least, to my mother, Dana Samson, who nurtured and loved me into being. Kyel Samson, you are the best brother a human could ask for. How did I get so lucky? Tu-Ha Agi. Ni draga-gan I degi-gan. Agents of Inevitability. Ta-an Aganai. And without further delay: Nancy Samson, "you met me at a very strange time in my life," and from the first moment (and the moments shared and to come) your mind's imprint on my own flitters from page to page. Our daily rituals give meaning to the rising sun and setting moon (despite us understanding its physics). Finally, a prayer to our future children as I envision them—they have the poise, beauty, and strength you grace me with every day. To the journey and the legacy that will endure.

In loving memory of LaVerne Weldon.

NOTES

PROLOGUE

1. N. A. Christakis, *Blueprint: The Evolutionary Origins of a Good Society* (New York: Little, Brown Spark, 2019).
2. M. Foucault, "Friendship as a Way of Life," in *Ethics: Subjectivity and Truth*, vol. 1 (1997), 135–140.

CHAPTER 1: THE TRIBE DRIVE

1. S. Pinker, *Enlightenment Now: The Case for Reason, Science, Humanism, and Progress* (New York: Penguin, 2018); A. H. Hastorf and H. Cantril, "They Saw a Game; a Case Study," *Journal of Abnormal and Social Psychology* 49, no. 1 (1954): 129; H. Mercier and D. Sperber, "Why Do Humans Reason? Arguments for an Argumentative Theory," *Behavioral and Brain Sciences* 34, no. 2 (2011): 57–74; discussion 74–111; C. G. Lord, L. Ross, and M. R. Lepper, "Biased Assimilation and Attitude Polarization: The Effects of Prior Theories on Subsequently Considered Evidence," *Journal of Personality and Social Psychology* 37, no. 11 (1979): 2098; D. M. Kahan, et al., "Motivated Numeracy and Enlightened Self-Government," *Behavioral Public Policy* 1, no. 1 (2017): 54–86; M. S. Nurse and W. J. Grant, "I'll See It When I Believe It: Motivated Numeracy in Perceptions of Climate Change Risk," *Environmental Communication* 14, no. 2 (2020): 184–201.
2. R. M. Sapolsky, *Behave: The Biology of Humans at Our Best and Worst* (New York: Penguin, 2017).
3. Sapolsky, *Behave.*
4. R. Wrangham, *The Goodness Paradox: The Strange Relationship Between Virtue and Violence in Human Evolution* (New York: Vintage, 2019).
5. C. Boehm, *Moral Origins: The Evolution of Virtue, Altruism, and Shame* (New York: Soft Skull Press, 2012).
6. Wrangham, *The Goodness Paradox.*

7. D. Eagleman, *Incognito (Enhanced Edition): The Secret Lives of the Brain* (New York: Knopf, 2011).
8. D. Hoffman, *The Case against Reality: Why Evolution Hid the Truth from Our Eyes* (New York: W. W. Norton & Company, 2019).
9. S. R. Cavanagh, *Hivemind: The New Science of Tribalism in Our Divided World* (New York: Grand Central Publishing, 2019).
10. D. Hoffman, *The Case against Reality*.
11. C. Sagan and A. Druyan, *Shadows of Forgotten Ancestors: A Search for Who We Are* (New York: Random House, 1992), xvi, 505p.
12. R. G. Foster and T. Roenneberg, "Human Responses to the Geophysical Daily, Annual and Lunar Cycles," *Current Biology* 18, no. 17 (2008): R784–R794.
13. C. Sagan and A. Druyan, *Shadows of Forgotten Ancestors*, xvi, 505p.
14. S. Harris, *Waking Up*. https://app.wakingup.com/, 2021.
15. S. R. Cavanagh, *Hivemind*.

CHAPTER 2: TRIBALISM MISMATCHED

1. M. A. Schlaepfer, M. C. Runge, and P. W. Sherman, "Ecological and Evolutionary Traps," *Trends in Ecology & Evolution* 1, no. 10 (2002): 474–480. D. S. Wilson, A. J. Basile, and J. B. Smith, *Evolutionary Mismatch and What to Do About It*, The Evolution Institute, 2019.
2. A. Hallam and P. B. Wignall, *Mass Extinctions and Their Aftermath* (London: Oxford University Press, 1997).
3. E. Lloyd, D. S. Wilson, and E. Sober, *Evolutionary Mismatch and What to Do About It: A Basic Tutorial*, The Evolution Institute, Wesley Chapel, Florida, 2011.
4. C. Sagan, *Broca's Brain: Reflections on the Romance of Science* (New York: Random House, 1979), xv, 347.
5. R. M. Sapolsky, *Behave: The Biology of Humans at Our Best and Worst* (New York: Penguin, 2017).
6. J. Henrich, S. J. Heine, and A. Norenzayan, "Beyond WEIRD: Toward a Broad-Based Behavioral Science," *Behavioral and Brain Sciences* 33, nos. 2–3 (2010): 111–135.
7. G. Horváth, et al., "Reducing the Maladaptive Attractiveness of Solar Panels to Polarotactic Insects," *Conservation Biology* 24, no. 6 (2010): 1644–1653.
8. J. Sorrentino, *How to Eliminate Going to the Dentist*, The Evolution Institute, Wesley Chapel, Florida, 2019.
9. A. Ströhle, A. Hahn, and A. Sebastian, "Latitude, Local Ecology, and Hunter-Gatherer Dietary Acid Load: Implications from Evolutionary Ecology," *American Journal of Clinical Nutrition* 92, no. 4 (2010): 940–945.
10. I. Spreadbury, *Humans: Smart Enough to Create Processed Foods, Daft Enough to Eat Them*, The Evolution Institute, 2019.
11. P. Bourrat and P. E. Griffiths, *The Idea of Mismatch in Evolutionary Medicine*, 2021.
12. Matt Schneiderman, "William Levitt: The King of Suburbia," The Real Deal, April 30, 2008, https://therealdeal.com/issues_articles/william-levitt-the-king-of-suburbia/.
13. D. Rushkoff, *Digital Capitalism*, in *Making Sense*, S. Harris, editor, 2019.
14. Colin Marshall, "Levittown, the Prototypical American Suburb—A History of Cities in 50 Buildings, Day 25," *The Guardian*, April 28, 2015, https://www.theguardian.com/cities/2015/apr/28/levittown-america-prototypical-suburb-history-cities.
15. Staff, "Up from the Potato Fields," *Time* LVI, no. 1, July 3, 1950.

16. S. Pinker, *Enlightenment Now: The Case for Reason, Science, Humanism, and Progress* (New York: Penguin, 2018).
17. J. F. Helliwell, H. Huang, and S. Wang, "The Distribution of World Happiness," in World Happiness Report 2016 (New York: UN Sustainable Development Solutions Network, 2016), 8–37.
18. S. Junger, *Tribe: On Homecoming and Belonging* (New York: Twelve, 2016).
19. S. Harris, *Making Sense*, 2020.
20. American College Health Association (ACHA), National College Health Assessment II, Fall 2018, Reference Group Executive Summary, Silver Spring, Maryland, accessed December 6, 2022, https://www.acha.org/documents/ncha/NCHA-II_Fall_2018_Reference_Group_Executive_Summary.pdf.
21. J. L. Ravelo and S. Jerving, "COVID-19—A Timeline of the Coronavirus Outbreak," Devex, accessed December 6, 2022, https://www.devex.com/news/covid-19-a-timeline-of-the-coronavirus-outbreak-96396.
22. J. Olds and R. S. Schwartz, *The Lonely American: Drifting Apart in the Twenty-First Century* (New York: Beacon Press, 2009).
23. R. D. Putnam, *Bowling Alone: The Collapse and Revival of American Community* (New York: Simon & Schuster, 2000).
24. M. McPherson, L. Smith-Lovin, and M. E. Brashears, "Social Isolation in America: Changes in Core Discussion Networks over Two Decades," *American Sociological Review* 71, no. 3 (2006): 353–375.
25. K. McGillivray, "Here's Where Single-Person Households Are Clustering in Toronto," CBC News, August 8, 2017, https://www.cbc.ca/news/canada/toronto/single-person-households-toronto-1.4236139.
26. C. Bennett, "Poll: Half of Manhattan Residents Live Alone," *New York Post*, October 30, 2009, https://nypost.com/2009/10/30/poll-half-of-manhattan-residents-live-alone/.
27. T. Sommers, *Why Honor Matters* (New York: Basic Books, 2018).
28. F. Zublin, "How Do You Stop a Plague of Loneliness?," Ozy, August 15, 2016, https://www.ozy.com/around-the-world/how-do-you-stop-a-plague-of-loneliness/70903/.
29. P. E. Slater, *The Pursuit of Loneliness: American Culture at the Breaking Point* (Boston: Beacon, 1990).
30. J. M. Twenge, et al., "Egos Inflating over Time: A Cross-Temporal Meta-Analysis of the Narcissistic Personality Inventory," *Journal of Personality* 76, no. 4 (2008): 875–902.
31. J. Hari, *Lost Connections: Why You're Depressed and How to Find Hope* (New York Bloomsbury, 2019).
32. Hari, *Lost Connections*; J. T. Cacioppo and W. Patrick, *Loneliness: Human Nature and the Need for Social Connection* (New York: W. W. Norton & Company, 2008).
33. Hari, *Lost Connections.*
34. J. Olds and R. S. Schwartz, *The Lonely American: Drifting Apart in the Twenty-First Century* (Boston: Beacon, 2009).
35. S. W. Cole, et al., "Social Regulation of Gene Expression in Human Leukocytes," *Genome Biology* 8, no. 9 (2007): R189.
36. S. Pinker, *The Village Effect: How Face-to-Face Contact Can Make Us Healthier and Happier* (Toronto: Vintage Canada, 2015).

37. D. Lester, "The Holinger/Easterlin Cohort Hypothesis about Youth Suicide and Homicide Rates," *Perceptual and Motor Skills* 79, no. 3 suppl. (1994): 1545–1546.
38. J. M. Twenge, et al., "If You Can't Join Them, Beat Them: Effects of Social Exclusion on Aggressive Behavior," *Journal of Personality and Social Psychology* 81, no. 6 (2001): 1058.
39. Junger, *Tribe.*
40. Junger, *Tribe.*
41. J. T. Cacioppo and S. Cacioppo, "Social Relationships and Health: The Toxic Effects of Perceived Social Isolation," *Social and Personality Psychology Compass* 8, no. 2: 58–72.
42. W. H. Martens and G. B. Palermo, "Loneliness and Associated Violent Antisocial Behavior: Analysis of the Case Reports of Jeffrey Dahmer and Dennis Nilsen," *International Journal of Offender Therapy and Comparative Criminology* 49, no. 3 (2005): 298–307.
43. A. M. Stranahan, D. Khalil, and E. Gould, "Social Isolation Delays the Positive Effects of Running on Adult Neurogenesis," *Nature Neuroscience* 9, no. 4 (2006): 526–533.
44. L. C. Hawkley and J. T. Cacioppo, "Loneliness and Pathways to Disease," *Brain, Behavior, and Immunity* 17, no. 1, Suppl. (2003): 98–105; B. M. Hagerty and A. Williams, "The Effects of Sense of Belonging, Social Support, Conflict, and Loneliness on Depression," *Nursing Research* 48, no. 4 (1999): 215–219.
45. J. T. Cacioppo and L. C. Hawkley, "Perceived Social Isolation and Cognition," *Trends in Cognitive Sciences* 13, no. 10 (2009): 447–454.
46. D. A. Lamis, E. D. Ballard, and A. B. Patel, "Loneliness and Suicidal Ideation in Drug-Using College Students," *Suicide and Life-Threatening Behavior* 44, no. 6 (2014): 629–640.
47. L. M. Kurina, et al., "Loneliness Is Associated with Sleep Fragmentation in a Communal Society," *Sleep* 34, no. 11 (2011): 1519–1526.
48. F. Zaidi, et al., "Postpartum Depression in Women: A Risk Factor Analysis," *Journal of Clinical and Diagnostic Research: JCDR* 11, no. 8 (2017): QC13–QC16.
49. Z. Zhou, et al., "The Association Between Loneliness and Cognitive Impairment among Older Men and Women in China: A Nationwide Longitudinal Study," *International Journal of Environmental Research and Public Health* 16, no. 16 (2019): 2877; K. A. Straits-Tröster, et al., "The Relationship Between Loneliness, Interpersonal Competence, and Immunologic Status in HIV-Infected Men," *Psychology and Health* 9, no. 3 (1994): 205–219; L. Dahlberg, et al., "Predictors of Loneliness among Older Women and Men in Sweden: A National Longitudinal Study," *Aging & Mental Health* 19, no. 5 (2015): 409–417.
50. R. W. Sussman, *Primate Ecology and Social Structure, Vol. 1, Lorises, Lemurs, and Tarsiers* (N.p: Pearson Custom, 2003).
51. Christakis, *Blueprint.*

CHAPTER 3: TRIBAL TRUSTS EVOLVE

1. M. Dash, "For 40 Years, This Russian Family Was Cut Off from All Human Contact, Unaware of World War II," *Smithsonian Magazine*, January 28, 2013, https://www.smithsonianmag.com/history/for-40-years-this-russian-family-was-cut-off-from-all-human-contact-unaware-of-world-war-ii-7354256/.
2. A. H. Westing, "Population: Perhaps the Basic Issue," in *From Environmental to*

Comprehensive Security, SpringerBriefs on Pioneers in Science and Practice: Texts and Protocols, vol. 13 (Cham, Switzerland: Springer, 2013), 133–145.

3. K. Aoki, "Avoidance and Prohibition of Brother-Sister Sex in Humans," *Population Ecology* 47, no. 1 (2005): 13–19.
4. N. B. Blurton Jones, *Demography and Evolutionary Ecology of Hadza Hunter-Gatherers* (Cambridge, UK: Cambridge University Press, 2016); F. Marlowe, *The Hadza: Hunter-Gatherers of Tanzania*, vol. 3, Origins of Human Behavior and Culture (Berkeley: University of California Press, 2010), x, 325p.
5. N. Howell, *Demography of the Dobe !Kung* (London: Routledge, 2017).
6. R. L. Holloway, D. C. Broadfield, and M. S. Yuan, *The Human Fossil Record* (New York: John Wiley & Sons, 2005).
7. P. V. Tobias, "The Brain of Homo Habilis: A New Level of Organization in Cerebral Evolution," *Journal of Human Evolution* 16, no. 7–8 (1987): 741–761.
8. R. Dunbar, C. Gamble, and J. A. Gowlett, *Lucy to Language: The Benchmark Papers* (Oxford: Oxford University Press, 2014).
9. C. Zimmer, "Down from the Trees, Humans Finally Got a Decent Night's Sleep," *New York Times*, December 17, 2015, https://www.nytimes.com/2015/12/22/science/down-from-the-trees-humans-finally-got-a-decent-nights-sleep.html.
10. C. Boehm, *Moral Origins: The Evolution of Virtue, Altruism, and Shame* (New York: Soft Skull Press, 2012); C. L. Apicella and A. N. Crittenden, "Hunter-Gatherer Families and Parenting," in *The Handbook of Evolutionary Psychology*, ed. D. M. Buss (Hoboken, NJ: John Wiley & Sons, 2015), 797–827; R. W. Wrangham, M. L. Wilson, and M. N. Muller, "Comparative Rates of Violence in Chimpanzees and Humans," *Primates* 47, no. 1 (2006): 14–26; M. A. Nowak, C. E. Tarnita, and E. O. Wilson, "The Evolution of Eusociality," *Nature* 466, no. 7310 (2010): 1057–1062.
11. J. N. Rosenquist, J. H. Fowler, and N. A. Christakis, "Social Network Determinants of Depression," *Molecular Psychiatry* 16, no. 3 (2011): 273.
12. D. R. Samson, "The Human Sleep Paradox: The Unexpected Sleeping Habits of Homo sapiens," *Annual Review of Anthropology* 50 (2021): 259–274.
13. Apicella and Crittenden, "Hunter-Gatherer Families and Parenting," 797–827.
14. Apicella and Crittenden, "Hunter-Gatherer Families and Parenting," 797–827.
15. A. N. Crittenden and F. W. Marlowe, "Allomaternal Care among the Hadza of Tanzania," *Human Nature* 19, no. 3 (2008): 249.
16. B. Chapais, "Monogamy, Strongly Bonded Groups, and the Evolution of Human Social Structure," *Evolutionary Anthropology: Issues, News, and Reviews* 22, no. 2 (2013): 52–65.
17. C. Boehm, *Moral Origins: The Evolution of Virtue, Altruism, and Shame* (New York: Soft Skull Press, 2012).
18. M. Gladwell, *The Tipping Point: How Little Things Can Make a Big Difference* (New York: Little, Brown, 2006).
19. G. A. Miller, "The Magical Number Seven, Plus or Minus Two: Some Limits on Our Capacity for Processing Information," *Psychological Review* 63, no. 2 (1956): 81.
20. R. A. Hill and R. I. Dunbar, "Social Network Size in Humans," Human Nature 14, no. 1 (2003): 53–72; A. Sutcliffe, et al., "Relationships and the Social Brain: Integrating Psychological and Evolutionary Perspectives," *British Journal of Psychology* 103, no. 2 (2012): 149–168; R. I. Dunbar, et al., "The Structure of Online Social Networks Mirrors Those in the Offline World," *Social Networks* 43 (2015): 39–47;

P. Mac Carron, K. Kaski, and R. Dunbar, "Calling Dunbar's Numbers," *Social Networks* 47 (2016): 151–155; R. I. Dunbar, "Do Online Social Media Cut Through the Constraints That Limit the Size of Offline Social Networks?," *Royal Society Open Science* 3, no. 1 (2016): 150292.

21. S. G. Roberts, et al., "Exploring Variation in Active Network Size: Constraints and Ego Characteristics," *Social Networks* 31, no. 2 (2009):138–146.
22. R. I. M. Dunbar, "The Social Brain Hypothesis," *Evolutionary Anthropology: Issues, News, and Reviews* 6, no. 5 (1998): 178–190.
23. M. J. Hamilton, et al., "The Complex Structure of Hunter-Gatherer Social Networks," Proceedings of the Royal Society B: Biological Sciences 274, no. 1622 (2007): 2195–2203; W.-X. Zhou, et al., "Discrete Hierarchical Organization of Social Group Sizes," Proceedings of the Royal Society B: Biological Sciences 272, no. 1561 (2005): 439–444; R. Dunbar, C. Gamble, and J. A. Gowlett, *Lucy to Language: The Benchmark Papers* (Oxford: Oxford University Press, 2014); Dunbar, "The Social Brain Hypothesis," 178–190.
24. C. J. Buys and K. L. Larson, "Human Sympathy Groups," *Psychological Reports* 45, no. 2 (1979): 547–553.
25. Dunbar, Gamble, and Gowlett, *Lucy to Language.*
26. B. Hare and V. Woods, *The Genius of Dogs: How Dogs Are Smarter Than You Think* (New York: Plume, 2013).
27. L. N. Trut, "Early Canid Domestication: The Farm-Fox Experiment: Foxes Bred for Tamability in a 40-Year Experiment Exhibit Remarkable Transformations That Suggest an Interplay Between Behavioral Genetics and Development," *American Scientist* 87, no. 2 (1999): 160–169.
28. L. Berkowitz, *Aggression: Its Causes, Consequences, and Control* (Philadelphia: Temple University Press, 1993).
29. R. Wrangham, *The Goodness Paradox: The Strange Relationship Between Virtue and Violence in Human Evolution* (New York: Vintage, 2019).
30. A. Brooks, *The Origin of Human Society*, in *Our Tribal Nature: Tribalism, Politics, and Evolution*. 2019. The Leakey Foundation.
31. F. Marlowe, *The Hadza: Hunter-Gatherers of Tanzania*, vol. 3, Origins of Human Behavior and Culture (Berkeley: University of California Press, 2010), x, 325p.
32. P. V. Marsden, "Core Discussion Networks of Americans," *American Sociological Review* 52, no. 1 (February 1987): 122–131; C. McCarty, "Structure in Personal Networks," Journal of Social Structure 3, no. 1 (2002): 20; J. Boissevain, "Conflict and Change: Establishment and Opposition in Malta," in *Choice and Change: Essays in Honor of Lucy Mair*, ed. J. Davis (London: Athlone, 1974), 17.
33. M. Domínguez-Rodrigo, et al., "The Meta-Group Social Network of Early Humans: A Temporal-Spatial Assessment of Group Size at FLK Zinj (Olduvai Gorge, Tanzania)," *Journal of Human Evolution* 127 (2019): 54–66.
34. S. A. West and A. Gardner, "Altruism, Spite, and Greenbeards," *Science* 327, no. 5971 (2010): 1341–1344.
35. S. Smukalla, et al., "FLO1 Is a Variable Green Beard Gene That Drives Biofilm-Like Cooperation in Budding Yeast," *Cell* 135, no. 4 (2008): 726–737.
36. Marlowe, *The Hadza: Hunter-Gatherers of Tanzania*, x, 325p.
37. I. Watts, M. Chazan, and J. Wilkins, "Early Evidence for Brilliant Ritualized Display: Specularite Use in the Northern Cape (South Africa) between~ 500 and~ 300

ka," *Current Anthropology* 57, no. 3 (2016): 287–310; J. Wilkins, et al., "Innovative Homo sapiens Behaviours 105,000 Years Ago in a Wetter Kalahari," *Nature* 592, no. 7853 (2021): 248–252.

38. S. Pinker, *The Language Instinct: How the Mind Creates Language* (New York: William Morrow, 1994).
39. J. E. Yellen, *Archaeological Approaches to the Present: Models for Reconstructing the Past*, vol. 1 (New York: Academic Press, 1977).
40. J. M. Miller and Y. V. Wang, "Ostrich Eggshell Beads Reveal 50,000-Year-Old Social Network in Africa," *Nature* 601, no. 7892 (2022): 234–239.
41. H. V. Merrick, F. Brown, and W. Nash, "Use and Movement of Obsidian in the Early and Middle Stone Ages of Kenya and Northern Tanzania," *Society, Culture, and Technology in Africa* 11, no. 6 (1994): 29–44.
42. Watts, Chazan, and Wilkins, "Early Evidence for Brilliant Ritualized Display," 287–310.
43. M. Singh and L. Glowacki, "Human Social Organization During the Late Pleistocene: Beyond the Nomadic-Egalitarian Model," *Evolution and Human Behavior* 43, no. 5 (September 2022): 418–431.
44. R. Wrangham, *The Goodness Paradox: The Strange Relationship Between Virtue and Violence in Human Evolution* (New York: Vintage, 2019).
45. P. Wiessner, "From Spears to M-16s: Testing the Imbalance of Power Hypothesis among the Enga," *Journal of Anthropological Research* 62, no. 2 (2006): 165–191.
46. R. M. Sapolsky, *Behave: The Biology of Humans at Our Best and Worst* (New York: Penguin, 2017).
47. F. Marlowe, "Why the Hadza Are Still Hunter-Gatherers," in *Ethnicity, Hunter-Gatherers, and The "Other": Association or Assimilation in Africa*, ed. S. Kent (Washington, DC: Smithosonian Institution Press, 2002), 247–281; R. B. Lee, *The !Kung San: Men, Women and Work in a Foraging Society* (Cambridge, UK: Cambridge University Press, 1979).
48. N. B. Tindale, "Ecology of Primitive Aboriginal Man in Australia," in *Biogeography and Ecology in Australia*, ed. A. Keast, R. L. Crocker, and C. S. Christian (Dordrecht: Springer, 1959), 36–51.
49. R. W. Wrangham, M. L. Wilson, and M. N. Muller, "Comparative Rates of Violence in Chimpanzees and Humans," Primates 47, no. 1 (2006): 14–26; S. A. Leblanc, "Prehistory of Warfare," *Archaeology* 56, no. 3 (2003): 18–25; B. M. Knauft, et al., "Violence and Sociality in Human Evolution [And Comments and Replies]," Current Anthropology 32, no. 4 (1991): 391–428.
50. T. C. Zeng, A. J. Aw, and M. W. Feldman, "Cultural Hitchhiking and Competition Between Patrilineal Kin Groups Explain the Post-Neolithic Y-Chromosome Bottleneck," *Nature Communications* 9, no. 1 (2018): 1–12.
51. T. Hobbes, *Leviathan, or The Matter, Form and Power of a Common Wealth Ecclesiastical and Civil* (1651).

CHAPTER 4: TRIBAL BEHAVIOR, TRUST IN ACTION

1. W. Schmidt, "Heavy-Metal Groups Shake Moscow," *New York Times*, September 29, 1991, https://www.nytimes.com/1991/09/29/world/heavy-metal-groups-shake-moscow.html.
2. D. Cadena, "Metal in Soviet Russia: Monsters of Rock 1991," StMU Research

Scholars, St. Mary's University, November 28, 2018, https://stmuscholars.org/metal-in-soviet-russia-monsters-of-rock-1991/.

3. Sapolsky, *Behave.*
4. M. Sherif, *Experimental Study of Positive and Negative Intergroup Attitudes Between Experimentally Produced Groups: Robbers Cave Study* (1954); M. Sherif, "Experiments in Group Conflict," *Scientific American* 195, no. 5 (1956): 54–59; M. Sherif, et al., *Intergroup Conflict and Cooperation: The Robbers Cave Experiment*, vol. 10 (Norman, OK: Institute of Group Relations, University of Oklahoma, 1961); M. Sherif, "Superordinate Goals in the Reduction of Intergroup Conflict," *American Journal of Sociology* 63, no. 4 (1958): 349–356.
5. M. Sherif, *Experimental Study of Positive and Negative Intergroup Attitudes Between Experimentally Produced Groups: Robbers Cave Study* (1954).
6. R. M. Sapolsky, *Behave.*
7. P. D. MacLean, *The Triune Brain in Evolution: Role in Paleocerebral Functions* (New York: Springer Science & Business Media, 1990).
8. S. R. Cavanagh, *Hivemind: The New Science of Tribalism in Our Divided World* (New York: Grand Central Publishing, 2019); O. Devinsky, M. J. Morrell, and B. A. Vogt, "Contributions of Anterior Cingulate Cortex to Behaviour," *Brain* 118, no. 1 (1995): 279–306.
9. G. Rizzolatti, et al., "Resonance Behaviors and Mirror Neurons," Archives italiennes de biologie 137, no. 2 (1999): 85–100; V. S. Ramachandran, *The Tell-Tale Brain: A Neuroscientist's Quest for What Makes Us Human* (New York: W. W. Norton & Company, 2012).
10. C. Lamm, C. D. Batson, and J. Decety, "The Neural Substrate of Human Empathy: Effects of Perspective-Taking and Cognitive Appraisal," *Journal of Cognitive Neuroscience* 19, no. 1 (2007): 42–58.
11. V. Caggiano, et al., "Mirror Neurons Differentially Encode the Peripersonal and Extrapersonal Space of Monkeys," *Science* 324, no. 5925 (2009): 403–406.
12. B. Duce, et al., "The AASM Recommended and Acceptable EEG Montages Are Comparable for the Staging of Sleep and Scoring of EEG Arousals," *Journal of Clinical Sleep Medicine* 10, no. 7 (2014): 803–809.
13. J. Decety and K. J. Michalska, "Neurodevelopmental Changes in the Circuits Underlying Empathy and Sympathy from Childhood to Adulthood," *Developmental Science* 13, no. 6 (2010): 886–899.
14. L. T. Harris and S. T. Fiske, "Dehumanizing the Lowest of the Low: Neuroimaging Responses to Extreme Out-Groups," *Psychological Science* 17, no. 10 (2006): 847–853.
15. J. W. Dalley, B. J. Everitt, and T. W. Robbins, "Impulsivity, Compulsivity, and Top-Down Cognitive Control," *Neuron* 69, no. 4 (2011): 680–694; J. N. Gutsell and M. Inzlicht, "Intergroup Differences in the Sharing of Emotive States: Neural Evidence of an Empathy Gap," *Social Cognitive and Affective Neuroscience* 7, no. 5 (2012): 596–603; H. Takahashi, et al., "When Your Gain Is My Pain and Your Pain Is My Gain: Neural Correlates of Envy and Schadenfreude," *Science* 323, no. 5916 (2009): 937–939.
16. C. N. DeWall, et al., "Depletion Makes the Heart Grow Less Helpful: Helping as a Function of Self-Regulatory Energy and Genetic Relatedness," *Personality and Social Psychology Bulletin* 34, no. 12 (2008): 1653–1662.

17. R. J. Nelson and B. C. Trainor, "Neural Mechanisms of Aggression," *Nature Reviews Neuroscience* 8, no. 7 (2007): 536–546.
18. C.-B. Zhong and K. Liljenquist, "Washing Away Your Sins: Threatened Morality and Physical Cleansing," *Science* 313, no. 5792 (2006): 1451–1452.
19. Cialdini, R., *Pre-suasion: A Revolutionary Way to Influence and Persuade* (New York: Simon & Schuster, 2016).
20. M. A. Nowak, C. E. Tarnita, and E. O. Wilson, "The Evolution of Eusocialit," *Nature* 466, no. 7310 (2010): 1057–1062; E. O. Wilson, *Sociobiology: The New Synthesis* (Cambridge, MA: Harvard University Press, 2000); E. O. Wilson, *The Social Conquest of Earth* (New York: W. W. Norton & Company, 2012).
21. S. R. Cavanagh, *Hivemind: The New Science of Tribalism in Our Divided World* (New York: Grand Central Publishing, 2019).
22. J. Haidt, *The Righteous Mind: Why Good People Are Divided by Politics and Religion* (New York: Vintage, 2012).
23. J. A. Coan and D. A. Sbarra, "Social Baseline Theory: The Social Regulation of Risk and Effort," *Current Opinion in Psychology* 1 (2015): 87–91.
24. E. B. Gross and S. E. Medina-DeVilliers, "Cognitive Processes Unfold in a Social Context: A Review and Extension of Social Baseline Theory," *Frontiers in Psychology* 11, no. 378 (2020).
25. L. Tomova, et al., "The Need to Connect: Acute Social Isolation Causes Neural Craving Responses Similar to Hunger," Nature Neuroscience 23 (2020): 1597–1605.
26. M. Levine, et al., "Identity and Emergency Intervention: How Social Group Membership and Inclusiveness of Group Boundaries Shape Helping Behavior," *Personality and Social Psychology Bulletin* 31, no. 4 (2005): 443–453.
27. J. A. Coan, H. S. Schaefer, and R. J. Davidson, "Lending a Hand: Social Regulation of the Neural Response to Threat," *Psychological Science* 17, no. 12 (2006): 1032–1039.
28. M. López-Solà, et al., "Brain Mechanisms of Social Touch-Induced Analgesia in Females," *Pain* 160, no. 9 (2019): 2072–2085.
29. R. M. Sapolsky, *Behave: The Biology of Humans at Our Best and Worst* (New York: Penguin, 2017).
30. L. W. Tsai and R. M. Sapolsky, "Rapid Stimulatory Effects of Testosterone upon Myotubule Metabolism and Sugar Transport, as Assessed by Silicon Microphysiometry," *Aggressive Behavior: Official Journal of the International Society for Research on Aggression* 22, no. 5 (1996): 357–364; A. Boissy and M. Bouissou, "Effects of Androgen Treatment on Behavioral and Physiological Responses of Heifers to Fear-Eliciting Situations," *Hormones and Behavior* 28, no. 1 (1994): 66–83.
31. R. I. Wood, "Reinforcing Aspects of Androgens," *Physiology & Behavior* 83, no. 2 (2004): 279–289; R. M. Sapolsky, *Behave.*
32. A. F. Dixson and C. M. Nevison, "The Socioendocrinology of Adolescent Development in Male Rhesus Monkeys (Macaca Mulatta)," *Hormones and Behavior* 31, no. 2 (1997): 126–135.
33. P. C. Bernhardt, et al., "Testosterone Changes During Vicarious Experiences of Winning and Losing among Fans at Sporting Events," *Physiology & Behavior* 65, no. 1 (1998): 59–62.

34. M. Wibral, et al., "Testosterone Administration Reduces Lying in Men," *PloS One* 7, no. 10 (2012): e46774.
35. C. D. Navarrete, et al., "Race Bias Tracks Conception Risk across the Menstrual Cycle," *Psychological Science* 20, no. 6 (2009): 661–665.
36. C. D. Navarrete, et al., "Fertility and Race Perception Predict Voter Preference for Barack Obama," Evolution and Human Behavior 31, no. 6 (2010): 394–399.
37. H. Wang, et al., "Histone Deacetylase Inhibitors Facilitate Partner Preference Formation in Female Prairie Voles," *Nature Neuroscience* 16, no. 7 (2013): 919.
38. M. Nagasawa, et al., "Oxytocin-Gaze Positive Loop and the Coevolution of Human-Dog Bonds," *Science* 348, no. 6232 (2015): 333–336.
39. R. Feldman, "The Neurobiology of Human Attachments," *Trends in Cognitive Sciences* 21, no. 2 (2017): 80–99; C. K. De Dreu, et al., "Oxytocin Promotes Human Ethnocentrism," *Proceedings of the National Academy of Sciences* 108, no. 4 (2011): 1262–1266.
40. T. Baumgartner, et al., "Oxytocin Shapes the Neural Circuitry of Trust and Trust Adaptation in Humans," *Neuron* 58, no. 4 (2008): 639–650.
41. C. K. De Dreu, "Oxytocin Modulates Cooperation within and Competition Between Groups: An Integrative Review and Research Agenda," *Hormones and Behavior* 61, no. 3 (2012): 419–428.
42. R. M. Sapolsky, *Behave*; D. J. Langford, et al., "Social Modulation of Pain as Evidence for Empathy in Mice," *Science* 312, no. 5782 (2006): 1967–1970.
43. McNeill, W.H., *Keeping Together in Time: Dance and Drill in Human History* (Cambridge, MA: Harvard University Press, 1997).
44. J. Altman and G. D. Das, "Autoradiographic and Histological Evidence of Postnatal Hippocampal Neurogenesis in Rats," *Journal of Comparative Neurology* 124, no. 3 (1965): 319–335.
45. G. Kempermann, "New Neurons for 'Survival of the Fittest,'" *Nature Reviews Neuroscience* 13, no. 10 (2012): 727–736; S. K. Droste, et al., "Effects of Long-Term Voluntary Exercise on the Mouse Hypothalamic-Pituitary-Adrenocortical Axis," *Endocrinology* 144, no. 7 (2003): 3012–3023.
46. H. Van Praag, "Neurogenesis and Exercise: Past and Future Directions," *Neuromolecular Medicine* 10, no. 2 (2008): 128–140.
47. H. Tajfel, "Experiments in Intergroup Discrimination," *Scientific American* 223, no. 5 (1970): 96–103; H. Tajfel, et al., "Social Categorization and Intergroup Behaviour," *European Journal of Social Psychology* 1, no. 2 (1971): 149–178.
48. M. Billig and H. Tajfel, "Social Categorization and Similarity in Intergroup Behaviour," *European Journal of Social Psychology* 3, no. 1 (1973): 27–52.
49. J. M. Rabbie and H. F. Lodewijkx, "Conflict and Aggression: An Individual-Group Continuum," *Advances in Group Processes* 11 (1994): 139–174.
50. T. Yamagishi, N. Jin, and T. Kiyonari, "Bounded Generalized Reciprocity: Ingroup Boasting and Ingroup Favoritism," *Advances in Group Processes* 16, no. 1 (1999): 161–197.
51. R. M. Sapolsky, *Behave*.
52. Y. Dunham, A. S. Baron, and S. Carey, "Consequences of 'Minimal' Group Affiliations in Children," *Child Development* 82, no. 3 (2011): 793–811.
53. Y. Bar-Haim, et al., "Nature and Nurture in Own-Race Face Processing," *Psychological Science* 17, no. 2 (2006): 159–163.

54. K. D. Kinzler, E. Dupoux, and E. S. Spelke, "The Native Language of Social Cognition," *Proceedings of the National Academy of Sciences* 104, no. 30 (2007): 12577–12580.
55. R. C. Knickmeyer, et al., "A Structural MRI Study of Human Brain Development from Birth to 2 Years," *Journal of Neuroscience* 28, no. 47 (2008): 12176–12182.
56. D. E. Brown, "Human Universals and Their Implications," in *Being Humans: Anthropological Universality and Particularity in Transdisciplinary Perspectives*, ed. N. Roughley (Berlin: Walter de Gruyter, 2000), 156–174.
57. H. R. Markus and S. Kitayama, "Culture and the Self: Implications for Cognition, Emotion, and Motivation," *Psychological Review* 98, no. 2 (1991): 224.
58. C. R. Ember and M. Ember, "Warfare, Aggression, and Resource Problems: Cross-Cultural Codes," *Behavior Science Research* 26, nos. 1–4 (1992): 169–226; R. B. Textor, *A Cross-Cultural Summary*, vol. 10. 1967: Human Relations Area Files; H. C. Peoples and F. W. Marlowe, "Subsistence and the Evolution of Religion," Human Nature 23, no. 3 (2012): 253–269.
59. T. Sommers, *Why Honor Matters* (New York: Basic Books, 2018).
60. R. E. Nisbett, *Culture of Honor: The Psychology of Violence in the South* (New York: Routledge, 2018).
61. D. Cohen, R. E. Nisbett, B. F. Bowdle, and N. Schwarz, "Insult, Aggression, and the Southern Culture of Honor: An 'Experimental Ethnography,'" *Journal of Personality and Social Psychology* 70, no. 5 (1996): 945–960, https://doi.org/10.1037/0022-3514.70.5.945
62. Cohen, Nisbett, Bowdle, and Schwarz, "Insult, Aggression, and the Southern Culture of Honor."
63. M. Gelfand, *Rule Makers, Rule Breakers: Tight and Loose Cultures and the Secret Signals That Direct Our Lives* (New York: Scribner, 2019).

CHAPTER 5: TRUST SIGNALS

1. M. A. Halleran, *The Better Angels of Our Nature: Freemasonry in the American Civil War* (Tuscaloosa: University of Alabama Press, 2010).
2. S. Pinker, *The Language Instinct: How the Mind Creates Language* (New York: William Morrow, 1994).
3. C. Handley and S. Mathew, "Human Large-Scale Cooperation as a Product of Competition Between Cultural Groups," *Nature Communications* 11, no. 1 (2020): 702.
4. L. Jussim, J. T. Crawford, and R. S. Rubinstein, "Stereotype (in) Accuracy in Perceptions of Groups and Individuals," *Current Directions in Psychological Science* 24, no. 6 (2015): 490–497.
5. R. M. Sapolsky, *Behave.*
6. A. Zahavi, "Mate Selection—A Selection for a Handicap," *Journal of Theoretical Biology* 53, no. 1 (1975): 205–214.
7. G. F. Miller, "Sexual Selection for Moral Virtues," *Quarterly Review of Biology* 82, no. 2 (2007): 97–125.
8. D. Lanska, et al., "Factors Influencing Anatomic Location of Fat Tissue in 52,953 Women," *International Journal of Obesity* 9, no. 1 (1985): 29–38; J. Arechiga, et al., "Women in Transition—Menopause and Body Composition in Different Populations," *Collegium Antropologicum* 25, no. 2 (2001): 443–448; A. Misra and N. K.

Vikram, "Clinical and Pathophysiological Consequences of Abdominal Adiposity and Abdominal Adipose Tissue Depots," *Nutrition* 19, no. 5 (2003): 457–466; M. Van Hooff, et al., "Insulin, Androgen, and Gonadotropin Concentrations, Body Mass Index, and Waist to Hip Ratio in the First Years after Menarche in Girls with Regular Menstrual Cycles, Irregular Menstrual Cycles, or Oligomenorrhea," *Journal of Clinical Endocrinology & Metabolism* 85, no. 4 (2000): 1394–1400; B. M. Zaadstra, et al., "Fat and Female Fecundity-Prospective-Study of Effect of Body-Fat Distribution on Conception Rates," *British Medical Journal* 306, no. 6876 (1993): 484–487; R. Pasquali, et al., "The Natural History of the Metabolic Syndrome in Young Women with the Polycystic Ovary Syndrome and the Effect of Long-Term Oestrogen-Progestagen Treatment," *Clinical Endocrinology* 50, no. 4 (1999): 517–527.

9. C. Moya, *Why Chimpanzees Don't Stereotype, We Do, and Whales Might*, 2019, https://evolution-institute.org/why-chimpanzees-dont-stereotype-we-do-and-whales-might/.
10. Moya, *Why Chimpanzees Don't Stereotype.*
11. Moya, *Why Chimpanzees Don't Stereotype.*
12. G. A. Fine and L. Holyfield, "Secrecy, Trust, and Dangerous Leisure: Generating Group Cohesion in Voluntary Organizations," *Social Psychology Quarterly* 59, no. 1 (1996): 22–38.
13. E. L. Bridges and S. K. Lothrop, "The Canoe Indians of Tierra del Fuego," in *A Reader in General Anthropology* (1948), 84–116.
14. W. L. Warner, *A Black Civilization: A Social Study of an Australian Tribe, Rev.* (1958).
15. J. Loenen, "Was Anaximander an Evolutionist?," *Mnemosyne* 7, no. 1 (1954): 215–232.
16. D. Hoffman, *The Case Against Reality: Why Evolution Hid the Truth from Our Eyes* (New York: W. W. Norton & Company, 2019).
17. J. T. Mark, B. B. Marion, and D. D. Hoffman, "Natural Selection and Veridical Perceptions," *Journal of Theoretical Biology* 266, no. 4 (2010): 504–515.
18. D. M. Kahan, H. Jenkins-Smith, and D. Braman, "Cultural Cognition of Scientific Consensus," *Journal of Risk Research* 14, no. 2 (2011): 147–174; D. M. Kahan, "Climate-Science Communication and the Measurement Problem," Political Psychology 36 (2015): 1–43.
19. S. Pinker, *Enlightenment Now: The Case for Reason, Science, Humanism, and Progress* (New York: Penguin, 2018).
20. D. M. Kahan, et al., *The Tragedy of the Risk-Perception Commons: Culture Conflict, Rationality Conflict, and Climate Change.* Temple University Legal Studies Research Paper, 2011-26, June 24, 2011.
21. T. Purnell, W. Idsardi, and J. Baugh, "Perceptual and Phonetic Experiments on American English Dialect Identification," *Journal of Language and Social Psychology* 18, no. 1 (1999): 10–30.
22. T. Rakić, M. C. Steffens, and A. Mummendey, "Blinded by the Accent! The Minor Role of Looks in Ethnic Categorization," *Journal of Personality and Social Psychology* 100, no. 1 (2011): 16.
23. S. Pinker, *The Language Instinct: How the Mind Creates Language* (New York: William Morrow, 1994).
24. A. Gupta, "Why Young Men of Color Are Joining White-Supremacist Groups," *The*

Daily Beast, September 6, 2018, https://www.thedailybeast.com/why-young-men-of-color-are-joining-white-supremacist-groups.

25. M. Baram, "Cult Murder Suspect's Mom: It's Not Her Fault," ABC News, August 12, 2008, ../customXml/item1.xml.
26. J. Velikovsky, "The Holon/Parton Theory of the Unit of Culture (or the Meme, and Narreme): In Science, Media, Entertainment, and the Arts," in *Technology Adoption and Social Issues: Concepts, Methodologies, Tools, and Applications* (IGI Global, 2018), 1590–1627.
27. H. L. Hoogland, "Infanticide in Prairie Dogs: Lactating Females Kill Offspring of Close Kin," *Science* 230, no. 4729 (1985): 1037–1040.
28. J. Chang and J. Halliday, *Mao: The Unknown Story* (New York: Anchor, 2011).
29. Y. Gao, *Born Red: A Chronicle of the Cultural Revolution* (Stanford, CA: Stanford University Press, 1987).
30. H. Whitehouse, et al., "Complex Societies Precede Moralizing Gods Throughout World History," *Nature* 568, no. 7751 (2019): 226–229.
31. K. Jaspers, *The Origin and Goal of History*, trans. M. Bullock (1949).
32. C. Sagan and A. Druyan, *Shadows of Forgotten Ancestors: A Search for Who We Are* (New York: Random House, 1992), xvi, 505p.
33. S. B. Schaefer and P. T. Furst, *People of the Peyote: Huichol Indian History, Religion & Survival* (Albuquerque: University of New Mexico Press, 1996).
34. B. G. Myerhoff, *Peyote Hunt: The Sacred Journey of the Huichol Indians* (Ithaca, NY: Cornell University Press, 1976).
35. P. E. Tetlock, "Thinking the Unthinkable: Sacred Values and Taboo Cognitions," *Trends in Cognitive Sciences* 7, no. 7 (2003): 320–324; S. Atran and R. Axelrod, "Reframing Sacred Values," *Negotiation Journal* 24, no. 3 (2008): 221–246.

CHAPTER 6: TRIBAL BENEFITS

1. M. Gladwell, *Outliers: The Story of Success* (New York: Little, Brown, 2008).
2. T. Miller, *Communes in America, 1975–2000* (Syracuse, NY: Syracuse University Press, 2019).
3. N. Fouriezos, "Why Is Making Friends So Hard?," OZY, April 10, 2021, https://www.ozy.com/the-new-and-the-next/why-is-making-friends-so-hard/428214/.
4. M. Curtin, "This 75-Year Harvard Study Found the 1 Secret to Leading a Fullfiling Life," Inc., accessed December 6, 2022, https://www.inc.com/melanie-curtin/want-a-life-of-fulfillment-a-75-year-harvard-study-says-to-prioritize-this-one-t.html.
5. L. Tomova, et al., "The Need to Connect: Acute Social Isolation Causes Neural Craving Responses Similar to Hunger," *Nature Neuroscience* 23 (2020): 1597–1605.
6. N. J. Donovan, et al., "Association of Higher Cortical Amyloid Burden with Loneliness in Cognitively Normal Older Adults," *JAMA Psychiatry* 73, no. 12 (2016): 1230–1237.
7. M. López-Solà, et al., "Brain Mechanisms of Social Touch-Induced Analgesia in Females," *Pain* 160, no. 9 (2019): 2072–2085.
8. D. A. Campbell, et al., "A Randomized Control Trial of Continuous Support in Labor by a Lay Doula," *Journal of Obstetric, Gynecologic & Neonatal Nursing* 35, no. 4 (2006): 456–464.
9. D. P. Weekes, et al., "The Phenomenon of Hand Holding as a Coping Strategy in

Adolescents Experiencing Treatment-Related Pain," *Journal of Pediatric Oncology Nursing* 10, no. 1 (1993): 19–25.

10. H. IJzerman, et al., "Caring for Sharing," *Social Psychology* 44, no. 2 (2013): 160–166.
11. D. M. Campagne, "Stress and Perceived Social Isolation (Loneliness)," *Archives of Gerontology and Geriatrics* 82 (2019): 192–199; R. Glaser, et al., "Stress, Loneliness, and Changes in Herpesvirus Latency," *Journal of Behavioral Medicine* 8, no. 3 (1985): 249–260.
12. A. Steptoe, et al., "Loneliness and Neuroendocrine, Cardiovascular, and Inflammatory Stress Responses in Middle-Aged Men and Women," *Psychoneuroendocrinology* 29, no. 5 (2004): 593–611.
13. J. K. Kiecolt-Glaser, et al., "Urinary Cortisol Levels, Cellular Immunocompetency, and Loneliness in Psychiatric Inpatients," *Psychosomatic Medicine* 46, no. 1 (1984): 15–23.
14. R. M. Sapolsky, *Behave: The Biology of Humans at Our Best and Worst* (New York: Penguin, 2017).
15. J. D. Clapp and J. Gayle Beck, "Understanding the Relationship Between PTSD and Social Support: The Role of Negative Network Orientation," *Behaviour Research and Therapy* 47, no. 3 (2009): 237–244.
16. L. C. Hawkley, K. J. Preacher, and J. T. Cacioppo, "Multilevel Modeling of Social Interactions and Mood in Lonely and Socially Connected Individuals: The Macarthur Social Neuroscience Studies," in *Oxford Handbook of Methods in Positive Psychology*, ed. A. D. Ong and M. H. M. van Dulmen (Oxford: Oxford University Press, 2007), 559–575.
17. A. M. Stranahan, D. Khalil, and E. Gould, "Social Isolation Delays the Positive Effects of Running on Adult Neurogenesis," *Nature Neuroscience* 9, no. 4 (2006): 526–533.
18. V. Slaughter, M. J. Dennis, and M. Pritchard, "Theory of Mind and Peer Acceptance in Preschool Children," *British Journal of Developmental Psychology* 20, no. 4 (2002): 545–564.
19. R. M. Sapolsky, *Behave*.
20. J. T. Cacioppo and L. C. Hawkley, "Perceived Social Isolation and Cognition," *Trends in Cognitive Sciences* 13, no. 10 (2009): 447–454.
21. S. W. Cole, et al., "Social Regulation of Gene Expression in Human Leukocytes," *Genome Biology* 8, no. 9 (2007): R189.
22. General Social Survey, accessed December 6, 2022, https://gss.norc.org/get-the-data.
23. General Social Survey.
24. L. Smith-Lovin, "On Point," National Public Radio, 2006.
25. J. Olds and R. S. Schwartz, *The Lonely American: Drifting Apart in the Twenty-First Century* (Boston: Beacon Press, 2009).
26. J. Olds, et al., "Part-Time Employment and Marital Well-Being: A Hypothesis and Pilot Study," Family Therapy: The Journal of the California Graduate School of *Family Psychology* 20, no. 1 (1993).
27. Olds and Schwartz, *The Lonely American*.
28. R. M. Seyfarth and D. L. Cheney, "The Evolutionary Origins of Friendship," *Annual Review of Psychology* 63 (2012): 153–177.

29. R. R. Bell, "Friendships of Women and of Men," *Psychology of Women Quarterly* 5, no. 3 (1981):402–417.
30. S. J. Oliker, *Best Friends and Marriage: Exchange Among Women* (Berkeley: University of California Press, 1989).
31. M. Heid, "Are You Headed for a Friendship Crisis?," *Men's Health*, June 17, 2013, https://www.menshealth.com/health/a19519239/are-you-headed-for-a-friendship-crisis/.
32. D. Jerrome, "The Significance of Friendship for Women in Later Life," *Ageing & Society* 1, no. 2 (1981): 175–197.
33. N. Gerstel and N. Sarkisian, "Marriage: The Good, the Bad, and the Greedy," *Contexts* 5, no. 4 (2006): 16–21.
34. Olds and Schwartz, *The Lonely American*.
35. Olds and Schwartz, *The Lonely American*.
36. E. H. Newberger, et al., "Child Abuse and Pediatric Social Illness: An Epidemiological Analysis and Ecological Reformulation," *American Journal of Orthopsychiatry* 56, no. 4 (1986): 589–601.
37. Oliker, *Best Friends and Marriage*.
38. B. F. Hutchens and J. Kearney, "Risk Factors for Postpartum Depression: An Umbrella Review," *Journal of Midwifery & Women's Health* 65, no. 1 (2020): 96–108.
39. J. H. Shaver, et al., "Church Attendance and Alloparenting: An Analysis of Fertility, Social Support and Child Development among English Mothers," *Philosophical Transactions of the Royal Society* B 375, no. 1805 (2020): 20190428.
40. Shaver, et al., "Church Attendance and Alloparenting."
41. T. Gleason and D. Narvaez, "Beyond Resilience to Thriving: Optimizing Child Wellbeing," *International Journal of Wellbeing* 9, no. 4 (2019).
42. C. Koverola, et al., "Longitudinal Investigation of the Relationship among Maternal Victimization, Depressive Symptoms, Social Support, and Children's Behavior and Development," *Journal of Interpersonal Violence* 20, no. 12 (2005): 1523–1546.
43. A. S. Masten, K. M. Best, and N. Garmezy, "Resilience and Development: Contributions from the Study of Children Who Overcome Adversity," *Development and Psychopathology* 2, no. 4 (1990): 425–444.
44. P. R. Amato, "The Impact of Family Formation Change on the Cognitive, Social, and Emotional Well-Being of the Next Generation," *Future of Children* 15, no. 2 (2005): 75–96.
45. D. Narvaez, ed., *Basic Needs, Wellbeing and Morality: Fulfilling Human Potential* (Cham, Switzerland: Springer, 2018); D. Narvaez, "Triune Ethics: The Neurobiological Roots of Our Multiple Moralities," New Ideas in Psychology 26, no. 1 (2008): 95–119; M. J. Bundick, et al., "Thriving across the Life Span," in *The Handbook of Life-Span Development, Vol. 1., Cognition, Biology, and Methods*, ed. W. F. Overton and R. M. Lerner (New York: John Wiley & Sons, 2010), 882–923.
46. Olds and Schwartz, *The Lonely American*.
47. G. M. Ehrle and H. Day, "Adjustment and Family Functioning of Grandmothers Rearing Their Grandchildren," *Contemporary Family Therapy* 16, no. 1 (1994): 67–82.
48. D. P. Waldrop and J. A. Weber, "From Grandparent to Caregiver: The Stress and Satisfaction of Raising Grandchildren," *Families* in Society 82, no. 5 (2001): 461–472.

49. M. J. Poulin and C. M. Haase, "Growing to Trust: Evidence That Trust Increases and Sustains Well-Being Across the Life Span," *Social Psychological and Personality Science* 6, no. 6 (2015): 614–621.
50. S. Galloway, "Making Sense," *Wealth & Happiness*, S. Harris, editor, 2019.
51. C. L. Apicella and A. N. Crittenden, "Hunter-Gatherer Families and Parenting," in *The Handbook of Evolutionary Psychology*, ed. D. M. Buss (Hoboken, NJ: John Wiley & Sons, 2015), 797–827; A. N. Crittenden and F. W. Marlowe, "Cooperative Child Care among the Hadza: Situating Multiple Attachment in Evolutionary Context," in *Attachment Reconsidered: Cultural Persepectives on a Western Theory*, ed. N. Quinn and J. M. Mageo (New York: Palgrave Macmillan, 2013), 67–83; A. N. Crittenden, "Ancestral Attachment," in *Ancestral Landscapes in Human Evolution: Culture, Childrearing and Social Wellbeing*, ed. D. Narvaez, et al. (Oxford: Oxford University Press, 2014), 282.
52. M. Kuhn, M. Schularick, and U. I. Steins, "Income and Wealth Inequality in America, 1949–2016," *Journal of Political Economy* 128, no. 9 (2017): 3469–3519.
53. D. Brooks, "The Nuclear Family Was a Mistake," *The Atlantic*, March 2020, https://www.theatlantic.com/magazine/archive/2020/03/the-nuclear-family-was-a-mistake/605536/.
54. N. E. Adler and J. M. Ostrove, "Socioeconomic Status and Health: What We Know and What We Don't," *Annals of the New York Academy of Sciences* 896, no. 1 (1999): 3–15; I. Kawachi and B. P. Kennedy, *Health of Nations* (New York: New Press, 2006); J. Lynch, et al., "Income Inequality, the Psychosocial Environment, and Health: Comparisons of Wealthy Nations," *The Lancet* 358, no. 9277 (2001): 194–200; G. A. Kaplan, et al., "Inequality in Income and Mortality in the United States: Analysis of Mortality and Potential Pathways," *BMJ* 312, no. 7037 (1996): 999–1003; J. R. Dunn, B. Burgess, and N. A. Ross, "Income Distribution, Public Services Expenditures, and All Cause Mortality in US States," *Journal of Epidemiology & Community Health* 59, no. 9 (2005): 768–774; C. R. Ronzio, E. Pamuk, and G. D. Squires, "The Politics of Preventable Deaths: Local Spending, Income Inequality, and Premature Mortality in US Cities," *Journal of Epidemiology & Community Health* 58, no. 3 (2004): 175–179.
55. N. E. Adler, et al., "Relationship of Subjective and Objective Social Status with Psychological and Physiological Functioning: Preliminary Data in Healthy, White Women," *Health Psychology* 19, no. 6 (2000): 586.
56. L. Ayalon, "Subjective Social Status as a Predictor of Loneliness: The Moderating Effect of the Type of Long-Term Care Setting," *Research* on Aging 41, no. 10 (2019): 915–935.
57. I. Kawachi and B. P. Kennedy, *Health of Nations* (New York: New Press, 2006).
58. T. Miller, *Communes in America, 1975–2000* (Syracuse, NY: Syracuse University Press, 2019).
59. Miller, *Communes in America.*
60. G. Meltzer, "Cohousing: Verifying the Importance of Community in the Application of Environmentalism," *Journal of Architectural and Planning Research* 17, no. 2(2000): 110–132.
61. B. Leung and K. Shen, *Quit Like a Millionaire: No Gimmicks, Luck, or Trust Fund Required* (New York: TarcherPerigree, 2019).

62. H. Vernon, "I Tested the Saving Technique That Promises Retirement at 40," *Vice*, February 26, 2019, https://www.vice.com/en/article/gya8bx/i-tested-the-saving-technique-that-promises-retirement-at-40.
63. J. L. Collins, "Community, Chautauqua, and AMA with JL Collins," Choose FI, 2017.
64. M. Dakhli and D. De Clercq, "Human Capital, Social Capital, and Innovation: A Multi-Country Study," *Entrepreneurship & Regional Development* 16, no. 2 (2004): 107–128.
65. J. S. Coleman, "Social Capital in the Creation of Human Capital," *American Journal of Sociology* 94 (1988): S95–S120.
66. R. S. Burt, *Structural Holes: The Social Structure of Competition* (Cambridge, MA: Harvard University Press, 2009).
67. J. Nahapiet and S. Ghoshal, "Social Capital, Intellectual Capital, and the Organizational Advantage," *Academy of Management Review* 23, no. 2 (1998): 242–266; M. Freel, "External Linkages and Product Innovation in Small Manufacturing Firms," *Entrepreneurship & Regional Development* 12, no. 3 (2000): 245–266.
68. M. Dakhli and D. De Clercq, "Human Capital, Social Capital, and Innovation: A Multi-Country Study," *Entrepreneurship & Regional Development* 16, no. 2 (2004): 107–128.
69. D. Bollier, "The Cornucopia of the Commons," Yes! Magazine 18 (2001).
70. J. M. Pennings, K. Lee, and A. V. Witteloostuijn, "Human Capital, Social Capital, and Firm Dissolution," *Academy of Management Journal* 41, no. 4 (1998): 425–440.
71. S. McChrystal, et al., *Team of Teams: New Rules of Engagement for a Complex World* (New York: Profolio, 2015).
72. E. Diener and M. E. P. Seligman, "Very Happy People," *Psychological Science* 13, no. 1 (2002): 81–84.
73. B. H. Hidaka, "Depression as a Disease of Modernity: Explanations for Increasing Prevalence," *Journal of Affective Disorders* 140, no. 3 (2012): 205–214.
74. J. Hari, *Lost Connections: Why You're Depressed and How to Find Hope* (London: Bloomsbury Publishing, 2019).
75. M. T. Gonzalez, et al., "Therapeutic Horticulture in Clinical Depression: A Prospective Study of Active Components," *Journal of Advanced Nursing* 66, no. 9 (2010): 2002–2013.
76. S. Turkle, *Alone Together: Why We Expect More from Technology and Less from Each Other* (New York: Basic Books, 2017).
77. M. J. Prince, et al., "A Prospective Population-Based Cohort Study of the Effects of Disablement and Social Milieu on the Onset and Maintenance of Late-Life Depression. The Gospel Oak Project VII," *Psychological Medicine* 28, no. 2 (1998): 337–350.
78. M. F. de Mello, et al., "A Systematic Review of Research Findings on the Efficacy of Interpersonal Therapy for Depressive Disorders," *European Archives of Psychiatry and Clinical Neuroscience* 255, no. 2 (2005): 75–82.
79. T. Kasser, *The High Price of Materialism* (Cambridge, MA: MIT Press, 2002).
80. T. Kasser, et al., "Changes in Materialism, Changes in Psychological Well-Being: Evidence from Three Longitudinal Studies and an Intervention Experiment," *Motivation and Emotion* 38, no. 1 (2014): 1–22.

81. Hari, *Lost Connections.*
82. Hari, *Lost Connections.*
83. M. G. Marmot, et al., "Employment Grade and Coronary Heart Disease in British Civil Servants," *Journal of Epidemiology & Community Health* 32, no. 4 (1978): 244–249; M. G. Marmot, et al., "Health Inequalities among British Civil Servants: The Whitehall II Study," *The Lancet* 337, no. 8754 (1991): 1387–1393; F. M. North, et al., "Psychosocial Work Environment and Sickness Absence among British Civil Servants: The Whitehall II Study," *American Journal of Public Health* 86, no. 3 (1996): 332–340.
84. P. P. Baard, E. L. Deci, and R. M. Ryan, "Intrinsic Need Satisfaction: A Motivational Basis of Performance and Well-Being in Two Work Settings," *Journal of Applied Social Psychology* 34, no. 10 (2004): 2045–2068.
85. Hari, *Lost Connections.*
86. M. Connolly, "Rumspringa: Amish Teens Venture into Modern Vices," Talk of the Nation, NPR, June 7, 2006, https://www.npr.org/templates/story/story.php?storyId=5455572.
87. K. Stollznow, *God Bless America: Strange and Unusual Religious Beliefs and Practices in the United States* (Durham, NC: Pitchstone, 2014).
88. J. A. Egeland and A. M. Hostetter, "Amish Study: I. Affective Disorders among the Amish, 1976–1980," *American Journal of Psychiatry* 140, no. 1 (1983): 46–51.
89. Hari, *Lost Connections.*
90. E. Gellner, *Conditions of Liberty: Civil Society and Its Rivals* (London: Hamish Hamilton, 1994).
91. R. Wrangham, *The Goodness Paradox: The Strange Relationship Between Virtue and Violence in Human Evolution* (New York: Vintage, 2019).
92. G. Meltzer, "Cohousing: Verifying the Importance of Community in the Application of Environmentalism," *Journal of Architectural and Planning Research* 17, no. 2(2000): 110–132.
93. M. Gladwell, *Outliers: The Story of Success* (New York: Little, Brown, 2008).

CHAPTER 7: TRIBAL FRIENDS

1. С. И. Руденко, *Frozen Tombs of Siberia: The Pazyryk Burials of Iron Age Horsemen* (Berkeley: University of California Press, 1970).
2. N. A. Christakis, *Blueprint: The Evolutionary Origins of a Good Society* (New York: Springer, 1996).
3. Christakis, *Blueprint.*
4. C. J. Buys, and K. L. Larson, "Human Sympathy Groups," *Psychological Reports* 45, no. 2 (1979): 547–553.
5. P. E. Tetlock, and D. Gardner, *Superforecasting: The Art and Science of Prediction* (New York: Random House, 2016).
6. S. Pinker, *Enlightenment Now: The Case for Reason, Science, Humanism, and Progress* (New York: Penguin, 2018).
7. J. H. Fowler, J. E. Settle, and N. A. Christakis, "Correlated Genotypes in Friendship Networks," *Proceedings of the National Academy of Sciences* 108, no. 5 (2011): 1993–1997.
8. Christakis, *Blueprint.*
9. D. Gilbert, *Stumbling on Happiness* (Toronto: Vintage Canada, 2009).

10. S. Harris, "Making Sense," in *The Map of Misunderstanding: A Conversation with Daniel Kahneman*. 2019.
11. Christakis, *Blueprint*.
12. Christakis, *Blueprint*.
13. B. J. Bigelow, "Children's Friendship Expectations: A Cognitive-Developmental Study," *Child Development* 48, no. 1 (March 1977): 246–253.
14. C. L. Apicella, et al., "Social Networks and Cooperation in Hunter-Gatherers," *Nature* 481, no. 7382 (2012): 497.
15. D. J. Hruschka, *Friendship: Development, Ecology, and Evolution of a Relationship*, vol. 5 (Berkeley: University of California Press, 2010).
16. E. A. Madsen, et al., "Kinship and Altruism: A Cross-Cultural Experimental Study," *British Journal of Psychology* 98, no. 2 (2007): 339–359.
17. M. J. Rantala and U. M. Marcinkowska, "The Role of Sexual Imprinting and the Westermarck Effect in Mate Choice in Humans," *Behavioral Ecology and Sociobiology* 65, no. 5 (2011): 859–873.
18. I. G. Sarason, et al., "A Brief Measure of Social Support: Practical and Theoretical Implications," *Journal of Social and Personal Relationships* 4, no. 4 (1987): 497–510.
19. M. McPherson, L. Smith-Lovin, and M. E. Brashears, "Social Isolation in America: Changes in Core Discussion Networks over Two Decades," *American Sociological Review* 71, no. 3 (2006): 353–375; P. V. Marsden, "Core Discussion Networks of Americans," *American Sociological Review* 52, no. 1 (February 1987): 122–131; A. J. O'Malley, et al., "Egocentric Social Network Structure, Health, and Pro-Social Behaviors in a National Panel Study of Americans," *PLOS One* 7, no. 5 (2012): e36250.
20. O'Malley, et al., "Egocentric Social Network Structure," e36250.
21. A. Levine and R. Heller, *Attached: The New Science of Adult Attachment and How It Can Help You Find—And Keep—Love* (New York: Penguin, 2012).
22. G. F. Miller, "Sexual Selection for Moral Virtues," *Quarterly Review of Biology* 82, no. 2 (2007): 97–125.
23. E. Perel, *Mating in Captivity: Unlocking Erotic Intelligence* (Harper Audio, 2006).
24. S. Coontz, *Marriage, a History: How Love Conquered Marriage* (New York: Penguin, 2006).
25. J. Olds and R. S. Schwartz, *The Lonely American: Drifting Apart in the Twenty-First Century* (Boston: Beacon Press, 2009).
26. E. Perel, *Mating in Captivity*.
27. C. E. Fritz, *Disasters and Mental Health: Therapeutic Principles Drawn from Disaster Studies* (Newark: Disaster Research Center, University of Delaware, 1996).
28. J. Drury, et al., "Facilitating Collective Psychosocial Resilience in the Public in Emergencies: Twelve Recommendations Based on the Social Identity Approach," Frontiers in Public Health 7 (2019): 141.
29. D. Brooks, "The Nuclear Family Was a Mistake," *The Atlantic*, March 2020, numbering.xml.
30. J. J. Clarke, *Creating Rituals: A New Way of Healing for Everyday Life* (Mahwah, NJ: Paulist Press, 2014).
31. D. Brooks, "There Should Be More Rituals!" *New York Times*, April 22, 2019, https://www.nytimes.com/2019/04/22/opinion/rituals-meaning.html.

32. S. Sagan, *For Small Creatures Such as We: Rituals for Finding Meaning in Our Unlikely World* (New York: Penguin Random House, 2019).

CHAPTER 8: CAMPCRAFTING (PART 1)

1. M. Rolland, "Our Children's Coming of Age," White Raven Center, November 28, 2017, https://www.whiteravencenter.org/2017/11/28/our-childrens-coming-of-age-by-dr-marianne-rolland/.
2. M. D'Avella (dir.), *Minimalism: A Documentary About the Important Things in Life*, 2016, https://www.youtube.com/watch?v=0Co1Iptd4p4.
3. J. Lane, "Choosing the Right Team," *Management Accounting*, 1987: 27–30; D. S. Kezsbom, "Reopening Pandora Box-Sources of Project Conflict in the 90s," *Industrial Engineering* 24, no. 5 (1992): 54–59.
4. A. Strickland, "Astronauts on Mars Mission Will Need to Be 'Conscientious' to Work Well Together," CNN, November 24, 2020, https://www.cnn.com/2020/11/24/world/mars-astronaut-traits-wellness-scn-trnd.
5. J. A. LePine, "Team Adaptation and Postchange Performance: Effects of Team Composition in Terms of Members' Cognitive Ability and Personality," *Journal of Applied Psychology* 88, no. 1 (2003): 27.
6. E. Molleman, A. Nauta, and K. A. Jehn, "Person-Job Fit Applied to Teamwork: A Multilevel Approach," *Small Group Research* 35, no. 5 (2004): 515–539.
7. G. A. Neuman, S. H. Wagner, and N. D. Christiansen, "The Relationship Between Work-Team Personality Composition and the Job Performance of Teams," *Group & Organization Management* 24, no. 1 (1999): 28–45.
8. A. Neal, et al., "Predicting the Form and Direction of Work Role Performance from the Big 5 Model of Personality Traits," *Journal of Organizational Behavior* 33, no. 2 (2012): 175–192.
9. S. Taggar, "Individual Creativity and Group Ability to Utilize Individual Creative Resources: A Multilevel Model," *Academy of Management Journal* 45, no. 2 (2002): 315–330; E. Molleman, A. Nauta, and K. A. Jehn, "Person-Job Fit Applied to Teamwork: A Multilevel Approach," *Small Group Research* 35, no. 5 (2004): 515–539; LePine, "Team Adaptation and Postchange Performance."
10. S. Mohammed and L. C. Angell, "Personality Heterogeneity in Teams: Which Differences Make a Difference for Team Performance?," Small Group Research 34, no. 6 (2003): 651–677.
11. S. L. Kichuk and W. H. Wiesner, "The Big Five Personality Factors and Team Performance: Implications for Selecting Successful Product Design Teams," *Journal of Engineering and Technology Management* 14, nos. 3–4 (1997): 195–221.
12. M. R. Barrick, et al., "Relating Member Ability and Personality to Work-Team Processes and Team Effectiveness," *Journal of Applied Psychology* 83, no. 3 (1998): 377; B. Barry and G. L. Stewart, "Composition, Process, and Performance in Self-Managed Groups: The Role of Personality," *Journal of Applied Psychology* 82, no. 1 (1997): 62.
13. P. T. Costa and R. R. McCrae, *Neo Personality Inventory-Revised (NEO PI-R)* (Odessa, FL: Psychological Assessment Resources, 1992); W. G. Graziano, L. A. Jensen-Campbell, and E. C. Hair, "Perceiving Interpersonal Conflict and Reacting to It: The Case for Agreeableness," *Journal of Personality and Social Psychology* 70, no. 4 (1996): 820.
14. A. E. Van Vianen and C. K. De Dreu, "Personality in Teams: Its Relationship to So-

cial Cohesion, Task Cohesion, and Team Performance," *European Journal of Work and Organizational Psychology* 10, no. 2 (2001): 97–120.

15. S. Mohammed, J. E. Mathieu, and A. 'Bart' Bartlett, "Technical-Administrative Task Performance, Leadership Task Performance, and Contextual Performance: Considering the Influence of Team—and Task—Related Composition Variables," *Journal of Organizational Behavior: The International Journal of Industrial, Occupational and Organizational Psychology and Behavior* 23, no. 7 (2002): 795–814.
16. R. Klimoski and S. Mohammed, "Team Mental Model: Construct or Metaphor?" *Journal of Management* 20, no. 2 (1994): 403–437.
17. D. M. Kahan, et al., "Motivated Numeracy and Enlightened Self-Government," Behavioural Public Policy 1, no. 1 (2017): 54–86; S. Harris, *Waking Up*, 2021, https://app.wakingup.com/.
18. M. A. Peeters, et al., "Personality and Team Performance: A Meta-Analysis," *European Journal of Personality* 20, no. 5 (2006): 377–396.
19. R. I. Damian, et al., "Sixteen Going on Sixty-Six: A Longitudinal Study of Personality Stability and Change across 50 Years," *Journal of Personality and Social Psychology* 117, no. 3 (2019): 674.
20. J. Rantanen, et al., "Long-Term Stability in the Big Five Personality Traits in Adulthood," *Scandinavian Journal of Psychology* 48, no. 6 (2007): 511–518.
21. J. G. Dunn and N. L. Holt, "A Qualitative Investigation of a Personal-Disclosure Mutual-Sharing Team Building Activity," *The Sport Psychologist* 18, no. 4 (2004): 363–380.
22. A. V. Carron, H. A. Hausenblas, and M. A. Eys, *Group Dynamics in Sport* (Morgantown, WV: Fitness Information Technology, 2005); R. P. Tett and D. D. Burnett, "A Personality Trait-Based Interactionist Model of Job Performance," *Journal of Applied Psychology* 88, no. 3 (2003): 500.
23. P. W. Atkins, D. S. Wilson, and S. C. Hayes, *Prosocial: Using Evolutionary Science to Build Productive, Equitable, and Collaborative Groups* (Oakland, CA: New Harbinger Publications, 2019).
24. L. Festinger, "Informal Social Communication," *Psychological Review* 57, no. 5 (1950): 271.
25. J. E. Mathieu, et al., "Modeling Reciprocal Team Cohesion-Performance Relationships, as Impacted by Shared Leadership and Members' Competence," *Journal of Applied Psychology* 100, no. 3 (2015): 713.
26. B. E. Ashforth and F. Mael, "Social Identity Theory and the Organization," *Academy of Management Review* 14, no. 1 (1989): 20–39; A. Chang and P. Bordia, "A Multidimensional Approach to the Group Cohesion-Group Performance Relationship," *Small Group Research* 32, no. 4 (2001): 379–405; D. De Cremer, "Respect and Cooperation in Social Dilemmas: The Importance of Feeling Included," *Personality and Social Psychology Bulletin* 28, no. 10 (2002): 1335–1341; D. De Cremer and J. Stouten, "When Do People Find Cooperation Most Justified? The Effect of Trust and Self-Other Merging in Social Dilemmas," *Social Justice Research* 16, no. 1 (2003): 41–52; M. Van Vugt and C. M. Hart, "Social Identity as Social Glue: The Origins of Group Loyalty," *Journal of Personality and Social Psychology* 86, no. 4 (2004): 585.
27. G. L. Stewart, S. H. Courtright, and M. R. Barrick, "Peer-Based Control in Self-Managing Teams: Linking Rational and Normative Influence with Individual and Group Performance," *Journal of Applied Psychology* 97, no. 2 (2012): 435.

28. C. C. Tossell, et al., "Spiritual over Physical Formidability Determines Willingness to Fight and Sacrifice Through Loyalty in Cross-Cultural Populations," *Proceedings of the National Academy of Sciences* 119, no. 6 (2022): e2113076119.
29. I. Spreadbury, *Humans: Smart Enough to Create Processed Foods, Daft Enough to Eat Them.* The Evolution Institute, 2019.
30. V. H. Fetvadjiev and J. He, "The Longitudinal Links of Personality Traits, Values, and Well-Being and Self-Esteem: A Five-Wave Study of a Nationally Representative Sample," *Journal of Personality and Social Psychology* 117, no. 2 (2019): 448.
31. T. Miller, *Communes in America, 1975–2000* (Syracuse, NY: Syracuse University Press, 2019).
32. C. Duhigg, "What Google Learned from Its Quest to Build the Perfect Team," *The New York Times Magazine,* February 25, 2016, https://www.nytimes.com/2016/02/28/magazine/what-google-learned-from-its-quest-to-build-the-perfect-team.html.
33. J. Jones, "U.S. Church Membership Falls Below Majority for First Time," Gallup, March 29, 2021, https://news.gallup.com/poll/341963/church-membership-falls-below-majority-first-time.aspx.
34. D. Masci and M. Lipka, "Americans May Be Getting Less Religious, but Feelings of Spirituality Are on the Rise," Pew Research Center, January 21, 2016, https://www.pewresearch.org/fact-tank/2016/01/21/americans-spirituality/.
35. A. Norman, *Mental Immunity: Infectious Ideas, Mind-Parasites, and the Search for a Better Way to Think* (New York: HarperCollins, 2021).
36. J. H. Shaver, et al., "Church Attendance and Alloparenting: An Analysis of Fertility, Social Support and Child Development among English Mothers," *Philosophical Transactions of the Royal Society* B 375, no. 1805 (2020): 20190428.
37. G. S. Paul, "Cross-National Correlations of Quantifiable Societal Health with Popular Religiosity and Secularism in the Prosperous Democracies: A First Look," *Journal of Religion and Society,* 7 (2005), https://dspace2.creighton.edu/xmlui/bitstream/handle/10504/64409/2005-11.pdf.
38. B. Vlaardingerbroek, "Paganism in 21st-century Europe—What's the Attraction?," MercatorNet, February 19, 2020, https://mercatornet.com/paganism-in-21st-century-europe-whats-the-attraction/47604/.
39. M. W. Johnson, et al., "Classic Psychedelics: An Integrative Review of Epidemiology, Therapeutics, Mystical Experience, and Brain Network Function," *Pharmacology & Therapeutics* 197 (2019): 83–102.
40. P. Lundborg, *Psychedelia: An Ancient Culture, a Modern Way of Life* (Stockholm: Lysergia, 2012).
41. B. C. Muraresku, *The Immortality Key: The Secret History of the Religion with No Name* (New York: St. Martin's Press, 2020).
42. D. J. McKenna, J. C. Callaway, and C. S. Grob, "10. The Scientific Investigation of Ayahuasca: A Review of Past and Current Research," *The Heffter Review of Psychedelic Research* 1 (January 1998): 65–77, https://www.researchgate.net/publication/237540713_10_The_Scientific_Investigation_of_Ayahuasca_A_Review_of_Past_and_Current_Research.
43. R. R. Griffiths, et al., "Psilocybin-Occasioned Mystical-Type Experience in Combination with Meditation and Other Spiritual Practices Produces Enduring Positive Changes in Psychological Functioning and in Trait Measures of Pro-

social Attitudes and Behaviors," *Journal of Psychopharmacology* 32, no. 1 (2018): 49–69.

44. P. Lundborg, *Psychedelia: An Ancient Culture, a Modern Way of Life* (Stockholm: Lysergia, 2012).
45. Lundborg, *Psychedelia*.
46. T. Kim, et al., "Work Group Rituals Enhance the Meaning of Work," *Organizational Behavior and Human Decision Processes* 165 (2021): 197–212.
47. M. Bagherniya, et al., "The Effect of Fasting or Calorie Restriction on Autophagy Induction: A Review of the Literature," *Ageing Research Reviews* 47 (2018): 183–197; K. K. Hoddy, et al., "Changes in Hunger and Fullness in Relation to Gut Peptides Before and After 8 Weeks of Alternate Day Fasting," *Clinical Nutrition* 35, no. 6 (2016): 1380–1385; R. E. Patterson and D. D. Sears, "Metabolic Effects of Intermittent Fasting," *Annual Review of Nutrition* 37, no. 1, (2017): 371–393.
48. H. Gould, et al., "Robot Death Care: A Study of Funerary Practice," *International Journal of Cultural Studies* 24, no. 4 (2021): 603–621.
49. M. D. S. Ainsworth, et al., *Patterns of Attachment: A Psychological Study of the Strange Situation* (New York: Psychology Press, 2015).
50. D. M. Raup, "Geometric Analysis of Shell Coiling: General Problems," *Journal of Paleontology* 40, no. 5 (September 1966): 1178–1190.
51. L. Cronk, *That Complex Whole. Culture and the Evolution of Behavior* (Boulder, CO: Westview, 1999).
52. N. A. Christakis, *Blueprint: The Evolutionary Origins of a Good Society* (New York: Springer, 1996).
53. R. I. Dunbar, et al., "The Structure of Online Social Networks Mirrors Those in the Offline World," *Social Networks* 43 (2015): 39–47; R. I. Dunbar and R. Sosis, "Optimising Human Community Sizes," *Evolution and Human Behavior* 39, no. 1 (2018): 106–111.
54. D. G. Rand, et al., "Static Network Structure Can Stabilize Human Cooperation," *Proceedings of the National Academy of Sciences* 111, no. 48 (2014): 17093–17098.
55. H. Shirado, et al., "Quality versus Quantity of Social Ties in Experimental Cooperative Networks," *Nature Communications* 4, no. 1 (2013): 1–8.
56. M. Szell, R. Lambiotte, and S. Thurner, "Multirelational Organization of Large-Scale Social Networks in an Online World," *Proceedings of the National Academy of Sciences* 107, no. 31 (2010): 13636–13641.
57. F. Marlowe, *The Hadza: Hunter-Gatherers of Tanzania*, vol. 3, Origins of Human Behavior and Culture (Berkeley: University of California Press, 2010), x, 325p.
58. A. Nishi, et al., "Inequality and Visibility of Wealth in Experimental Social Networks," *Nature* 526, no. 7573 (2015): 426–429.
59. M. L. Burton, et al., "Regions Based on Social Structure," *Current Anthropology* 37, no. 1 (1996): 87–123.
60. B. F. Skinner, *Walden Two* (Indianapolis: Hackett Publishing, 1969).
61. J. L. Ravelo and S. Jerving, "COVID-19—A Timeline of the Coronavirus Outbreak," Devex, accessed December 6, 2022, https://www.devex.com/news/covid-19-a-timeline-of-the-coronavirus-outbreak-96396.
62. R. Sosis and B. J. Ruffle, "Ideology, Religion, and the Evolution of Cooperation: Field Experiments on Israeli Kibbutzim," in *Socioeconomic Aspects of Human Behavioral Ecology*, ed. M. Alvard (Bingley, UK: Emerald Group, 2004), 89–118.

63. H. Kuhlman, *Living Walden Two: BF Skinner's Behaviorist Utopia and Experimental Communities* (Champaign: Illinois University Press, 2010).
64. I. Komar, *Living the Dream: A Documentary Study of the Twin Oaks Community*, vol. 1 (Norwood Editions, 1983).
65. C. Los Horcones, "News from Now-Here, 1986: A Response to 'News from Nowhere, 1984.'" *The Behavior Analyst* 9, no. 1 (1986): 129.

CHAPTER 9: CAMPCRAFTING (PART 2)

1. D. Grover, "What Is the Farthest Distance That a Normal Human Scream or Sound Can Travel Considering It Is Night and Normal Climatic Conditions?," Quora, 2015, accessed December 6, 2022, https://www.quora.com/What-is-the-farthest-distance-that-a-normal-human-scream-or-sound-can-travel-considering-it-is-night-and-normal-climatic-conditions.
2. J. Stratmann, L. W. Ferreiro, and R. Narayan, *Toward Sustainability—Analysis of Collaborative Behaviour in Urban Cohousing*, 2013.
3. J. Olds and R. S. Schwartz, *The Lonely American: Drifting Apart in the Twenty-First Century* (New York: Beacon Press, 2009).
4. T. Miller, *Communes in America, 1975–2000* (Syracuse, NY: Syracuse University Press, 2019).
5. F. Barone, *Urban Firewalls: Place, Space and New Technologies in Figueres, Catalonia* (PhD thesis, University of Kent, 2010).
6. S. Mallett, "Understanding Home: A Critical Review of the Literature," *The Sociological Review* 52, no. 1 (2004): 62–89.
7. F. Barone, "Home Truths: An Anthropology of House and Home," Human Relations Area Files, December 12, 2019, https://hraf.yale.edu/home-truths-an-anthropology-of-house-and-home/.
8. J. E. Arnold, et al., *Life at Home in the Twenty-First Century: 32 Families Open Their Doors* (Los Angeles: Cotsen Institute of Archaeology Press, UCLA, 2012).
9. B. M. Brown, "Population Estimation from Floor Area: A Restudy of 'Naroll's Constant,'" *Behavior Science Research* 21, nos. 1–4 (1987): 1–49; M. Porčić, "House Floor Area as a Correlate of Marital Residence Pattern: A Logistic Regression Approach," *Cross-Cultural Research* 44, no. 4 (2010): 405–424.
10. Miller, *Communes in America, 1975–2000*.
11. K. McCammant and C. Durret, *Cohousing: A Contemporary Approach to Housing Ourselves* (Berkeley: Ten Speed Press, 1994).
12. D. Milman, "Where It All Began: Cohousing in Denmark," The History of Cohousing, Canadian Cohousing Network, accessed December 6, 2022, https://cohousing.ca/about-cohousing/history-of-cohousing/.
13. McCammant and Durret, *Cohousing*.
14. A. Glass, "An Evaluation of an Elder Cohousing Community," *Gerontologist* 51 (2011).
15. D. Fromm, "American Cohousing: The First Five Years," *Journal of Architectural and Planning Research* 17, no. 2 (Summer 2000): 94–109.
16. M. O'Connor, "What Co-Living Is Like: An Insider Describes His Life in an 'Adult Dorm.'" Brick Underground, December 21, 2018, https://www.brickunderground.com/rent/what-it%27s-like-to-live-in-a-co-living-building.
17. A. Yu, "These Families Just Co-Purchased a $900.000 Semi in Mount Dennis. Now

They're Learning How to Live Together," *Toronto Life*, March 8, 2021, https://torontolife.com/real-estate/these-families-just-co-purchased-a-900000-semi-in-mount-dennis-now-theyre-learning-how-to-live-together/.

18. P. W. Atkins, D. S. Wilson, and S. C. Hayes, *Prosocial: Using Evolutionary Science to Build Productive, Equitable, and Collaborative Groups* (Oakland, CA: New Harbinger Publications, 2019).
19. M. R. Beauchamp, et al., "Leadership Behaviors and Multidimensional Role Ambiguity Perceptions in Team Sports," *Small Group Research* 36, no. 1 (2005): 5–20.
20. G. Fusco and A. Minelli, *The Biology of Reproduction* (Cambridge, UK: Cambridge University Press, 2019); A. M. Holub and T. K. Shackelford, *Sex in the Animal Kingdom*, 2021.
21. M. D. Breed and J. Moore, *Animal Behavior* (New York: Academic Press, 2015); R. J. Haier, et al., "The Neuroanatomy of General Intelligence: Sex Matter," *Neuroimage* 25, no. 1 (2005): 320–327.
22. K. L. Nadal, *The SAGE Encyclopedia of Psychology and Gender* (Thousand Oaks, CA: SAGE, 2017); M. L. Miville and A. D. Ferguson, *Handbook of Race-Ethnicity and Gender in Psychology* (New York: Springer, 2014).
23. L. Mealey, *Sex Differences: Developmental and Evolutionary Strategies* (New York: Academic Press, 2000).
24. A. Treuer, *The Assassination of Hole in the Day* (Nepean, Ontario: Borealis Books, 2011); A. Matzner, *An Oral History from Hawai'i.*
25. G. S. Elder, "Third Sex, Third Gender: Beyond Sexual Dimorphism in Culture and History," *Gender, Place and Culture* 6, no. 3 (1999): 291.
26. P. T. Costa Jr., A. Terracciano, and R. R. McCrae, "Gender Differences in Personality Traits across Cultures: Robust and Surprising Findings," *Journal of Personality and Social Psychology* 81, no. 2 (2001): 322.
27. S. Junger, *Tribe: On Homecoming and Belonging* (New York: Twelve, 2016).
28. R. Weil and F. Dunsworth, "Psychiatric Aspects of Disaster—A Case History: Some Experiences During the Springhill, NS Mining Disaster," *Canadian Psychiatric Association Journal* 3, no. 1 (1958): 11–17.
29. F. A. Dunsworth, "Springhill Disaster (Psychological Findings in the Surviving Miners," *Nova Scotia Medical Bulletin* 37 (1958): 111–114.
30. Junger, *Tribe.*
31. Q. M. Roberson, "Disentangling the Meanings of Diversity and Inclusion in Organizations," *Group & Organization Management* 31, no. 2 (2006): 212–236.
32. P. W. Atkins, D. S. Wilson, and S. C. Hayes, *Prosocial: Using Evolutionary Science to Build Productive, Equitable, and Collaborative Groups* (Oakland, CA: New Harbinger Publications, 2019).
33. A. M. O'Leary-Kelly, J. J. Martocchio, and D. D. Frink, "A Review of the Influence of Group Goals on Group Performance," *Academy of Management Journal* 37, no. 5 (1994): 1285–1301.
34. W. N. Widmeyer and K. Ducharme, "Team Building Through Team Goal Setting," *Journal of Applied Sport Psychology* 9, no. 1 (1997): 97–113.
35. A. Caspi, B. W. Roberts, and R. L. Shiner, "Personality Development: Stability and Change," *Annual Review Psychology* 56 (2005): 453–484.

36. M. Gelfand, *Rule Makers, Rule Breakers: Tight and Loose Cultures and the Secret Signals That Direct Our Live* (New York: Scribner, 2019).
37. Gelfand, *Rule Makers, Rule Breakers.*
38. N. A. Christakis, *Blueprint: The Evolutionary Origins of a Good Society* (New York: Springer, 1996).
39. H. E. Gardner, *Leading Minds: An Anatomy of Leadership* (New York: Basic Books, 2011).
40. F. Marlowe, *The Hadza: Hunter-Gatherers of Tanzania*, vol. 3, Origins of Human Behavior and Culture (Berkeley: University of California Press, 2010), x, 325p.
41. C. Boehm, *Moral Origins: The Evolution of Virtue, Altruism, and Shame* (New York: Soft Skull Press, 2012).
42. H. G. Barnett, "The Nature of the Potlatch," *American Anthropologist* 40, no. 3 (1938): 349–358.
43. S. Sinek, *Leaders Eat Last: Why Some Teams Pull Together and Others Don't* (New York: Penguin, 2014).
44. S. McChrystal, et al., *Team of Teams: New Rules of Engagement for a Complex World* (New York: Penguin, 2015).
45. J. Willink and L. Babin, *Extreme Ownership: How US Navy SEALS Lead and Win* (New York: St. Martin's Press, 2017).
46. A. M. O'Leary-Kelly, J. J. Martocchio, and D. D. Frink, "A Review of the Influence of Group Goals on Group Performance," *Academy of Management Journal* 37, no. 5 (1994): 1285–1301.
47. S. Sagan, *For Small Creatures Such as We: Rituals for Finding Meaning in Our Unlikely World* (New York: Penguin Random House, 2019).
48. Sagan, *For Small Creatures Such as We.*

CHAPTER 10: THE TRIBE VIRUS

1. A. Kuhrt, *The Ancient Near East, c. 3000–330 BC*, vol. 2 (London: Taylor & Francis, 1995).
2. A. Kuhrt, *The Persian Empire: A Corpus of Sources from the Achaemenid Period* (London: Routledge, 2013).
3. A. Norman, *Mental Immunity: Infectious Ideas, Mind-Parasites, and the Search for a Better Way to Think* (New York: HarperCollins, 2021).
4. E. O. Wilson, *The Social Conquest of Earth* (New York: W. W. Norton & Company, 2012).
5. "Democracy Index 2020: In Sickness and in Health?," Economist Intelligence Unit, *The Economist*, accessed December 6, 2022, https://www.eiu.com/n/campaigns/democracy-index-2020/.
6. S. Pinker, *Enlightenment Now: The Case for Reason, Science, Humanism, and Progress* (New York: Penguin, 2018).
7. D. Ariely and S. Jones, *Predictably Irrational* (New York: Harper Audio, 2008).
8. R. I. Dunbar, "Do Online Social Media Cut Through the Constraints That Limit the Size of Offline Social Networks?," *Royal Society Open Science* 3, no. 1 (2016): 150292.
9. C. J. Buys and K. L. Larson, "Human Sympathy Group," *Psychological Reports* 45, no. 2 (1979): 547–553.
10. J. Bowlby, *Attachment and Loss, vol. II: Separation, Anxiety and Anger* (London:

The Hogarth Press and the Institute of Psycho-Analysis, 1973), 1–429; I. Bretherton, "The Origins of Attachment Theory: John Bowlby and Mary Ainsworth," *Developmental Psychology* 28, no. 5 (1992): 759.

11. R. Topolski, et al., "Choosing Between the Emotional Dog and the Rational Pal: A Moral Dilemma with a Tail," *Anthrozoös* 26, no. 2 (2013): 253–263.
12. J. Tooby and L. Cosmides, "Friendship and the Banker's Paradox: Other Pathways to the Evolution of Adaptations for Altruism," *Proceedings of the British Academy* 88 (1996).
13. J. M. Rabbie and H. F. Lodewijkx, "Conflict and Aggression: An Individual-Group Continuum," *Advances in Group Processes* 11 (1994): 139–174.
14. Christakis, *Blueprint*; D. J. Hruschka, *Friendship: Development, Ecology, and Evolution of a Relationship*, vol. 5 (Berkeley: University of California Press, 2010).
15. J. Ross, "Global Deaths: This Is How COVID-19 Compares to Other Diseases," in World Economic Forum, 2020.
16. A. Avenanti, et al., "Transcranial Magnetic Stimulation Highlights the Sensorimotor Side of Empathy for Pain," *Nature Neuroscience* 8, no. 7 (2005): 955–960; X. Xu, et al., "Do You Feel My Pain? Racial Group Membership Modulates Empathic Neural Responses," *Journal of Neuroscience* 29, no. 26 (2009): 8525–8529; V. A. Mathur, et al., "Neural Basis of Extraordinary Empathy and Altruistic Motivation," *Neuroimage* 51, no. 4 (2010): 1468–1475.
17. A. Cohn, E. Fehr, and M. A. Maréchal, "Business Culture and Dishonesty in the Banking Industry," Nature 516, no. 7529 (2014): 86–89.
18. J. D. Greene, *Moral Tribes: Emotion, Reason, and the Gap Between Us and Them* (New York: Penguin, 2013).
19. J. Donovan, *The Way of Men* (N.p.: Self-published via Dissonant Hum, 2012); J. Donovan, *Becoming a Barbarian* (N.p.: Self-published via Dissonant Hum, 2016).
20. J. Rawls, "Kantian Constructivism in Moral Theory," *Journal of Philosophy* 77, no. 9 (1980): 515–572.
21. S. R. Cavanagh, *Hivemind: The New Science of Tribalism in Our Divided World* (New York: Grand Central Publishing, 2019).
22. S. Harris, *The Moral Landscape: How Science Can Determine Human Values* (New York: Simon & Schuster, 2011).
23. Plato, *The Republic*, in *The Dialogues of Plato*, trans. B. Jowett (Oxford: Clarendon Press, 1953), 1–499.
24. A. Norman, *Mental Immunity: Infectious Ideas, Mind-Parasites, and the Search for a Better Way to Think* (New York: HarperCollins, 2021).
25. C. Andreou, "Phillipa Foot, Natural Goodness (Oxford: Clarendon Press 2001), 125; *Utilitas* 17, no. 3 (2005): 359–361.
26. D. N. Perkins, "Reasoning as It Is and Could Be: An Empirical Perspective," in *Thinking Across Cultures: The Third International Conference on Thinking*, ed. D. M. Topping, D. C. Crowell, and V. N. Kobayashi (New York: Routledge, 1989), 175–194.
27. S. Pinker, *Enlightenment Now: The Case for Reason, Science, Humanism, and Progress* (New York: Penguin, 2018).
28. A. Chua, *Political Tribes: Group Instinct and the Fate of Nations* (New York: Penguin Books, 2019).
29. L. Zummo, B. Donovan, and K. Busch, "Complex Influences of Mechanistic

Knowledge, Worldview, and Quantitative Reasoning on Climate Change Discourse: Evidence for Ideologically Motivated Reasoning among Youth," *Journal of Research in Science Teaching* 58, no. 1 (2021): 95–127.
30. E. Peterson and S. Iyengar, "Partisan Gaps in Political Information and Information-Seeking Behavior: Motivated Reasoning or Cheerleading?," *American Journal of Political Science* 65, no. 1 (2021): 133–147.
31. C. Puryear, et al., "Bridging Political Divides by Correcting the Basic Morality Bias," PsyArXiv, January 11, 2022, doi:10.31234/osf.io/fk8g6.
32. J. Galak and C. R. Critcher, "Who Sees Which Political Falsehoods as More Acceptable and Why: A New Look at In-Group Loyalty and Trustworthiness," *Journal of Personality and Social Psychology* (June 2022).
33. https://www.edge.org/response-detail/27181
34. M. Fuoco, "What's in a President's Signature? Pittsburgh Expert Says 'A Lot!'" *Pittsburgh Post-Gazette*, 2018.
35. J. M. Curtis and M. J. Curtis, "Factors Related to Susceptibility and Recruitment by Cults," *Psychological Reports* 73, no. 2 (1993): 451–460.
36. https://journals.sagepub.com/doi/10.1177/1470594X211065080
37. D. Kahan, E. Peters, E. Dawson, and P. Slovic, "Motivated Numeracy and Enlightened Self-Government," *Behavioural Public Policy* 1, no. 1 (2017): 54–86. doi:10.1017/bpp.2016.2
38. A. Norman, *Mental Immunity: Infectious Ideas, Mind-Parasites, and the Search for a Better Way to Think* (New York: HarperCollins, 2021).
39. W. Storr, *The Status Game: On Human Life and How to Play It* (London: William Collins, 2021).
40. R. D. Hare, *Without Conscience: The Disturbing World of the Psychopaths among Us* (New York: Guilford, 1999).
41. C. Boehm, *Moral Origins: The Evolution of Virtue, Altruism, and Shame* (New York: Soft Skull Press, 2012).
42. S. Dadfarnia, M.-S. Sadeghi, and L. Panaghi, "Investigating Couples' Interactive Patterns at Three Stages of the Family Life Cycle," *International Journal of Behavioral Sciences* 14, no. 4 (2021): 191–198.
43. S. Atran, R. Axelrod, and R. Davis, "Sacred Barriers to Conflict Resolution," *Science* 317 (2007): 1039–1040.
44. R. M. Sapolsky, *Behave: The Biology of Humans at Our Best and Worst* (New York: Penguin, 2017).
45. A. Mattes, "Techniques for Communicating with the Public Where Hazard Is Low, but Public Concern Is High," *Air Quality and Climate Change* 54, no. 1 (2020): 13–16.
46. D. Huff, "Decentralization: Freedom by Diffusion," Foundation for Economic Education, November 1, 1988, https://fee.org/articles/decentralization-freedom-by-diffusion/.

CHAPTER 11: OUR TRIBAL FUTURE

1. B. G. Myerhoff, *Peyote Hunt: The Sacred Journey of the Huichol Indians* (Ithaca, NY: Cornell University Press, 1976).
2. "Democracy Index 2020: In Sickness and in Health?," Economist Intelligence Unit, *The Economist*, accessed December 6, 2022, https://www.eiu.com/n/campaigns/democracy-index-2020/.

3. K. Sale, *Human Scale* (New York: Coward, McCann & Geoghegan, 1980).
4. A. De Tocqueville, *Democracy in America*, vol. 2, trans. Henry Reeve (London: Saunders & Otley, 1838).
5. T. Jefferson, *Thomas Jefferson to Joseph C. Cabell*, Founders Online, 1816.
6. R. Chetty, et al., "The Opportunity Atlas: Mapping the Childhood Roots of Social Mobility," Working Paper 25147, National Bureau of Economic Research, October 2018, revised February 2020,
7. D. Brooks, "Who Is Driving Inequality? You Are," *New York Times*, April 23, 2020, https://www.nytimes.com/2020/04/23/opinion/income-inequality.html.
8. A. W. Crosby, *The Measure of Reality: Quantification in Western Europe, 1250–1600* (Cambridge, UK: Cambridge University Press, 1997); A. Norman, *Mental Immunity: Infectious Ideas, Mind-Parasites, and the Search for a Better Way to Think* (New York: HarperCollins, 2021).
9. J. D. Davidson and W. Rees-Mogg, *The Sovereign Individual: Mastering the Transition to the Information Age* (New York: Touchstone, 1997).
10. S. Nakamoto, "Bitcoin: A Peer-to-Peer Electronic Cash System," Whitepaper, *Decentralized Business Review*, October 31, 2008, https://www.debr.io/article/21260-bitcoin-a-peer-to-peer-electronic-cash-system.
11. J. Gevers, "The Four Pillars of a Decentralized Society," TedxZug, 2014.
12. M. Mullenweg, "The Future of Work," in *Making Sense*, S. Harris, editor, 2021.
13. T. Sommers, *Why Honor Matters* (New York: Basic Books, 2018).
14. Gevers, "The Four Pillars of a Decentralized Society."
15. B. Srinivasan, "Summary: The Network State in One Sentence, One Paragraph, and One Thousand Words," November 11, 2021, https://1729.com/summary; B. Srinivasan, *The Network State: How to Start a New Country*, accessed December 6, 2022, https://thenetworkstate.com/.
16. J. Lowery, "Software: Bitcoin and the Future of National Strategy Defense," in Design and Management Program, MIT, 2022.
17. Norman, *Mental Immunity*.
18. Genesis, A.i., *Bill Nye Debates Ken Ham*, 2014.
19. G. Pennycook, et al., "On the Belief That Beliefs Should Change According to Evidence: Implications for Conspiratorial, Moral, Paranormal, Political, Religious, and Science Beliefs," *Judgment and Decision Making* 15, no. 4 (2020): 476.
20. R. F. Baumeister and M. R. Leary, "The Need to Belong: Desire for Interpersonal Attachments as a Fundamental Human Motivation," *Psychological Bulletin* 117, no. 3 (1995): 497.
21. C. Sagan and A. Druyan, *Shadows of Forgotten Ancestors: A Search for Who We Are* (New York: Random House, 1992), xvi, 505p.
22. D. Rushkoff, *Team Human* (New York: W. W. Norton & Company, 2019).
23. The Leakey Foundation, *Our Tribal Nature: Tribalism, Politics, and Evolution*, 2019.
24. P. W. Atkins, D. S. Wilson, and S. C. Hayes, *Prosocial: Using Evolutionary Science to Build Productive, Equitable, and Collaborative Groups* (Oakland, CA: New Harbinger Publications, 2019).
25. The Human Energy Project, accessed December 6, 2022, https://humanenergy.io/.
26. A. Maass, S. G. West, and R. B. Cialdini, "Minority Influence and Conversion," in *Group Processes: Review of Personality and Social Psychology*, vol. 8, ed. C. A. Hendrick (Newbury Park, CA: SAGE, 1987), 55–79; T. B. Kashdan, *The Art of*

Insubordination: How to Dissent and Defy Effectively (New York: Avery, 2022, Kindle), 220.

27. Norman, *Mental Immunity*.
28. Norman, *Mental Immunity*.

EPILOGUE

1. P. Schulttheiss, S. S. Nooten, R. Wang, and G. Benoit, "The Abundance, Biomass, and Distribution of Ants on Earth," *Proceedings of the National Academy of Sciences* 119, no. 40 (2022): e2201550199.
2. C. Darwin, *The Descent of Man* (London: John Murray, 1871).

APPENDIX

1. A. Zahavi, "Mate Selection—A Selection for a Handicap," *Journal of Theoretical Biology* 53, no. 1 (1975): 205–214.
2. M. A. Nowak, C. E. Tarnita, and E. O. Wilson, "The Evolution of Eusociality," Nature 466, no. 7310 (2010): 1057–1062; P. W. Atkins, D. S. Wilson, and S. C. Hayes, *Prosocial: Using Evolutionary Science to Build Productive, Equitable, and Collaborative Groups* (Oakland, CA: New Harbinger Publications, 2019).
3. J. Haidt, *The Righteous Mind: Why Good People Are Divided by Politics and Religion* (New York: Vintage, 2012); R. Kurzban, J. Tooby, and L. Cosmides, "Can Race Be Erased? Coalitional Computation and Social Categorization," *Proceedings of the National Academy of Sciences* 98, no. 26 (2001): 15387–15392.
4. G. F. Miller, "Sexual Selection for Moral Virtues," *Quarterly Review of Biology* 82, no. 2 (2007): 97–125.
5. N. A. Christakis, *Blueprint: The Evolutionary Origins of a Good Society* (New York: Springer, 1996); J. N. Rosenquist, J. H. Fowler, and N. A. Christakis, "Social Network Determinants of Depression," *Molecular Psychiatry* 16, no. 3 (2011): 273; N. A. Christakis and J. H. Fowler, "Friendship and Natural Selection," *Proceedings of the National Academy of Sciences* 111, Suppl. 3 (2014): 10796–10801.
6. R. Wrangham, *The Goodness Paradox: The Strange Relationship Between Virtue and Violence in Human Evolution* (New York: Vintage, 2019); B. Hare, V. Wobber, and R. Wrangham, "The Self-Domestication Hypothesis: Evolution of Bonobo Psychology Is Due to Selection Against Aggression," *Animal Behaviour* 83, no. 3 (2012): 573–585.
7. J. Tooby, "Coalitional Instincts," Edge.org, accessed December 6, 2022, https://www.edge.org/response-detail/27168.

INDEX